Contents

Urban Regeneration in the UK

'*Urban Regenerat*
ation. It is well w
practitioners alike
history and discus

'This is a clearly a
century. It provid
while also offering
changing econom
policy as well as p
tioners and politicia

Exploring the streets of
notice the striking tran
regeneration of urban are
ant issue among governme
impacts of the decline of ci
ing of global competition. A
taken over many decades to at

This text provides an accessi
porating key policies, approach
place the historical and contem
conceptual and policy framework
that have been adopted by central
development of cities, including ea
community-focused urban policies of the late 1960s, entrepreneurial property-led regeneration of the 1980s and competition for urban funds in the 1990s. The penultimate section illustrates the key thematic policies and strategies that have been pursued by cities themselves, focusing particularly on improving economic competitiveness and tackling social disadvantage. These approaches are contextualised by discussions covering, for example, urban competitiveness policies and the focus on sustainable urban regeneration. The final section summarises key issues and debates facing urban regeneration, and speculates upon future directions.

Urban Regeneration in the UK blends the approaches taken by central government programmes and cities themselves in the regeneration process. The latest ideas and examples from across disciplines and across the UK's urban areas are illustrated. This book provides a comprehensive and up-to-date synthesis that will fill a significant gap in the current literature on regeneration and will be a tool for students as well as a seminal read for practitioners and researchers.

Andrew Tallon is Senior Lecturer in Urban Policy at the School of the Built and Natural Environment, University of the West of England, UK. His research interests include the fields of urban geography, urban policy and urban regeneration.

URBAN REGENERATION IN THE UK

ANDREW TALLON

LONDON AND NEW YORK

First published 2010
by Routledge
2 Park Square, Milton Park, Abingdon, Oxon, OX14 4RN

Simultaneously published in the USA and Canada
by Routledge
270 Madison Ave, New York, NY 10016

Routledge is an imprint of the Taylor & Francis Group, an informa business

Typeset in Times and Bell Gothic by RefineCatch Limited, Bungay, Suffolk
Printed and bound in Great Britain by T. J. International Ltd, Padstow, Cornwall

British Library Cataloguing in Publication Data
A catalogue record for this book is available from the British Library

Library of Congress Cataloging-in-Publication Data
Tallon, Andrew, 1972–
Urban regeneration in the UK / Andrew Tallon.
p. cm.
1. City planning—Great Britain. 2. Urban policy—Great Britain. I.Title.
HT169.G7T35 2009
307.3'4160941—dc22
2009006586

ISBN10: 0–415–42596–4 (hbk)
ISBN10: 0–415–42597–2 (pbk)
ISBN10: 0–203–87259–2 (ebk)

ISBN13: 978–0–415–42596–4 (hbk)
ISBN13: 978–0–415–42597–1 (pbk)
ISBN13: 978–0–203–87259–8 (ebk)

Acknowledgements

Thanks go to former colleagues and PhD supervisors Rosemary Bromley and Colin Thomas at the Department of Geography at Swansea University. Thanks also to all of those colleagues in the School of the Built and Natural Environment at the University of the West of England who have contributed indirectly to the development of this book. Particular thanks go to Peter Malpass, Ron Griffiths, Vincent Nadin, Christine Lambert and Martin Boddy for their inspiration and support over the last five years. I am grateful for the assistance and support from the team at Routledge, particularly Andrew Mould, Michael P. Jones and Jennifer Page. Thanks also to three anonymous referees for their constructive and insightful comments on the manuscript.

The author and publishers would like to thank the following for granting permission to reproduce the following illustrations in this work: Routledge for Figures 1.4, 2.1, 2.2 and 5.1 and Sage for Figure 3.1. Thanks to Nicola James of the Department of Geography, Swansea University for producing Figure 12.1.

Every effort has been made to contact copyright holders for their permission to reprint material in this book. The publisher would be grateful to hear from any copyright holder who is not here acknowledged and will undertake to rectify any errors or omissions in any future editions of this book.

Illustrations

FIGURES

TABLES

PLATES

SECTION I

The Context for Urban Regeneration

1 Introduction

The decline and rise of UK cities

CHAPTER AIMS

The aims of this opening chapter are to outline the structure of the textbook and to introduce the key contemporary debates and urban changes framing urban regeneration in the UK. This provides the academic and policy context for the subsequent two substantive sections of the textbook. The chapter also identifies the nature of urban regeneration and explores why it is an important field of study. Key challenges that are currently being faced in the UK's urban areas are explored, including an overview of dimensions of the urban problem linked to urban economic decline and urban social polarisation. The chapter then introduces the seemingly recent rejuvenation of post-modern cities associated with urban regeneration policies and approaches, which is developed later in the textbook.

AIMS AND ORGANISATION OF THE TEXTBOOK

Urban regeneration has been given an increased public profile within the UK policy agenda since the late 1990s and is of considerable contemporary public interest and debate. The central objective of this textbook is to assess the historical and contemporary approaches to urban regeneration and to place these approaches within their wider economic, social and political contexts. The textbook aims to present and evaluate existing research and to add new insights into the subject.

Policy initiatives pursued by public and private sector bodies of different kinds have played a central role in shaping the transformations taking place in cities. A key aim of this textbook is to present a balance between, first, policies that have been adopted by government at the national level to influence the social, economic and physical development of cities, and, second, the policies and strategies that have been pursued by cities themselves, focusing particularly on measures concerned with improving economic competitiveness, and tackling problems of social disadvantage and exclusion. The textbook therefore scrutinises central government urban policy initiatives, in addition to engaging with wider issues surrounding the transition of cities from places of industry to post-industrial landscapes dominated by consumption.

The key themes within each chapter are informed and driven by the literature, each covering a distinctive and important aspect of urban regeneration. The early chapters detail the origins and general context for urban regeneration and later chapters move on to looking at recent research on thematic approaches to urban regeneration. This approach is warranted to draw together and synthesise the many strands of research, which are currently fragmented in the form of parts of textbooks, book chapters, journal articles, official reports, policy documents and websites. Case study examples are illustrated and integrated throughout to link theory with real-world practice. Examples are drawn from a variety of towns and cities across England, Scotland, Wales and Northern Ireland. Where appropriate, international comparisons are made with European, North American and other advanced cities, as many of the trends and patterns observed are international in extent. The textbook is interdisciplinary in scope, drawing upon the urban studies, geography, town and country planning, economics, local government and sociology literatures. Following this chapter, Section II outlines and traces central government urban interventions since the mid-1940s, and Section III focuses on cities and their strategies in the context of Section II, providing a focus on a number of key thematic contemporary issues. Section IV offers a conclusion and possible future directions for urban regeneration in the UK.

DEFINING URBAN REGENERATION AND URBAN POLICY

It is pertinent to begin with a discussion of what urban regeneration means and encompasses. At the most general level, regeneration has come to be associated with any development that is taking place in towns and cities. An initial question surrounds the definition of what 'the urban' is. This, as with 'the rural', is a somewhat contested question, but without getting too embroiled in definitional arguments, this textbook will take 'the urban' to mean 'relating to towns and cities'. Urban places are 'spatial concentrations of human economic, social, cultural and political activities distinguished from non-urban/rural places by both physical aspects such as population density or administrative definition and lifestyle characteristics' (Pacione 2005, p. 676). Definitions concerning the development, diversity and key themes within 'urban geography', a field of study similar to urban studies, can be further examined in Johnston et al. (2000).

Urban regeneration is a significant component of wider 'urban policy', which is not necessarily exclusively concerned with 'regeneration'. Urban policy has evolved in its own distinct way during the post-war period in the UK as demonstrated in Section II. Urban policy can be viewed as spatial in that it relates to urban areas and urban processes, and to the populations who live in urban areas, and particularly the resolution of urban problems. At a basic level, policy is a course of action adopted and pursued by government; it is an approach, method, practice and code of conduct (Roberts 2000). Urban areas are complex and dynamic systems and reflect the many processes that drive economic, social, physical and environmental transition (Roberts 2000, p. 9). Urban regeneration can be seen as the outcome of the interplay between these many processes, and is also a response to the opportunities and challenges that are presented by urban degeneration (Roberts 2000).

Urban regeneration, revitalisation, renewal and renaissance are amongst the plethora of phrases and buzzwords have come to characterise the key themes of this textbook and tend to be used by the media, government and even academics as interchangeable terms essentially relating to the same process. However, as Lees (2003b) identified, there are subtleties relating to the use of these terms by academics and policy makers. For example, urban renewal in the 1960s was public sector-driven and primarily concerned with the large-scale redevelopment of overcrowded inner city slum areas (Couch 1990). By contrast, urban regeneration in the 1980s focused on economic growth and used public funds to lever in largely undirected market investment, as exemplified by London's Docklands (Brownill 1990; 1999). Current policies seek to combine both the public and private sectors in partnership to achieve urban regeneration, with a more heightened environmental awareness than before (Hall 2006).

Lees (2003a) argued that the urban regeneration metaphor was being replaced by the urban renaissance metaphor under New Labour in the late 1990s and early 2000s. However, since the mid-2000s it appears that renaissance has been subsumed by the wider sustainable communities agenda (ODPM 2003a; Raco 2005b; Raco and Henderson 2006). However, regeneration remains the most recognised and widely used term by professionals and academics alike. All of the terms have similar meanings and connotations relating to rebirth, revival and reconstitution. Regeneration has its roots in religion, social theory and medicine as a word infused with religious hope (Furbey 1999; Lees 2003a). New Labour's concept of urban renaissance has gone beyond purely physical objectives to include concerns for social inclusion, wealth creation, sustainable development, urban governance, health and welfare, crime prevention, educational opportunity, freedom of movement, environmental quality and good design (ODPM 2003a; www.communities.gov.uk).

A definition of urban regeneration put forward by Roberts (2000, p. 17) is

> a comprehensive and integrated vision and action which leads to the resolution of urban problems and which seeks to bring about a lasting improvement in the economic, physical, social and environmental conditions of an area that has been subject to change.

However, Turok (2005) qualified this statement by noting that regeneration is rarely, if ever, comprehensive, and that it can be the case that the urban problems addressed are not resolved in practice because they can be amongst the most intractable or 'wicked' problems in society (see Harrison 2000).

Turok (2005, p. 57) identified three distinctive features of contemporary urban regeneration:

1. It is intended to change the nature of a place and in the process to involve the community and other actors with a stake in its future.
2. It embraces multiple objectives and activities that cut across the main functional responsibilities of central government, depending on the area's particular problems and potential.
3. It usually involves some form of partnership working amongst different stakeholders, although the form of partnership can vary.

Urban regeneration can be classified in a number of ways, but for the purpose of this textbook, Turok's (2005) categorisation of 'people', 'business' and 'place' appears appropriate. In terms of people, regeneration aims to enhance skills, capacities and aspirations to enable them to participate in and benefit from opportunities. Regeneration also aims to improve economic competitiveness in terms of business performance to create more local jobs and prosperity. To attract both people and business, regeneration aims to improve the general appeal of a place. The theory is that in balance all three elements combine to secure the upward trajectory of a locality in a long-term and sustainable manner. Dimensions of urban regeneration can be broadly described as economic, social and cultural, physical and environmental, and governance-related in nature (Table 1.1).

These four dimensions are not mutually exclusive, but interconnected. Successful urban regeneration should recognise the linked nature of economic, social and physical environmental problems in the context of local geographies. Related to the definition and scope of regeneration, Lichfield (1992, p. 19) pointed to a need for 'a better understanding of the processes of decline'. Hausner (1993, p. 526) emphasised the inherent weaknesses of approaches to regeneration that are 'short term, fragmented, ad hoc and project-based without an overall strategic framework for city-wide development'.

Urban regeneration is an activity rooted in practice, and there is a high degree of similarity between theory and practice in urban regeneration. There is a great deal of pragmatism

TABLE 1.1 APPROACHES TO URBAN REGENERATION

Dimension	*Concerns*
Economic	Job creation, income, employment, skills, employability, development (see Chapter 6)
Social/cultural	Quality of life, health, education, crime, housing, quality of public services (see Chapter 5)
Physical/environmental	Infrastructure, built and natural environment, transport and communications (see Chapter 6)
Governance	Nature of local decision-making, engagement of local community, involvement of other groups, style of leadership (see Chapters 7 and 8)

and experimentation in implementing regeneration, and substantial evidence has emerged regarding which policies are effective (Wilks-Heeg 1996; Turok 2005).

According to Roberts (2000, p. 22), key features of urban regeneration are that it is:

- an interventionist activity;
- an activity that straddles the public, private and voluntary and community sectors;
- an activity that is likely to experience considerable changes in its institutional structures over time in response to changing economic, social, environmental and political circumstances;
- a means of mobilising collective effort and providing the basis for the negotiation of appropriate solutions;
- a means of determining policies and actions designed to improve the condition of urban areas and developing institutional structures necessary to support the preparation of specific proposals.

Due to its nature and practice, urban regeneration is far from being a completely fixed set of guiding principles and practices, and does not have a proven or well-established track record of success. However, Turok (2005) argued that some ambiguity about the scope and purpose of urban regeneration can be helpful in terms of flexibility and modification. Urban regeneration strategies can be aware of the difficulties arising from 'one size fits all' approaches and of complicated local scenarios and geographies.

Leading on from a discussion of the definition and characteristics of regeneration, a pertinent question is, why is it important to regenerate urban areas? Regeneration of urban areas matters as 'the tragedy of the inner city affects everyone' (Stegman 1995, p. 1602). Cities matter, and effective urban regeneration is of fundamental importance to a wide range of actors and stakeholders including local communities; city, regional and national government; property owners and investors; businesses; environmental organisations; residents; and visitors at all levels from the local to the global (Roberts 2000). Government and society have made the value judgement that cities should be maintained as the focus of urban life in the UK. Urban regeneration is required to keep pace with the consequences of continued processes of urban change.

Urban areas and regeneration are of importance from an empirical or statistical as well as a normative or value perspective. The twenty-first century can be seen as the era of the city as it is estimated that 60 per cent of the world's population will be living in urban areas 2030 (Hamnett 2005), and economic, political and administrative power is concentrated in urban areas. In the USA, metropolitan areas house 75 per cent of the population on just 1.5 per cent of the total land. In England and Wales, which are smaller and more densely urbanised than most countries, 89 per cent of the population lives in urban areas on 7.7 per cent of the land. In the highly urbanised South-East of England, the urban coverage is approximately 20 per cent of the total land and this figure is growing (see Pacione 2005; Hamnett 2005).

THE CONTEMPORARY URBAN REGENERATION AGENDA IN THE UK

In general, in the late 2000s, three approaches to urban regeneration are apparent. Each of these is related to policy approaches that have evolved over successive decades since the 1960s (see Section II). Turok (2005) summarised these under the urban renaissance, social inclusion and economic competitiveness umbrellas.

The urban renaissance agenda, which has now to a large extent been subsumed within the sustainable communities programme (ODPM 2003a; see Chapters 5, 9 and 11) has been concerned with physical and environmental conditions, linked with the trend towards brownfield redevelopment (see *Local Environment* 2006, Vol. 11, no. 5; Dixon et al. 2007) and issues surrounding greenfield development. It promotes high quality urban design (Urban Task Force 1999), mixed-use environments (Coupland 1997) and sustainable cities (Hall 2006).

The social inclusion agenda focuses more on social conditions within deprived neighbourhoods. It encourages the development of social capital (see Putnam 2000; 2002; Kearns 2003) and community participation (see Taylor 2003) to bring about the regeneration of neighbourhood and community.

The economic competitiveness agenda is concerned with improving economic performance and employment by increasing output, productivity and innovation (see Begg 2002a; Boddy and Parkinson 2004; Buck et al. 2005; ODPM 2006a).

Actors involved in urban regeneration have changed relative to each other and according to the situation over many decades as is apparent from Section II, but usually comprise partnerships which combine members of the local community, non-profit voluntary sector organisations, public sector agencies and private sector business (see Chapters 7 and 8).

The co-ordination of the actors involved in urban regeneration has been a central but problematic element. This is required because of the multiple aims of many urban regeneration initiatives and because of the many organisations involved in delivering these. The most complex co-ordination takes the form of area-wide partnerships that cover a particular neighbourhood that include representatives from the public, private, community and voluntary sectors (see Chapters 5 and 7; Bailey et al. 1995). Partnership arrangements have emerged as a central feature of urban regeneration strategies in the UK, starting with a number of local authority and private sector-led initiatives in the 1980s (see Chapter 3), and gradually leading to the incorporation of partnership into central government policy from the 1990s with policies such as City Challenge and Single Regeneration Budget (see Chapter 4), and area-based initiatives from the late 1990s (see Chapter 5). These tend to be voluntary agreements that operate by consensus and persuasion rather than being strongly controlled hierarchical institutions (Turok 2005). Partnerships can have a lifespan between five and ten years and aim to start to address the significant structural issues in an area.

At the other end of the spectrum, Turok (2005) argued that urban regeneration could be co-ordinated by tight contractual arrangements involving a similar number of partners. This arrangement is more common in economic and physical regeneration schemes where the private sector is involved and substantial amounts of money are injected.

In between these extreme alternatives, are special purpose locally based organisations with a narrower range of responsibilities than large multi-agency partnerships. These can be established to deliver business advice services, local employment and training programmes, or small-scale property development. Their internal structure is usually hierarchical and they are responsible for delivering particular projects.

As Sections II and III make clear, there has been a massive variety of policies and approaches to urban regeneration, all of which strive to achieve similar goals (Hall 2006, p. 58). Before moving on to take a historical perspective regarding the decline and rise of UK cities, it is worth summarising the contemporary general concerns of the urban regeneration agenda:

1. Physical environment: urban regeneration has attempted to improve the built environment, concerns which have now embraced environmental sustainability.
2. Quality of life: urban regeneration has sought to improve the physical living conditions, or local cultural activities, or facilities for particular social groups.
3. Social welfare: urban regeneration has endeavoured to improve the provision of basic social services in certain areas and for certain populations.
4. Economic prospects: urban regeneration has sought to enhance the employment prospects for deprived groups and areas through job creation or through education and training programmes.
5. Governance: there has been a shift from government to governance within urban regeneration, and public policy more generally, which is highlighted by the rise in importance of partnership, community engagement and multiple stakeholders in the process and delivery of urban regeneration (see Chapters 7 and 8).

The remainder of this chapter begins to present a chronology of the events and processes that were to have a fundamental impact on urban areas and that led to the emergence of urban problems which urban regeneration policies seek to ameliorate. The origins of what is often termed the 'urban problem' is discussed and its history is tracked from the latter half of the nineteenth century, when the problems of the urban poor were first recognised, to the emergence of an 'urban crisis' in post-war UK cities. Also explored are some of the differing approaches to understanding these problems by academics and policy-makers, many of which are developed in Section II of the textbook. The chapter then considers the concepts of poverty, deprivation, the underclass and social exclusion, and questions whether these definitions will contribute in any way to the quest for a more equal and inclusive society. The chapter finishes on a more positive note in discussing the seeming rise in fortunes of many urban areas since the late twentieth century associated with the apparent transition towards the post-modern city.

NINETEENTH- AND EARLY TWENTIETH-CENTURY INDUSTRIALISATION, URBANISATION AND MUNICIPAL INTERVENTION

The beginnings of urban decline in the UK were linked to the massive increase in urban populations. The forces that transformed the UK economy from an agricultural to industrial base during the eighteenth and nineteenth centuries also caused a striking redistribution of population and laid the foundations for an urban society (Herbert 2000). By the mid-nineteenth century more people were living in towns than in the country. The shape of the urban UK was influenced by pre-existing settlements to which many new towns and cities were added. Advances in transport technology allowed urban areas to widen their influence and underpinned processes of concentration and centralisation. Processes of urbanisation and industrialisation in the UK intensified and continued apace, which brought about their own problems and challenges.

Intervention in urban areas can be traced back to the late nineteenth century, when the idea of regulating capitalist industrial cities emerged (Hall 2006; Atkinson and Moon 1994a). This was in response to the catastrophic consequences arising from unregulated urban growth associated with the ongoing process of industrial capitalism and industrialisation. The period from the 1880s witnessed the emergence of urban slums, dereliction, municipal corruption, and 'moral dangers' associated with the perceived threat of the working-class mob during this period of urban history (Atkinson and Moon 1994a). At this time the poor were divided into the 'deserving working poor' who were seen as a proud and hard-working group in society, and the 'useless poor' characterising those living in poverty, dependent on the limited state, and seen as not contributing to society (Atkinson and Moon 1994a). Politicians and social reformers recognised that intervention was required to address such problems in large cities of the UK, and indeed of continental Europe and North America. It was during this time that formal planning systems began to emerge in urban areas to regulate the development of places (Cullingworth and Nadin 2006; Hall 2006). Most action at this time was focused on the physical living conditions of the urban poor through bricks and mortar, sanitation improvements and sunlight laws. However, social reformers could not keep pace with the sheer rapidity of change. For example, Birmingham saw an increase in its population from 71,000 in 1800 to 765,000 by 1901 (Hall 2001, p. 7). At this time, most urban intervention was associated with town planning rather than with urban regeneration as such (see Cullingworth and Nadin 2006).

Industrial cities were created through the processes of urbanisation linked to the industrial revolution. This influenced the internal geographies of many cities along with the economic, political and physical links between them (Hall 2006). These foundations greatly influenced future urbanisation in the UK. Famous models of urban form were devised by the Chicago School of academics in the early twentieth century, such as the concentric rings model of Burgess in 1925 (Figure 1.1; see Pacione 2005, Chapter 7; Hall 2006, Chapter 2).

The Burgess model consists of a city based on the historic core of the central business district of medieval age in the UK and dating from the eighteenth and nineteenth centuries in the US. This was dominated by banking and financial businesses, which evolved into a growth pole for financial services comprising tertiary and quaternary sector activities. Burgess's model depicted the central business district as the centre of the city, the financial services economy, the tertiary sector economy, and the place with the highest property

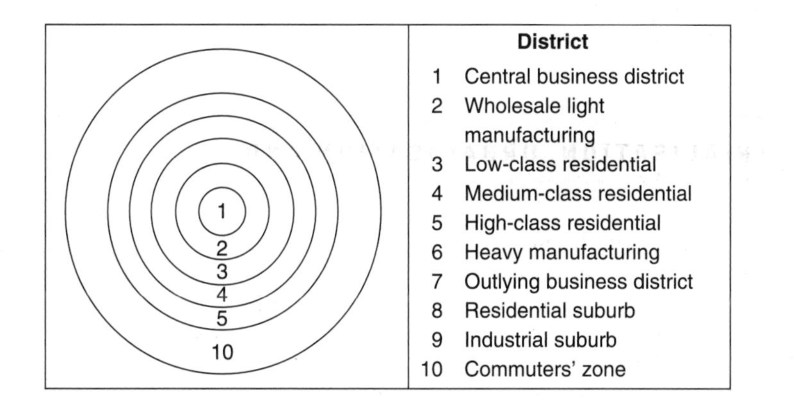

Figure 1.1 Burgess's concentric zone model
Ley (1983, p. 73).

prices, high status banks, boutique shops, the focus of the transport system, and the focus of local government. Agglomeration economies were generated in the city core from the reputation of area, access to market information, business with other firms, skilled labour in the area, and proximity to support services in the zone of transition. Disadvantages that subsequently emerged included high property prices and traffic congestion. The Burgess model next included a ring around the centre termed the 'zone of transition', which was dominated by residential and warehousing uses with small terraced houses for workers near to warehousing and docklands. Big Georgian houses were built for merchants and were within walking distance of the centre but usually spatially separate. Zones of transition were characterised by a mix of residential, office, retail, leisure and infrastructure uses, often containing the 'support economy' for the central business district. Over time, the zone of transition in both UK and US cities has witnessed a transformation to a gentrified landscape of refurbished housing, arts and leisure functions. The sequence of rings progressively moving outwards then encompassed the suburban fringe comprised of the early inner suburbs which shifted outwards over time, and finally the rural-urban fringe where the city blends into the country (see Chapter 13).

By contrast, the Harris Ullman model of 1945 addressed the simplicity of the Burgess model (Pacione 2005, p. 145). This was based on the Los Angeles experience and depicted urban form as based on the progressive integration of a number of separate nuclei or 'cells' of use rather than around a single central business district. The model illustrated that different areas have different agglomeration economies and that uses will 'cluster' around different parts of the city. Drivers of this model were increasing car ownership, zoned planning of land uses and large-scale public transport systems.

Models of urban form evolved over time for a variety of reasons, key amongst these being the emergence of mass car ownership, the development of the motorway network, the de-industrialisation process, the growth in the tertiary and quaternary economy, social and demographic changes, and due to urban sprawl as cities reach their limits. Modifications to models of urban form are reflected in the subsequent variants. A further famous model is the sector model devised by Hoyt in 1939 (Figure 1.2).

In addition, Mann created a model of the typical UK city in 1965 (Figure 1.3) (see Hall 2006, Chapter 2; Pacione 2005, Chapters 3 and 7).

Others illustrated by Pacione (2005, pp. 146–150) were Kearsley's modified Burgess model, Vance's urban-realms model, and White's model of the twenty-first-century city. These models all represented a radical change in the form and function of cities in terms of transport infrastructure, the economy and demographics. Industrialisation in the UK

District

1 Central business district
2 Wholesale light manufacturing
3 Low-class residential
4 Medium-class residential
5 High-class residential

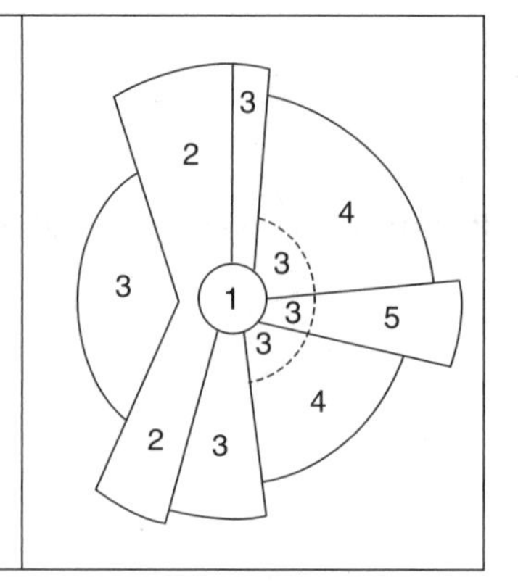

Figure 1.2 Hoyt's sector model
Ley (1983, p. 73).

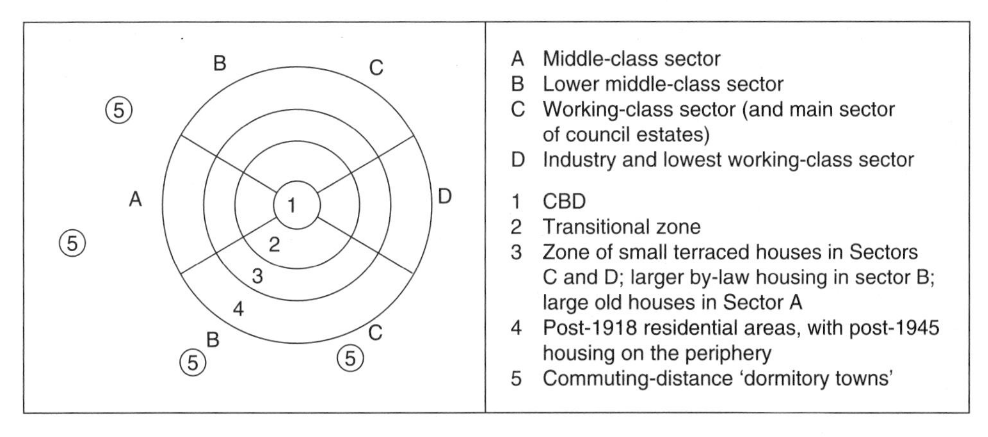

Figure 1.3 Mann's model of a UK city
Knowles and Wareing (1976, p. 245).

was rapid as cities shifted from being religious or mercantile centres towards foci of manufacturing industry, and subsequently new spatial patterns emerged associated with de-industrialisation and edge-city development.

SUBURBANISATION, COUNTER-URBANISATION AND THE DECLINE OF THE INNER CITY

From around the 1920s, more flexible transport systems and the rise of the private car facilitated the deconcentration and decentralisation of people and capital from urban areas, and this began to lead to the decline of cities. In this sense, transport can be described as both the 'maker and breaker' of cities (Herbert and Thomas 1997; Herbert 2000). The dominant flows in the mid-twentieth century were outwards from the city centre towards the peripheries (see Pacione 2005, Chapter 4). The process of counter-urbanisation from the early 1960s in the UK occurred when areas situated at distance from metropolitan influences began to grow at a faster rate than the main conurbations and their dependent regions (Pacione 2005, pp. 86–87). This process continued in subsequent decades.

Accompanying the counter-urbanisation process were growing regional-scale shifts in population and economic activity. Regional economic planning sought to contain urban growth and direct activities to the declining north and west of the UK, which were increasingly affected by the de-industrialisation process. Containment was affected by post-war town and country planning controls, green belts, and major new initiatives such as New Towns and urban expansion schemes (see Cullingworth and Nadin 2006; Pacione 2005; Chapter 2). Redistribution policies proved more difficult as a succession of regional economic policies testified (see Prestwich and Taylor 1990; Healey and Ilbery 1990).

Associated with the counter-urbanisation process, suburbanisation ran strongly during the post-war decades contributing to the growth of the outer rings of cities at the expense of the central city (see Pacione 2005, Chapter 4; Boyle et al. 1998; Chapter 13). Essentially, counter-urbanisation and suburbanisation were indistinguishable and produced the ongoing dispersal process. Suburban growth took place in a piecemeal and epochal fashion over the course of the twentieth century (Hall 2006, Chapter 7), a process that continues to an extent in the 2000s (see Chapter 13).

THE SOCIAL AND ECONOMIC CONSEQUENCES OF WARTIME BOMBING AND POST-WAR RECONSTRUCTION PROJECTS

Post-war urban reconstruction and Keynesian (redistributive) economics stimulated economic growth and brought a new confidence to the economy and a general air of optimism after the Second World War. From 1945, it is possible to argue that the UK experienced its first form of urban regeneration, with post-war reconstruction projects. However, there was a growing unease about a 'threat from within', rather than with the earlier threat from foreign invaders. Despite economic growth, persistent poverty remained and refused to be shifted (Atkinson and Moon 1994a).

Various research projects on dimensions of the welfare state began to question the assumption that the post-war boom had eradicated poverty. People with incomes below National Assistance rates had trebled between 1953 and 1960. Research reports included the 1958 Milner Holland Report on housing in Greater London, the 1966 Plowden Report on children and their primary schools, and the 1968 Seebohm Report on local authority and allied personal services (see Cullingworth and Nadin 2006, pp. 360–361). These reports concluded that services were failing to reach those living in small pockets of deprivation in older cities, and there was consensus on developing positive discrimination on an area basis (Atkinson and Moon 1994a; Mullins and Murie 2006). There were major concerns about helping the poor and about the growing fear of racial unrest. Greater selectivity was now the goal for delivering benefit, focusing on certain groups and certain areas.

Both the housing and economic markets ensured that poor immigrants from the former empire were concentrated in inner city areas. It was at this time that poverty became an area-based problem issue (Atkinson and Moon 1994a). High levels of migrant workers coming to the UK had a weak bargaining position forcing them to take low paid, unskilled jobs and find cheap accommodation in the inner city. The Notting Hill riots in 1958 underlined the growing tension felt about the level of immigration. The Labour Party returned to power in 1964 but lost the previously safe seat of Smethwick in the West Midlands largely due to the immigration issue and, with a parliamentary majority of just three, they bowed to public pressure on the issue and introduced immigration controls. In 1968, Enoch Powell, the then Conservative shadow minister, made his infamous 'Rivers of Blood' speech. Here, he quoted Virgil's warning of the River Tiber foaming with blood and preached both a halt to immigration and a policy of voluntary repatriation. The 'rivers of blood' speech was seized upon by the press and sealed his political fate, but he nevertheless drew attention to this emotive issue. Partly in response, in May 1968, Harold Wilson announced the Urban Programme, which really symbolised the emergence of a distinct urban policy in the UK. It was at this time urban regeneration as an explicit strategy was imported from the US, and this marked the beginning of three decades of continuous central government intervention in urban affairs (Johnstone and Whitehead 2004b) (see Chapter 2).

LATE TWENTIETH-CENTURY DE-INDUSTRIALISATION AND THE CHANGING GEOGRAPHIES OF PRODUCTION

Forces of de-industrialisation had severe and long-lasting effects on industrial cities since the 1970s in particular (Pacione 2005; Hall 2006). Many cities of North America and Europe were founded on the expansion of industrial capital from the 1850s onwards, and the rise of industry within cities represents a major phase in their history. During the 1970s and 1980s the older industrial cities of the UK experienced a sharp downturn in manufacturing. Hall (1985) observed that it took over 100 years from 1851 to 1951 for technology and foreign competition to halve the numbers employed in agriculture, but it took only 13 years from 1971 to 1983 to cut manufacturing jobs by a third.

The decline in manufacturing industry had a number of implications. The emergence of long-term unemployment saw significant numbers of people out of work for over one year. Unemployment has been concentrated in manufacturing, which had once dominated the national economy with a fall in employment from 7.4 million workers in 1975 to 3.3 million in 2004 (Hall 2006, p. 38). Between 1971 and 2001, the UK's biggest 20 cities lost 2.8 million manufacturing jobs and gained 1.9 million service sector jobs (Moore and Begg 2004), representing what has been termed a 'jobs gap' (Turok and Edge 1999). Regions such as the North of England and South Wales suffered severe economic problems, and particularly the cities of such regions that had been the focus of these industries, especially the inner cities (Lawless 1981). The worst impact of unemployment has been felt within certain social groups such as the young, the late middle-aged, males, and ethnic minorities (Hall 2006, p. 38).

De-industrialisation has been attributed to three principal factors as illustrated by Hall (2001, p. 37). A key cause was factory closure linked to poor and inadequate sites and intense global competition, which contributed to every major urban area in the UK losing between one quarter and one half of its employment between 1960 and 1982. There were transfers of firms, forced moves due to urban regeneration and closure of uneconomic units.

The migration of jobs to suburban and rural locations and overseas occurred due to the changing nature and requirements of manufacturing industries in the post-Fordist era (see Healey and Ilbery 1990). Firms were more successful outside inner cities and the expansion of information technology and service sector employment was striking along the M4 corridor in particular.

Technological advances witnessed automated systems of production replacing the human labour force (Hall 2006). Some of the worst affected cities in the UK have been Glasgow, Newcastle, Liverpool, Manchester, Sheffield, Birmingham and London. These processes began in the 1940s, but deepened after 1960 and massive disinvestment in the industrial capacity of the inner city continued through until the mid-1980s.

Within the UK, the geographical pattern of spaces of economic dynamism and economic depression have become increasingly polarised and the north–south divide in the UK persists at a general level (IPPR 2007). Some former manufacturing cities have been physically transformed through investment in convention centres, offices, hotels, and retail and leisure developments (Hall 2006; see Section III). However, other cities have failed to capture significant economic inward investment and have continued to decline. Cities have responded by attempting to combat the negative impacts of economic change with diverse responses and these have wide-ranging implications. Urban regeneration and urban policy in the UK have had a range of impacts on city landscapes, economies, images and social geographies, as Sections II and III illustrate.

SOCIAL EXCLUSION AND SOCIAL POLARISATION IN CITIES

Economic decline in urban areas over the last 30 or so years has been paralleled by striking social and spatial polarisation, and deprivation in many urban areas. Reflections of this are highly visible today in many inner suburban and inner city areas, and in peripheral local authority estates. A number of processes have led to this polarisation, deprivation and even exclusion of certain groups in the city. These processes are related principally and traditionally to divisions related to employment, social status and ethnicity, but also to variables such as age, gender, sexuality, and able-bodiedness (see Pacione 1997).

There has been a long-standing concern of the issue of 'haves' and 'have-nots' in cities, especially so with the 'have-nots'. This has been an intractable problem, however, the language for talking about and ways of conceptualising the issue has changed over ti
Victorians talked about the 'deserving' and 'undeserving' poor (Atkinson and Moon

Twentieth-century terms included poverty, disadvantage and deprivation. In the latter part of the twentieth century the term underclass came to prominence with echoes of the Victorian notion of the undeserving poor. More recently the term social exclusion has become common parlance in European Union and in UK government policies, exemplified here with the setting up of the Social Exclusion Unit in 1998 (SEU 7 1998; Chapter 5).

Underlying the different language and different concepts is a long-standing tension – whether people are poor through no fault of their own (due to the system) or whether they are responsible in some way (original sin). Key questions relating to social exclusion and polarisation are of relevance to urban regeneration and the remainder of this textbook. These include the meaning of the term social exclusion and whether it is conceptually different from terms like poverty, deprivation or the underclass; how social exclusion is measured and identified; and what evidence is available about the incidence and extent of poverty and social exclusion in UK cities. A final issue is whether or not cities in the UK are becoming more unequal places, despite urban policy and regeneration.

Social exclusion as a concept has multiple meanings. These different meanings are significant, because they point to and emphasise different kinds of explanation and lead onto different kinds of policy solutions. Poverty denotes lack of money, which constrains people in all sorts of ways and affects their opportunities. There is a debate about how to measure poverty and whether it is absolute or relative. In policy terms it implies a minimum level of income for all. Deprivation emerged as a development of the idea of poverty and was based on the observation that those in poverty frequently experience complex problems of poor housing, poor levels of education and skills, poor health, and so on. The idea is that problems of poverty are multi-dimensional and require a multi-dimensional approach (Madanipour et al. 1998).

The concept of the underclass came to prominence in the 1980s, especially in the US. It was originally coined to describe the chronically unemployed who were associated with the collapse of manufacturing industry in US cities. As a term it was taken up and used more widely, and academic and popular circles shifted its meaning. The term became associated with assumptions about the behaviour and characteristics of the affected groups – not the circumstances in which people found themselves, but the choices they made. This behavioural definition became the dominant one and it took on a moralistic and blaming the victim tone. In the US it became a racialized term applied to the black population of cities – problems of the ghetto were associated with unwillingness to adhere to the American work ethic. Conservative commentators in the 1980s took up the term and also argued that the provision of welfare was part of the problem – it encouraged dependency and undermined the work ethic (see Chapter 3). They argued the solution was to cut back welfare and encourage more self-reliance, and the concept has now largely been discredited.

Social exclusion is a more recent term and implies a multi-dimensional and dynamic process (Gaffikin and Morrissey 1999), and takes on a wider definition than poverty, which is defined more narrowly in terms of material wealth. The key dimension of social exclusion is the idea of detachment. As the SEU (1998, p. 1) quoted, 'Over the last 20 years hundreds of poor neighbourhoods have seen their basic quality of life become increasingly detached from the mainstream.' Someone or something is acting to push some individuals and groups out of mainstream society, causing them to become excluded. Social exclusion is not so much about how much money or resources individuals have, but what the mechanisms or causal processes are that lead them into that situation. The idea is that these processes or mechanisms of exclusion generate outcomes in terms of poverty or deprivation.

Social exclusion is multi-dimensional, social and spatial in its form. As Madanipour et al. (1998, p. 22) suggested, 'Social exclusion is defined as a multi-dimensional process, in which various forms of exclusion are combined: participation in decision making and political processes, access to employment and material resources, and integration into common cultural processes. When combined, they create acute forms of exclusion that find a spatial manifestation in particular neighbourhoods.' Solutions to the problem must focus on the processes or mechanisms that cut people off from the mainstream, rather than dealing with the state of poverty itself. Crudely, if someone is poor then giving them money provides only temporary relief. To deal with problems in the long term it is necessary to examine why those who are excluded experience the problems they do and tackle those issues. In practice there is still a lot of confusion and ambiguity about the term. This revolves around the same kind of ambiguity that affected the term underclass and led to its demise.

It is understood that social exclusion points towards processes and mechanisms, but there is a debate as to quite what these are. Levitas (2005) identified three ways in which the term social exclusion is used in current UK policy and politics (see Chapter 5). The first is termed a 'redistributive discourse', which views social exclusion as a consequence of poverty and essentially a lack of money. In other words the poor are unable to participate in society fully, and this discourse very much reflects the 'Old' Labour view. A second understanding of social exclusion is an 'integration discourse' which sees employment or paid work as the primary way in which individuals are integrated into society, and that social exclusion results from a lack of work. A third 'moral discourse' emphasises cultural and behavioural factors and that exclusion stems from lack of morals. The argument here is that some groups who are not integrated, that is, they are not in the labour market, are in that position because of their behaviour or attitudes, which need to be changed for reintegration to take place. The case put forward by Levitas (2005) is that much current policy in relation to social exclusion emphasises either integration into the workforce or altering values and morals, and less so redistribution of wealth (see Chapter 5).

While the concept of social exclusion emphasises processes and causal mechanisms, when it comes to measuring it, the focus is actually on outcomes. Patterns of income are measured to identify households with very low incomes or proxies for this such as unemployment. This is similar to older ideas of multiple deprivation, which attempt to measure factors such as low educational performance and poor health. In practice the concept of social exclusion does incorporate poverty but it also adds in other aspects of disadvantage or deprivation.

The Indices of Multiple Deprivation were initially developed by the New Labour government in 1998 and were revisited in 2001. A revised set of seven domains emerged in 2004 (see Chapter 5). These were income (for example, households dependent on benefits of various kinds); employment (for example, unemployment, long-term unemployment); health and disability (for example, mortality rates, low birth weights); education, skills and training (for example, percentage getting five A–C GCSE grades); housing and services (for example, overcrowded housing, distance to certain services); living environment (for example, condition of housing, air quality); and crime (for example, rates of certain types of crime). Areas were scored according to these aspects and the findings are used to identify areas or neighbourhoods with high levels of deprivation, in other words, with high levels of social exclusion (ODPM 2004j). The Indices of Multiple Deprivation were more recently revised again in 2007 covering the seven domains of income; employment; health and disability; education, skills and training; barriers to housing and services; living environment; and crime (CLG 2007c). Due to changes in method, geographical areas and data used, these indices cannot be used to understand changes in deprivation longitudinally (Smith et al. 2007a, p. 225).

Patterns of social and spatial exclusion in the UK are strongly tied to those of deindustrialisation described earlier. Some patterns of inequality were further accentuated by government policy. In terms of the social patterns of change, inequality increased significantly in the period after the mid-1970s as revealed by the work of the Joseph Rowntree Foundation in the 1990s (Hills 1995). This research showed that incomes for the richest group in society increased by 65 per cent in the period 1979 to 1995 whereas the incomes of the poorest group fell in real terms. There was also a large growth in the proportion of the population with low incomes relative to the average; this proportion doubled from 10 per cent in the mid-1960s to 20 per cent in the mid-1990s. The proportion of children growing up in households with very low incomes increased from 10 per cent in 1979 to 32 per cent in the mid-1990s. For those in work, wages and salaries became more dispersed during this period. This was demonstrated by a widening polarisation of income between richer and poorer groups since the 1960s. Evidence suggests that this trend has continued since the late 1990s as inequality has persisted, although the poorest in society are arguably better off in general because of policies and benefits targeted at these groups (Dorling et al. 2007; Centre for Cities 2008; see Chapter 5).

Looking at spatial patterns of social exclusion in the UK, there is a strong urban-rural dimension to the Indices of Multiple Deprivation. Generally speaking more poverty and deprivation is found in towns and cities as measured by the percentage of population with low incomes, unemployed, living in poor housing, low educational attainment, and so on. Areas with the highest concentration of deprivation match those most severely affected by the de-industrialisation process and include inner London boroughs, large metropolitan cities and old industrial cities (see, for example, Boddy 2003c; Hamnett 2003a; Pacione 2005).

Within cities, the areas with the highest levels of deprivation are a mixture of inner city areas and certain council estates. The example of Bristol highlights these patterns of striking spatial polarisation (Bassett 2001; Tallon 2007a). There is also evidence that cities are becoming more polarised places – the richer neighbourhoods are becoming better off and poorer neighbourhoods are getting poorer (see Centre for Cities 2008). Rates of unemployment in different neighbourhoods are becoming more divergent and some council estates appear to be the areas where deprivation is becoming more concentrated. Council housing is more closely associated with people on very low incomes and who lack choice (Hall 1997b; see Mullins and Murie 2006).

The evidence that deprivation or problems of social exclusion are concentrated in particular areas of cities has led to a debate about whether this makes the problems of individuals worse. It could be the case that living in a poor area compounds the problems of individuals and make it harder to escape the area because of 'neighbourhood' or 'area' effects (see Lupton 2003). However, research on 'area effects' (for example, Buck and Gordon 2004) finds very small neighbourhood effects on social and economic outcomes. Individual and family characteristics have far more impact on outcomes than neighbourhood characteristics.

Overall, the concept of social exclusion has been seen as useful insofar as it focuses on the processes and causal mechanisms, which lead to poverty and deprivation. From the point of view of developing solutions in terms of community regeneration, it seems a sensible approach. However, the exact nature of these processes and causal mechanisms remains a matter of some debate. In the late 2000s the discussion continues to be tied up with a political and ideological debate about the extent to which governments should try to redistribute income through the taxation system.

THE RISE OF THE POSTMODERN CITY AND NEW URBAN SPACES

Much has been written in recent years about the apparent transformation of the form and type of cities in Europe and North America. Much of this debate has focused on the emergence of the 'post-industrial' or 'post-modern' city and this is significantly different from the industrial or modern city as depicted by models such as Burgess's (Figure 1.1). Contrasts between the modern and postmodern city can be assessed in terms of urban form and structure, urban planning, the urban economy, society and culture, architecture and the urban landscape, and urban government (Table 1.2). Each of these themes is closely

TABLE 1.2 CONTRASTING THE MODERN AND POST-MODERN CITY

Modern	Post-modern
Urban form	
Modern	**Post-modern**
Homogeneous functional zoning	Chaotic multi-nucleated structure
Dominant commercial core	Highly spectacular centres
Steady decline in land values away from centre	Large 'seas' of poverty
	Post-suburban edge-city developments
Urban planning	
Modern	**Post-modern**
Cities planned in totality	Spatial 'fragments' designed for aesthetic ends
Urban space shaped for social ends	
The urban economy	
Modern	**Post-modern**
Industrial	Service sector-based
Mass production	Flexible production aimed at niche markets
Economies of scale	Economies of scope
Production-based	Globalised
	Telecommunications and information-based
	Finance
	Consumption-oriented
	Jobs in new edge-of-city zones
Society and culture	
Modern	**Post-modern**
Class divisions	Highly fragmented
Large degree of internal homogeneity within class groups	Lifestyle divisions
	High degree of social polarisation
	Groups distinguished by consumption patterns
Architecture and the urban landscape	
Modern	**Post-modern**
Functional modernist architecture	Eclectic 'collage' of styles
Mass production of styles	Spectacular
	Playful
	Ironic
	Use of heritage
	Produced for specialist markets
Urban government	
Modern	**Post-modern**
Government	Governance
Managerial – redistribution of resources for social purposes	Entrepreneurial – use of resources to attract mobile international capital and investment
Public provision of essential services	Place promotion by city authorities
	Public and private sectors working in partnership
	Private provision of services

Source: **Hall (2006, p. 100).**

interlinked and the changes have taken place in parallel. These transitions have been closely linked with urban regeneration strategies in the UK and cities globally (see Section III).

URBAN FORM

The homogenous, functional zoning of the city with a dominant city centre and a steady decline in land values moving away from the centre, as represented by the concentric rings of the Burgess model, has been overtaken by the post-modern urban form. This new 'centre-less' urban form is chaotic, multi-nucleated and disarticulated (Dear 2000; Lees 2002; Figure 1.4).

Post-modern urban form is so unpredictable that Dear (2000) argued that it is difficult to represent graphically. The post-modern city is characterised by spatial polarisation with islands of regeneration existing in seas of poverty, the emergence of high-tech industrial corridors and ex-urban or edge-city developments.

URBAN PLANNING

Closely allied with the form of urban areas, cities have witnessed changes in approaches to planning. Modern cities were planned in totality as a single entity and urban space was shaped for social ends. Post-modern urban planning sees fragments of space designed for aesthetic rather than social purposes (Hall 2006).

THE POST-MODERN URBAN ECONOMY

The key change has been in the nature of the urban economy. The modern economy was essentially industrial, and was dominated by mass production or 'Fordism', economies of scale in the production process, and was production-based (see Mackinnon and Cumbers

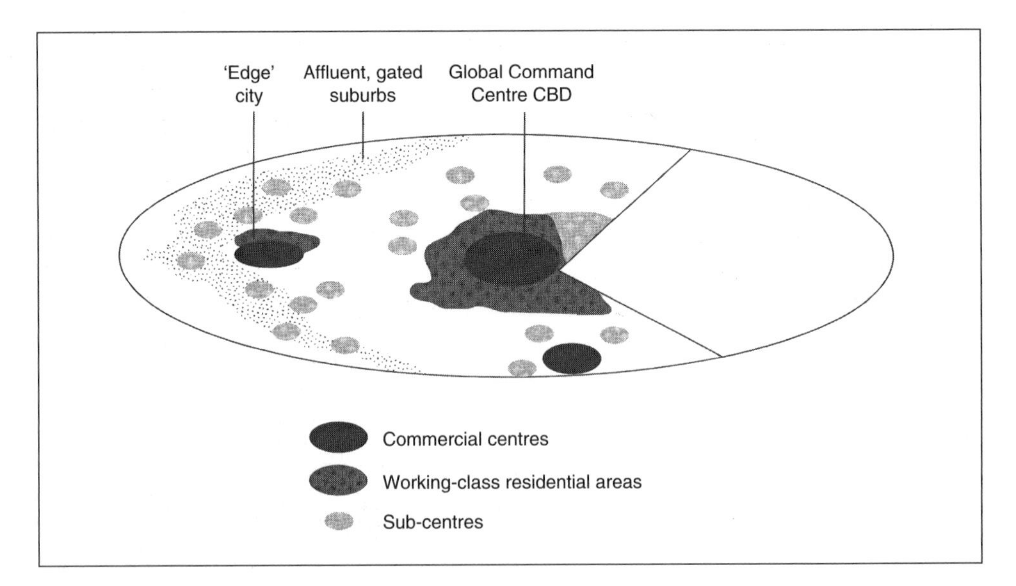

Figure 1.4 The post-industrial 'global' metropolis
Graham and Marvin (1996, p. 334).

2007). By contrast, the post-modern economy can be described as post-industrial. Urban economies have become increasingly service sector-based with flexible production methods aimed at niche markets. There has also been the emergence of an information and consumption-oriented economy, in the context of globalisation and the growth of telecommunications (see Dicken 2007).

The emergence of a post-industrial or post-modern city has seen the materialisation of a number of new urban economies, which have risen to take the place of the space left by the decline of the manufacturing sector. The major growth has been that of the service sector since the 1980s and this now employs over 70 per cent of employees in the UK (see Hall 2006, Chapter 4). However, the rise in service sector employment has not fully compensated for the loss of manufacturing jobs and the shift has an uneven regional dimension. The worst affected areas of unemployment continue to be cities in former manufacturing heartlands such as South Wales, Central Scotland, and the North East, North West and West Midlands in England.

The urban economy is being reorganised around new or growing sectors, which have been expressed in a number of ways. There is a growing role of corporate headquarters, which are increasingly concentrated in a small number of the world's largest cities termed 'world' or 'global cities' such as London, Tokyo and New York (see Sassen 1994). Producer service economies have also risen and offer legal, financial, advertising, consultancy and accountancy services to companies. There has also been the rise of research and development units which are highly significant in urban development and are sites of innovation where products are modified or developed and often appear as science and technology parks, and business parks. New industrial spaces have emerged within cities and are based on new technologies, which include computer hardware and software, telecommunications, virtual reality technology and biotechnology (Hall 2006).

Each of these growing sectors of the new economy requires the need for access to a highly qualified workforce, an environment that facilitates innovation, and good infrastructural and telecommunications linkages (see ODPM 2004b). These places are not usually the old industrial spaces, which experienced severe de-industrialisation. New industrial spaces appear to be dominated by a polarised income distribution requiring a small number of highly paid specialists and a large number of low paid workers (Short and Kim 1999).

The world economy is becoming increasingly interdependent, in that the fortunes of individual places are increasingly bound up with the fortunes of other places. This is the result of the emergence of multi-national corporations as major shapers of international economic flows, and the concentration of international command centres in global cities (see Short and Kim 1999; Dicken 2007). Spaces are becoming closer together, not that their geographical locations are changing but that interactions between them are becoming ever more instantaneous. Advances in telecommunications such as telephones, fax, e-mail, computer networks and the Internet have reduced or eliminated the time delay in communications between distant cities. However, this 'electronic economy' is predominantly located in a small number of major global cities.

SOCIETY AND CULTURE

Closely related to the economic changes that have been experienced by urban areas over the last 30 years have been social and cultural transitions. Modern cities exhibited class divisions and a large degree of internal homogeneity within class groups (Pacione 2005). However, the post-modern city reveals highly fragmented patterns of wealth, high degrees of social polarisation, divisions along lifestyles, and social groups being increasingly distinguished by consumption patterns. Divided or dual cities are argued to be characteristic of post-modern society (see, for example, Mollenkopf and Castells 1991; Pacione 1997).

One sector that has been particularly dynamic in the regeneration of declining city centre areas has been the cultural or creative industries such as fashion, media, film and video production, design, creative technology-based industries, and music (see Chapter 12). These industries not only have the potential to regenerate run-down city centre areas both economically and culturally, but also to combine this with community-oriented regeneration projects based around new media, the Internet or the creative arts such as music (Bassett et al. 2003; see Chapter 12). Cultural quarters or spaces have emerged in many cities throughout Europe and North America as key components of wider regeneration strategies (Montgomery 2004; Tallon et al. 2006; McCarthy 2007b).

ARCHITECTURE AND THE URBAN LANDSCAPE

The physical landscape of urban centres has radically altered since the 1970s. The city and its centre in particular had by the 1980s become a much-maligned place dominated by homogenous functionalist architecture and zoning of activities (see Chapter 10). They were blighted by a legacy of poor post-war planning and design, and were typified by poor physical environments, pedestrian-unfriendly traffic systems, downgraded retail environments, and economies dominated by offices that relegated cities to cultural deserts after the early evening (see Chapter 12). Since then, however, the urban landscape of cities and their centres has been the focus of an astounding degree of investment and development (Plate 1.1).

The 1980s saw a widespread and comprehensive 're-imagination' of city centres, and this process has involved a combination of physical enhancement and cultural animation

Plate 1.1 A post-modern urban landscape in Bristol city centre
Photograph by author.

processes which can be viewed in parallel to the transformation of the images of cities (Short and Kim 1999). This re-imagination of the city centre resulting from urban regeneration policies is expressed in a number of post-modern landscape elements that have emerged in city centres since the 1980s. These include spectacular and flagship developments, festival retailing, landscapes of heritage and nostalgia, cultural animation, gentrification and the rise of city centre living, and an emerging reurbanisation process (Hall 2006, Chapter 7; see Section III). Post-modern architectural styles have been evident in these new landscape uses, illustrated by an eclectic variety of styles, spectacular architecture, the use of heritage and pastiche, playful and ironic styles, and orientation towards niche markets.

URBAN GOVERNMENT

The management of cities has also shown changes reflecting the shift from the modern to post-modern city (see Chapter 7). Modern government was characterised by managerialism where resources were redistributed for social purposes, and the public sector provided essential services for the residents of urban areas. However, since the 1970s, urban areas have been managed in an entrepreneurial style with the primary aim being to use resources to attract mobile international capital and investment to boost cities (see Short and Kim 1999). Public–private partnerships have emerged since the 1980s as the usual mode of delivering many public services, and there has been the rise of the market provision of services (Barnekov et al. 1989; Fincher 1992; Bailey et al. 1995).

KEY CHALLENGES AND DEBATES

The processes of de-industrialisation and related social exclusion documented earlier have not affected all cities equally. Those cities with diverse economies without a significant manufacturing component have enjoyed very different economic fortunes. There has been a fall in the total urban populations of large cities in the UK associated with out-migration and the counter-urbanisation and suburbanisation processes, particularly in northern UK cities such as Manchester and Liverpool (Couch 1999; Centre for Cities 2008). However, since the 1990s some cities, such as London, Bristol and Southampton, have shown signs of a recovering population with a modest reurbanisation, most strikingly in central cities (Lever 1993; Couch 1999; DETR 2000b; Pacione 2005, pp. 84–85; Nathan and Urwin 2006; Bromley et al. 2007b; Centre for Cities 2008; see Chapter 11).

The shift from a modern to post-modern city has had uneven impacts on urban areas. It has in addition been argued that attempts at urban regeneration have so far been uneven and partial (Carley 2000; Chapter 14). Despite the seeming rise of cities in the post-industrial, post-modern age, there are still many deep and intractable social and economic problems entrenched in many cities associated with de-industrialisation and decline. At a general level, advances in urban areas have been confined to a small number of world cities, and to more favourably located cities and areas within cities.

It is essential to have an understanding of the wider urban geographical context for urban regeneration in which to situate urban interventions, and to appreciate that urban regeneration does not happen in isolation from wider changes affecting cities and their relationships with each other.

The rest of the textbook examines public, private, and voluntary and community sector policy responses to the problems resulting from these long-running processes, many of which are still being witnessed today, and remain as 'wicked' and as relevant as ever, despite four decades of a distinct government urban policy in the UK.

URBAN REGENERATION: SIGNIFICANT TEXTS

This textbook combines themes that until now have been to a large extent divided into those books focusing on urban policy and those on more general approaches to urban regeneration. Many of these are now out of date because of the rapid policy changes in the field but remain extremely useful in assessing past approaches and for historic case study examples. They are also more narrowly focused towards particular sets of issues rather than covering a breadth of urban regeneration topics. The former category of textbooks are usually in the form of critiques of particular policy periods such as the immediate post-war town and country planning policies (for example Couch 1990; Cullingworth and Nadin 2006), the property-led approaches characteristic of the 1980s (for example Imrie and Thomas 1993a; 1999; Healey et al. 1992), competition and urban policy in the 1990s (for example Oatley 1998a), or New Labour's brand of urban policy since 1997 (for example Imrie and Raco 2003a; Johnstone and Whitehead 2004a). In addition, Atkinson and Moon (1994a) traced urban policy from the late 1960s to the early 1990s within the context of theoretical approaches to policy analysis. Blackman (1995) and Hambleton and Thomas (1995) provided additional overviews of urban policy.

Key examples of the latter group of books on general approaches to urban regeneration include those of Colquhoun (1995), Roberts and Sykes (2000), Burwood and Roberts (2002) and Evans's (1997) book on regenerating town centres. These books look at the practice of urban regeneration drawing upon national and international experiences. Other accessible texts that cover urban regeneration issues to varying degrees include urban geography texts by Pacione (2005) and Hall (2006), and planning texts such as those by Couch (1990) on the theory and practice of urban renewal, Rydin (1998), which summarises UK urban policy and socio-economic processes affecting UK cities from the perspective of land-use planning and physical regeneration, Ward's (2004) account of urban planning, and Cullingworth and Nadin (2006) on town and country planning in the UK. Short (1996) offers a comprehensive, international overview of the generic urban studies literature illustrated with case studies and further reading, which links with the themes covered in Section III. Other useful textbooks deal in depth with one or a limited number of topics within urban regeneration such as Diamond and Liddle's (2005) work on the management of regeneration, Trott's (2002) text on best practice in regeneration and partnership working, McCarthy's (2007a) book on partnership, collaborative planning and urban regeneration, Purdue et al.'s (2000) and Chanan's (2003) work on community and regeneration, and Smith et al.'s (2007a) collection looking at neighbourhood governance and urban policy under New Labour. Judge et al. (1995) and Stoker (2000; 2004), for example, give useful detail on changes in the structures and practices of urban governance in the UK and its implications for policy delivery, including regeneration policy, which connect with Sections II and III of the textbook. Books that are international in scope include Couch et al.'s (2003) collection on urban regeneration in Europe and Colquhoun's (1995) international perspective looking at approaches in the United States and Europe.

Other books should be consulted to appreciate the wider processes of urban change that are intrinsically linked with urban regeneration. Lawless (1989; 1996), Robson (1988) and MacGregor and Pimlott (1990) set the broader scene of the nature of urban problems that remain relevant. In addition, Healey et al. (1995) looked at the links between local problems and global economic forces.

In addition to these key texts, international journals have devoted whole editions to central government urban regeneration policy. Significant amongst these since the early 1990s have been editions of *Local Economy* (1994, volume 9, number 3), *Cities* (1995, volume 12, number 4), *Planning Practice and Research* (1995, volume 10, numbers 3/4) and *Local Economy* (2000, volume 15, number 2).

This textbook does not aim to give an in-depth account of every theme covered but draws together key ideas and examples cross-cutting the subject, and directs the reader to a variety of more specific follow-up references and resources. This textbook is aimed primarily at undergraduate and postgraduate students involved in the study of urban regeneration as part of a geography, planning or regeneration course, and will also be of interest to policy-makers and a range of practitioners interested in transforming urban areas and regeneration.

SUMMARY

This textbook seeks to provide a comprehensive and critical synthesis of urban regeneration in the UK.

The textbook blends the approaches taken by central government programmes (Section II) with those undertaken by cities themselves in the urban regeneration process (Section III). It provides recent ideas and examples from across disciplines and across the UK's urban areas.

Long-running processes of urban change are expressed in an uneven pattern and have resulted in a range of distinctly urban problems.

Urban economic decline has largely stemmed from the long-running de-industrialisation process.

Patterns of redevelopment and regeneration have been uneven and partial, favouring fortunately located cities and particular areas within these cities.

Social inequalities are reflected in patterns of regeneration – new opportunities are socially selective, and the problem of social exclusion and polarisation has been intractable.

Many characteristics of urban regeneration are reflected in the transition from the modern to the post-modern city.

STUDY QUESTIONS

1. Discuss the origins of the 'urban problem' in the UK.
2. To what extent are cities of the UK continuing to struggle with the legacies of de-industrialisation?
3. Critically evaluate the claim that the twenty-first-century metropolis is post-modern.
4. To what extent is the post-modern city unequal?
5. Does living in an area of concentrated poverty add an extra dimension to deprivation? Assess the arguments and evidence relating to this assertion.
6. How is 'social exclusion' different from 'poverty'? Does this represent a useful addition to our understanding of urban social problems and how to tackle them?
7. Assess the symbols of urban regeneration and also of continued decline in a UK town or city with which you are familiar.

KEY READINGS ON URBAN REGENERATION AND URBAN POLICY

Atkinson and Moon (1994a) *Urban Policy in Britain: The City, the State and the Market* [this excellent text traces urban policy from its origins in the late 1960s through successive stages until the early 1990s in a theoretical context]

Blackman (1995) *Urban Policy in Practice* [a practical and critical guide to urban policy with case studies up until the early 1990s]

Burwood and Roberts (2002) *Learning from the Past: The BURA Guide to Effective and Lasting Regeneration* [this is more of a practitioner-oriented book and is a good guide to implementing successful regeneration schemes]

Cochrane (2007) *Understanding Urban Policy: A Critical Introduction* [looks at why clusters become identified as specifically urban problems suitable for intervention through urban policies at one time while they may be understood quite differently at different times and in different places]

Colquhoun (1995) *Urban Regeneration: An International Perspective* [a basic text but outlines a variety of UK and international examples of urban regeneration projects]

Gotham (ed.) (2001) *Critical Perspectives on Urban Redevelopment* [a collection of critical reviews of urban redevelopment approaches]

Hambleton and Thomas (eds.) (1995) *Urban Policy Evaluation: Challenge and Change* [includes evaluations of early 1990s urban policies]

Healey, Davoudi and Tavsanoglu (eds.) (1992) *Rebuilding the City: Property-Led Urban Regeneration* [a good collection of critiques of 1980s-style property-led regeneration]

Imrie and Raco (eds.) (2003) *Urban Renaissance? New Labour, Community and Urban Policy* [this edited collection analyses New Labour's approach to urban policy, focusing particularly on community and the urban renaissance]

Imrie and Thomas (eds.) (1999) *British Urban Policy: An Evaluation of the Urban Development Corporations* [this volume brings together eight detailed examples of 1980s/1990s Urban Development Corporations throughout the UK as well as a critical overview, updating the 1993 edition]

Johnstone and Whitehead (eds.) (2004a) *New Horizons in British Urban Policy: Perspectives on New Labour's Urban Renaissance* [this edited volume is an early critique of New Labour's urban renaissance policy]

Oatley (ed.) (1998a) *Cities, Economic Competition and Urban Policy* [this text is a key resource on 1990s-style urban policy characterised by competition and bidding for regeneration funding]

Pierson and Smith (eds.) (2001) *Rebuilding Community: Policy and Practice in Urban Regeneration* [an international overview of urban regeneration policy and initiatives]

Roberts and Sykes (eds.) (2000) *Urban Regeneration: A Handbook* [this is a good account of dimensions of the regeneration process up until the 2000s]

KEY WEBSITES ON URBAN REGENERATION AND URBAN POLICY

British Urban Regeneration Association (BURA): www.bura.org.uk [independent organisation promoting best practice in urban regeneration across the UK]

Centre for the Analysis of Social Exclusion (CASE): http://sticerd.lse.ac.uk/case/ [located within the London School of Economics, CASE produces research reports on social and economic change as well as Census briefings]

Communities and Local Government (CLG): www.communities.gov.uk [the government department responsible for urban regeneration and policy from 2006]

Economic and Social Research Council (ESRC): www.esrc.ac.uk [an independent, government-funded organisation which funds research and training in social and economic issues, including urban regeneration]

English Partnerships: www.englishpartnerships.co.uk [the national regeneration agency helping the government to support sustainable growth in England]

Institute for Public Policy Research: www.ippr.org [progressive left-of-centre independent public policy think tank; see especially their Centre for Cities]

New Economics Foundation: www.neweconomics.org [an independent 'think and do' tank with relevance to urban regeneration in the UK]

Online Planning Resources: http://planning.rudi.net [an excellent and up-to-date resource containing extensive bibliographies for numerous urban regeneration topics]

Planning Portal: www.planningportal.gov.uk [the UK government's online planning and building regulations resource]

Regeneration-UK.com: www.regeneration-uk.com [useful regeneration information portal for the UK]

Regen.net: www.regen.net [latest UK regeneration news]

Regional Development Agencies (RDAs): www.englandsrdas.com/home.aspx [information on England's nine RDAs which aim to increase prosperity in England's regions]

Social Market Foundation: www.smf.co.uk [an independent think tank, exploring ideas that link the state and the market to create a more just, free and prosperous society]
State of the English Cities: www.communities.gov.uk/index.asp?id=1162938 [2006 report giving a broad overview of the state of English cities and possible directions for urban policy]

SECTION II

Central Government Urban Regeneration Policy

2 The early years
Town and country planning and area-based policies

INTRODUCTION: ANALYSING URBAN POLICY PERIODS

This chapter is the first of four that set out and evaluate central government approaches to urban regeneration and urban policy in the post-war period. These chapters also attempt to elucidate lessons for current and future regeneration initiatives in towns and cities of the UK. There have been a number of distinct thematic shifts in urban policy since the Second World War, which are illustrated generally by Table 2.1. This shows the periodisation used to frame successive periods in this textbook, which illustrates continuities and changes in UK urban policy over more than 60 years. This set of chapters will chart urban policy chronologically over the decades through to the present day, stopping on the way to analyse particular policies in detail, and focusing on case studies in urban areas of the UK. It should be cautioned that these historical periods are not mutually exclusive. As Hall (2006, p. 62) warns, 'While it is tempting to characterise urban regeneration in a series of distinct periods, this tends to underemphasise the degree of continuity between periods.' Despite this reservation, such a classification of urban policy periods is a useful instructive device. For alternative, although similar, categorisations of the evolution of urban policy, see those of Roberts and Sykes (2000, p. 14), Pacione (2005, pp. 176–177) and Hall (2006, pp. 64–65).

This first chapter of Section II provides a brief background to the origins of urban regeneration to set the context for the subsequent evolution of urban regeneration policy. The chapter examines responses to the perceived problems of urban areas driven by central government in the immediate post-war decades. The period 1945 to 1968 was characterised by 'bricks and mortar' and physical redevelopment policies more associated with town and country planning than urban regeneration as such. The periods stretching from 1968 to 1977 and 1977 to 1979 witnessed the introduction of area-based community initiatives and indicated the importation of a specific urban regeneration policy.

URBAN POLICY AS 'EXPERIMENTATION'

As the chapters move through the consecutive periods of policy, readers should bear in mind that urban policy has resulted from an interaction between an intellectual process (based on research reports) and an institutional process (policy development). This means

TABLE 2.1 PERIODS OF UK URBAN POLICY, 1945–PRESENT

Period	*Problem construction*	*Policy response*
1945–1968	Housing problems, urban sprawl and ribbon development, need to comprehensively redevelop city centres, and to relocate jobs and populations out of congested city centres.	'Housing' and 'town and country planning' rather than 'urban' policy. Physical approach. City centre redevelopment.
1968–1977	Social pathology approach. Limited to small areas of towns.	Small-scale area-based initiatives. Largely experimental reflecting lack of knowledge.
1977–1979	Structural approach contained in Urban White Paper 1977 'Policy for the Inner Cities'. Identified four key problems: 1. Economic decline 2. Physical decline 3. Concentration of poverty 4. Racial discrimination. Assumed problem of urban decline lay in 'societal forces', those experiencing it seen as 'victims'.	Urban White Paper and Inner Urban Areas Act 1978. Attempted to develop an integrated approach, the formation of partnerships, new role for the private sector, and reference made to voluntary and community sectors. These partnerships were attempts to create vertical and horizontal coordination within the state. Still small area-based.
1979–1991	Mixture of structural and social pathology approaches. Urban problems seen as a product of: 1. Too much state intervention 2. Individual and group dependency 3. Restriction of the free market.	1. Roll back the state 2. Encourage self-help 3. Free up the market. Produced property-led urban regeneration – physical renewal in profitable locations. Public–private partnerships (e.g. Urban Development Corporations (UDCs)). Multiplicity of initiatives lacking coordination. Emphasis on better management of programmes. Local government marginalised as part of the problem.
1991–present	Retained elements of previous period but recognised failings, particularly the fact that deprived (socially excluded) communities were largely bypassed. Key problems: how to ensure that excluded communities benefit from policies, incoherence of area-based initiatives and levels of governance.	Development of new multi-sectoral partnerships (public, private, voluntary and community sectors). First in City Challenge, then Single Regeneration Budget (SRB), more recently New Deal for Communities (NDC). Creation of Regional Development Agencies (RDAs), creation of Neighbourhood Renewal Fund (NRF). Setting up of Social Exclusion Unit (SEU). Attempt to streamline and coordinate multiple initiatives and levels of governance – achieve greater policy coherence and synergy. Better use and targeting of resources. Strategic 'mainstreaming' approach. Joined-up approach.

Sources: **Based on Atkinson and Moon (1994a); Oatley (1998a); Imrie and Raco (2003a); Johnstone and Whitehead (2004a).**

that urban policy can be highly subjective and subject to powerful political processes – it is essentially based on experiment. Consequently, policies come and policies go, policies are taken up and discarded subject to whim, and policies often come 'full circle' (see Wilks-Heeg 1996). There are various advantages to seeing urban policy programmes as experiments, as well as reasons for the seemingly constant change (Wilks-Heeg 1996, p. 1264):

- Politicians can present the programmes as 'radical solutions to urban problems' without having to make any long-term commitment to them or specific urban areas due to their 'experimental' nature.
- The programmes can be cancelled at short notice if officials do not like the way they are being run.
- Ministers can make great claims for the projects even though generally they are bound to fail because of the constraints governing the programmes.
- Changes in government and therefore ministers cause policy programmes to change rapidly.
- The lack of effective evaluation of previous initiatives meant that lessons were not being learned and applied to future policies to ensure longer and more successful runs.

In short, it looks more impressive if a government appears to be doing a great deal to address urban problems that have proved intractable and insoluble.

The chapters that make up Section II identify the schematic shifts in urban policies. Related to these shifts, questions include the extent to which understandings of the urban problem have moved from social pathology as a perceived cause of poverty to structural–economic approaches to explaining the urban problem, often linked with the governing political party. As examined in Chapter 1, there is a debate about the nature of the urban problem and the extent to which it is caused by the problems of the people themselves, or the run-down state of urban areas and pockets of poverty, or de-industrialisation affecting the urban fabric, or the urban–rural shift (Atkinson and Moon 1994a). Other questions to consider within the evaluation of urban policy periods relate to how the policies were set up, and whether they were appropriately conceived, funded and managed.

URBAN PROBLEMS AT THE END OF THE SECOND WORLD WAR

The most obvious and striking urban problem in 1945 was the legacy of the war damage, particularly along the south coast, in London, and in many other major cities across the UK (Atkinson and Moon 1994a). There was a strong belief after the war to go on and 'win the peace', that was to eradicate poverty and regulate the economy to eliminate cycles of 'boom and slump'. This context created a more co-operative culture during the immediate post-war years between the political parties. This co-operation paved the way for the nationalisation of industry, the creation of the welfare state, and the nationalisation of development rights (see Cullingworth and Nadin 2006; Mullins and Murie 2006). These fundamental changes to the governing of society and the economy were all viewed as too radical before the war, and set the scene for six decades of urban policy.

Attitudes towards poverty on the part of government, as debated in Chapter 1, can be broadly divided into two halves for this urban policy period, from 1945 to 1965, and 1965 to 1979. In the first half of the period there was a strong belief that poverty had been all but eradicated, and that there was an urgent need to bring workers from the Empire to fill job vacancies in the rejuvenated post-war economy (Atkinson and Moon 1994a). In the second half of the period, there was a growing awareness that there were pockets where poverty still existed, that the gap between the rich and the poor had widened, and that racial tensions were beginning to surface.

TOWN AND COUNTRY PLANNING DEVELOPMENTS, 1945–1965

Urban problems that were prevalent in the period 1945 to 1965 were linked to physical housing conditions and unrestricted urban growth, which was leading to sprawl and ribbon development. The housing problem comprised a lack of housing and the poor quality of much of the remaining stock (Mullins and Murie 2006). Allied with the fact that nearly four million houses in the UK had been damaged or destroyed during the war (Mullins and Murie 2006, p. 28), there were also rising birth rates and thus an increasing population, coupled with changing social expectations which combined to ensure a short supply in the housing stock. This was an emerging problem before the war, which served to exacerbate the problem.

The fear of urban sprawl and ribbon development was associated with the growth in the road network in the pre-war period. There had been a concern that allowing the private sector to manage the housing shortage would increase the risk of sprawl and ribbon development, which would have severe implications for the UK. In the pre-war period, government attempted to limit ribbon development and sprawl with the Restriction of Ribbon Development Act (1935) (Cullingworth and Nadin 2006). Subsequently the issue was addressed following the Second World War by the Town and Country Planning Act (1947), which meant that the problem became less serious as the policy period evolved. New major issues needing attention were the need to comprehensively redevelop city centres and inner city housing, and these strategies were largely local authority-led and resulted in problems such as mono-tenured housing estates, and poorly planned and developed city centres (see Chapters 10 and 11).

Immediately at the end of the Second World War, central government sought to ameliorate the two major problems of urban sprawl and sub-standard housing with three central policies. These were strongly linked to town and country planning, regional development and housing policy. The three primary planning tools used to deal with physical urban problems from 1945 were New Towns, green belts and housing redevelopment.

NEW TOWNS

New Towns were conceived by the Reith Committee and were inspired by the long-established traditions in the UK of anti-urbanism and the Garden City Movement of Ebenezer Howard and Patrick Abercrombie (Atkinson and Moon 1994a; Cullingworth and Nadin 2006; CLG 2006c). A number of central principles underpinned the foundation of New Towns in the UK. The towns were planned for populations of 20,000 to 60,000 people, and they were sited around large, densely built-up urban areas to help reduce their populations (Pacione 2005, pp. 190–191). They were normally built on greenfield (although not on best agricultural land) sites, and were developed by New Town Development Corporations (central government appointed) or sponsored by the local planning authority (Atkinson and Moon 1994a; Pacione 2005). They were therefore public rather than privately funded and developed, and all the land in the designated areas would be vested in the development corporations and compulsory purchase powers were granted if necessary. The towns were self-contained and exhibited a mix of jobs, workers and social amenities (CLG 2006c). They also intended to mix social classes within neighbourhoods to reflect the community spirit developed during the war. The Reith Committee report's recommendations were generally accepted and the New Towns Act (1946) enabled the programme to go ahead. It passed through Parliament with very little objection consistent with the post-war political consensus.

In practice, 28 New Towns were created between 1946 and 1970 in two main phases of construction (Figure 2.1; Table 2.2).

Figure 2.1 New Towns in the UK
Pacione (2005, p. 191).

The first phase of 14 New Towns was built immediately after the war and primarily consisted of satellite settlements around London to meet the objectives pursued by Howard and Abercrombie (Pacione 2005). Some were also built in depressed regions like South Wales and Scotland to bolster regional growth along the lines advocated by the 1940 Barlow Committee Report on regional development. This seminal report argued that there should be a balanced distribution of industry and the industrial population throughout the regions of the UK, and that there should be an appropriate diversification of industries in those regions (see Cullingworth and Nadin 2006, pp. 19–21). A small phase was undertaken by the anti-interventionist Conservative government of the 1950s and were essentially 'expanded towns' where some local authorities agreed to accept people transferred from larger urban areas (Atkinson and Moon 1994a, p. 27). A second extensive phase of building was undertaken by Wilson's Labour government of the 1960s. This revitalised the programme and although the aim of decentralisation was retained, regional economic development became a strong directing force (Atkinson and Moon 1994a; Pacione 2005, p. 192).

Although officials of the time were positive regarding the proactive and innovative approach taken under the New Towns programme, evaluation of the scheme raised concerns with this style of physical new-build of housing. There was a national shortage of building materials in the early stages of New Town development. This, coupled with the massive amount of infrastructure planning that needed to take place to create New Towns,

TABLE 2.2 NEW TOWNS IN THE UK

Phase I New Towns	*Date of designation*
Stevenage	1946
Crawley	1947
Hemel Hempstead	1947
Harlow	1947
Aycliffe	1947
East Kilbride	1947
Peterlee	1948
Hatfield	1948
Welwyn	1948
Glenrothes	1948
Basildon	1949
Bracknell	1949
Cwmbran	1949
Corby	1950
Phase II New Towns	
Cumbernauld	1955
Skelmersdale	1961
Livingston	1962
Redditch	1964
Runcorn	1964
Washington	1964
Irvine	1966
Milton Keynes	1967
Peterborough	1967
Newtown	1967
Northampton	1968
Warrington	1968
Telford	1968
Central Lancashire New Town	1970

Source: **Pacione (2005, p. 191).**

meant that build rates were very slow. Eventually the policy looked expensive and bureaucratically cumbersome. The policy seemed autocratic and dictatorial to those people living in existing settlements that were designated to become expanded New Towns. There were also problems with relocating large numbers of young working-class people from existing urban areas to environments that initially had little in the way of entertainment provision and job opportunities. This reflected the fact that the policy was unashamedly physical, concomitant with wider approaches to town and country planning during this period. New Towns policy largely overlooked social and economic factors that caused problems in urban areas (Atkinson and Moon 1994a). However, as CLG (2006c) noted, there is little research-based material that sought to evaluate the New Towns programme as a whole in an objective way, and the programme was never reviewed systematically. As with many past urban policy initiatives, New Towns policy can offer both positive and negative transferable lessons, in this case for late 2000s 'growth areas' (CLG 2006c; see Chapter 5).

GREEN BELTS

The second key policy of this period was green belt policy, which emerged as part of wider land-use planning policy in the UK. A green belt was a boundary set around an urban area where growth and expansion could not take place thus containing urban areas, and within the green belt, agricultural land would be the norm along with some land for recreation. Green belts were envisaged to promote compact cities, restrict chaotic ribbon development and sprawl, reduce service costs, and preserve open and agricultural land, rural communities and the natural environment. Although not an original aim of green belts, they have more recently additionally been a way of promoting urban regeneration objectives.

One of the first modern proponents of green belts was Ebenezer Howard and the Garden Cities movement, who suggested that they be used to help support new proposed garden communities. The first official proposal to provide a reserve supply of public open space and of recreational areas and to establish a green belt of open space was made by the Greater London Regional Planning Committee in 1935. In policy terms, London was the first place to have a true green belt in 1938 (Ward 2004). This was supported by the famous London planner Patrick Abercrombie, who proposed a green belt to limit the growth of London and encourage development in New Towns as part of the 1944 Greater London Plan (see Pacione 2005, pp. 190–191). The Town and Country Planning Act (1947) allowed local authorities to include green belt proposals in their development plans, and in 1955 the government set out a green belt policy, asking local authorities to consider protecting any land acquired around their towns and cities by the formal designation of clearly-defined green belts. Duncan Sandys (father of the Civic Amenities Act and Civil Trust) lobbied heavily for green belts to become an essential part of planning in the UK. Concerns had arisen with house building programmes and the manner in which they were eating up land around the country.

Although there was strong criticism from senior civil servants who felt the powerful House Builders Federation would scupper the concept, green belts proved to be popular with the general public and they provided a focal point for the environmental lobby. In addition, planners were in favour with the Town Planning Institute president Desmond Heap commenting shortly after the publication of the Circular 4 in 1955 that green belts had become the 'raison d'être' of town planning (see Cullingworth and Nadin 2006, pp. 233–239).

Green belts today are not as scarce as is often popularly portrayed and cover 13 per cent of England, representing about 1.6 million hectares. There are 14 separate green belts in England (Figure 2.2), varying in size from 512,900 hectares around London to just 700 hectares around Burton-on-Trent and Swadlincote (Cullingworth and Nadin 2006).

Others are located in Tyne and Wear, York, South and West Yorkshire, the North West, Stoke-on-Trent, Nottingham and Derby, the West Midlands, Cambridge, Gloucester and Cheltenham, Oxford, Avon, and South West Hampshire and South East Dorset. The green belts surrounding London, Birmingham and Sheffield were among the first to be established in the 1930s. Wales has one green belt, between Cardiff and Newport, while Scotland has six. Green belt in England is protected both by planning controls and against 'inappropriate development' within its boundaries.

There are numerous problems associated with green belts despite their popularity. The main difficulty with green belts at this time was the severe limits placed on urban growth. The debate on developing green belt land continues to rage in the late 2000s in the context of the predicted need in 2007 for an additional 3 million homes to be developed by 2020 (see Chapters 5, 9 and 11). It has been argued that this makes building on green belts, in England in particular, inevitable.

Figure 2.2 Green belts in the UK
Cullingworth and Nadin (2006, p. 235).

HOUSING REDEVELOPMENT

The third major group of policy initiatives of the period 1945 to 1965 encompassed the related processes of housing development, slum clearance and rebuilding. Public sector interventions that immediately followed the Second World War, which were allied with town and country planning developments, were aimed at redeveloping severely damaged urban areas. There was a major problem with housing stock in particular in the immediate post-war period. To illustrate the scale of the problem, between 1955 and 1974, around 1.2 million dwellings were demolished through slum clearance schemes involving the need to re-house 3.1 million people. This resulted in the breaking apart and displacement of working class communities, and most of these communities did not offer much in the way of protest in the face of the promise of rehousing in better quality premises. This process generally resulted in decreasing housing densities due to a desire to limit overcrowding and implement new space requirements for housing.

By 1960, space standards and urban containment policies had seen most major cities almost exhaust the supply of land in their areas. Therefore higher densities were promoted, which meant methods of building up rather than out were imposed. As discussed in Chapter 1, modernist philosophies strongly influenced post-war urban planning and policy. Modernist architectural styles, functional zoning, the development of inner city and peripheral tower blocks and estates, and new shopping centre development characterised this period (see Chapters 1 and 10). Ensuing problems are well known, and have included the physical and social unsustainability of modernist developments. The push for high-rise dwellings in particular brought about the well-documented series of social and architectural problems exemplified by the partial collapse of the Ronan Point tower block in Newham, East London in 1968 following a minor gas explosion (see www.open2.net/modernity; Mullins and Murie 2006, p. 32).

To summarise the sub-period 1945 to 1965, urban problems being experienced were lack of housing and poor quality of existing housing, urban blight, and urban sprawl and the need to control decentralisation. These were tackled through physical solutions of New Towns, green belts and house building.

THE URBAN PROGRAMME, COMMUNITY DEVELOPMENT PROJECTS AND INNER AREA STUDIES, 1965–1979

The 1965–1979 sub-period of the early years, which set the context for later urban policies, witnessed a shift away from the physical approach observed in the immediate post-war period. However, in the beginning, there was still a strong tie with the 'social pathology' perspective, which attributed the cause of residual poverty to the 'pathological' behaviour of the people or communities who remained in poverty. This theory directed attention away from systemic failures and structural economic inequalities to the more restricted issue of dealing with the actual individuals and groups living in poverty (Atkinson and Moon 1994a, p. 33).

Urban problems during this period were still dominated by the major issue of pockets of poverty despite earlier physical approaches. The gap between rich and poor had widened significantly, and large areas of inner city poverty were juxtaposed with the wealth of suburbia. An additional issue was that racial tensions were beginning to simmer. In looking for people to attribute blame for social ills and poverty, immigrants became an easy target. Rioting over poor conditions in slums added to a tense atmosphere as shown by Table 2.3. Three key urban policies between 1965 and 1979 that warrant attention are the Urban Programme, Community Development Projects and Inner Area Studies.

THE URBAN PROGRAMME

The Urban Programme marked the beginning of a distinct 'urban policy' in the UK (Johnstone and Whitehead 2004b), developing beyond town and country planning and housing initiatives. The policy began as 'Urban Aid' in 1968 and became the Urban Programme after the Urban Areas Act (1978) was passed. It was designed to ease social tensions in the inner city, and the Labour government needed to be seen to be actively addressing the urban problems of the time. The programme was designed to fill in the gaps in existing provision and typically funded social schemes related to educational or youth-oriented projects. This was later widened to take in voluntary organisations, and to cover industrial, environmental and recreational provision (Cullingworth and Nadin 2006). There

TABLE 2.3 MAJOR EVENTS AND EFFECTS, 1968–1979

Year	Event	Effect
1968	Enoch Powell's 'rivers of blood' speech	Highlights growing racial unease
1968	Collapse of Ronan Point tower block	Death knell for system-built high-rises, and a growing reaction against wholesale redevelopment and clearance of properties and communities
1973/1974	Oil prices quadruple and industrial unrest surfaces	The bubble bursts; the chronic underlying weaknesses of the UK economy exposed
1973/1974	Coal miners' strike	Further worsening of the energy crisis
1974	Three-day weeks and blackouts	Deepens the sense of crisis
1976	UK receives a loan from the International Monetary Fund	Causes further heavy cuts in public expenditure
1977	Urban White Paper published	Urban problems addressed
1978/1979	Labour strikes in hospitals, refuse collection and other essential services – 'Winter of Discontent'	Loss of public confidence in the Labour government
1979	Vote of no confidence brings down the government	First occurrence of this since 1924; end of 'social democracy' and the beginning of neo-liberalism

was an emphasis on experimentation, self-help and co-ordinating existing services. The Urban Programme was specifically area-based for 'special need'. As will become clear over successive policies, the Urban Programme was beleaguered by various faults. A difficulty was that there was no clear definition of what constituted deprivation and little direction was given by government about what the proposed projects should specifically address. Ultimately, the emphasis of the Urban Programme was on small-scale community projects and social schemes (see Atkinson and Moon 1994a; Cullingworth and Nadin 2006, p. 361; Chapter 8).

COMMUNITY DEVELOPMENT PROJECTS

Community Development Projects were a second key area-based policy and were a subset of the Urban Programme. These were action-based research projects that began in 1969 and focused on small area projects dedicated to spend limited financial resources on 'deviant' populations. The initial aim was to overcome the sense of disintegration and depersonalisation felt by residents of deprived areas. The first neighbourhood-based projects ran in Coventry, Liverpool and Southwark, and, in all, 12 were established, adding the industrial locations of Birmingham, Cumbria, Glamorgan, Newcastle, Newham, Oldham, Paisley, Tynemouth and West Yorkshire to the list.

Community Development Projects were team-based and combined an action team drawn from the local area and a research team taken from a local university. In theory the action team was to encourage people in the area to improve their own situation through 'pump-priming' local initiatives. The research team was to monitor progress and report back to central government. In practice, however, this arrangement did not work. The Home Office was keen to apply a strictly social pathological approach to any findings of the research. The researchers were far less keen on this, however, and instead attributed troubles to structural inequalities and encouraged local groups to come up with radical, often quasi-Marxist, solutions to local urban problems. As Cullingworth and Nadin (2006, p. 361) observed, the Community Development Projects produced 'a veritable spate of publications ranging from carefully researched analyses to neo-Marxist denunciations of the basic structural weakness of capitalist society'. Needless to say, central government did not appreciate this analysis, and the programme was wound up in 1978 after less than 10 years.

A positive legacy of the Community Development Projects was the promotion of a new way of looking at poverty. Rather than the social pathology approach, poverty and deprivation could be viewed as the result of changes within the economic, educational and housing markets allied to the weak bargaining position of the poor (Atkinson and Moon 1994a). Cullingworth and Nadin (2006) noted that in addition in 1974, a small number of Comprehensive Community Programmes were introduced in areas of intense urban deprivation. These also resulted in a plethora of studies.

INNER AREA STUDIES

A third urban policy of this era was the Inner Area Studies, set up by the Secretary of State for the Environment, Peter Walker in 1972. There were three Inner Area Studies in Liverpool (Toxteth), Birmingham (Small Heath) and London's Lambeth, as these were districts within cities that were known to suffer from multiple deprivation. Oldham, Rotherham and Sunderland were 'complete industrial towns' which attracted 'Making Towns Better' status, which were slightly different to the Inner Area Studies and focused more on corporate working and public service delivery. The Inner Area Studies were designed to bring a 'total approach' to the multiplicity of urban problems being experienced. They were organised jointly between the Department of the Environment (DoE) and local planning authorities, and they also involved private sector consultants. Despite the fact that these reports used consultants rather than academics as was the case for Community Development Projects, their analyses were extremely similar. The Inner Area Studies ascribed the root cause of deprivation to basic poverty. Low personal incomes were caused by inadequate social security benefits for disadvantaged groups, as well as by rising unemployment due to changes in local and national labour markets. In other words, there was a resounding rejection of the social pathology approach to understanding urban poverty. By the late 1970s, a number of problems were culminating and resulted in the need for new approaches to the urban problem. Table 2.3 illustrates the evolving nature of social and economic problems in the UK and their influences on policy from the late 1960s to the late 1970s.

THE 1970s FOCUS ON REFURBISHMENT OF PRIVATE SECTOR HOUSING

The 1970s also witnessed central government promotion of the refurbishment and renewal of private sector housing, in association with local authorities. Housing improvement grant legislation during the late 1960s and 1970s enabled the upgrading of deteriorating property, often resulting in the gentrification of housing areas. This can be seen as an example of central and local government using housing renovation to regenerate urban areas (see Chapter 11). The intention in the 1970s was to improve conditions for sitting tenants, but speculative owners, developers and the housing market saw the potential to transform the whole character and marketability of these areas. Large areas of housing were improved and sold at much-inflated prices, which had the effect of displacing former low-income tenants (Herbert 2000). Cullingworth and Nadin (2006, pp. 351–354) trace the course from clearance of inadequate housing towards renewal of housing to regenerate inner city communities in particular.

THE 1977 URBAN WHITE PAPER – 'A POLICY FOR THE INNER CITIES'

The Inner Area Studies fed into the 1977 Urban White Paper: 'A Policy for the Inner Cities' (DoE 1977), which in turn resulted in the Inner Urban Areas Act (1978). This emerged in the context of a severe and deepening economic crisis in the UK (Table 2.3). It represented significant changes in urban policy and was the first explicit 'Urban' White Paper, although being very much 'inner city' focused. After the publication of the Urban White Paper, the government's view of urban problems shifted (Johnstone and Whitehead 2004b). Urban decay and deprivation were now acknowledged to result from a number of factors including low economic well-being and the physical decay of urban areas, low pay and unemployment, population shifts from urban to rural areas, discrimination for mortgages and amenities for inner city residents, and race.

The remedy proposed by the Urban White Paper was fourfold. There was a need to seek economic improvement, to improve the physical environment, to improve social conditions, and to accomplish a new balance between population and jobs in the UK.

The Urban White Paper proposed to increase the funding of the Urban Programme from £30 million to £125 million per annum by 1979/1980. It also proposed giving additional powers to local authorities with severe inner city problems so that they could operate more effectively in the economic development of their areas. This was also the first time that central government acknowledged that local authorities were unable to address the plethora of urban problems themselves, and would therefore have to work in partnership with other groups including the private sector and local communities.

However, many of the proposed changes did not come to fruition, and therefore it is difficult to know whether the Urban Programme in its new late 1970s form would have worked. In the end, the Urban Programme limped along until it was subsumed under the Single Regeneration Budget in 1994 (see Chapter 4), and it never achieved or fulfilled the aims set out by the 1977 Urban White Paper. Later in its life, the Urban Programme increased its emphasis on economic development and became an important source of funding for thousands of projects and organisations. At its height, Cullingworth and Nadin (2006, p. 361) show that 10,000 projects in the 57 Urban Programme areas were funded each year. This cost the government £236 million in 1992/1993. Towards its latter years in the late 1980s and early 1990s, almost half of the Urban Programme's expenditure was allocated to economic objectives, and the remainder was shared roughly equally between social and environmental objectives.

Little explicitly resulted from the Urban White Paper, as its release came at a time when there was increasing speculation as to the role of government spending and administration in solving societal problems. This was because Margaret Thatcher's Conservative government came to power in 1979 in the UK, and the Conservatives were to reign for 18 years. The view of poverty and the role of government in resolving urban problems through urban policy changed once again.

The year 1979 represented a sea change in the approach to urban regeneration and urban policy in the UK. This was strongly linked to urban problems resulting from an economic crisis and the global crisis of the welfare state at the end of the 1970s as listed in Table 2.3.

The next chapter moves on to the era of entrepreneurialism and property-led regeneration approaches associated with Thatcherite ideologies, which characterised urban regeneration policies throughout the 1980s. It will give an overview of the period 1979 to 1991 and will present case study examples of some of the initiatives within this significant urban policy period.

SUMMARY

This chapter focused on the successive policy periods which stretched from the end of the Second World War in 1945 until 1979 when Margaret Thatcher's Conservative Party was elected.

The period between 1945 and 1965 was dominated by town and country planning and physical approaches to urban problems of housing shortages and poor quality of housing, and urban sprawl.

Key policies to ameliorate these problems were New Towns, green belts and housing redevelopment.

During the period 1965 to 1979, the problem of pockets of poverty was recognised, and racial tensions were growing.

Urban policies shifted to an area-based focus exemplified by the Urban Programme, Community Development Projects, Inner Area Studies and the refurbishment of private housing areas.

Explanation of the urban problem moved from social pathology of individuals and communities to wider structural economic changes.

The 1977 Urban White Paper took the structural approach, but its impact was cut short by the change in government in 1979, and the changing nature of social and economic problems.

STUDY QUESTIONS

1. Characterise and assess approaches to urban intervention in the period between 1945 and 1965.
2. What has been the legacy of New Towns and green belts in the UK?
3. Critically evaluate late 1960s and 1970s area-based social welfare policies. What can contemporary urban regeneration initiatives learn from this experience?

KEY READINGS ON THE EARLY YEARS OF URBAN REGENERATION

Atkinson and Moon (1994a) *Urban Policy in Britain: The City, the State and the Market* [Chapter 2 covers in greater detail the nature of post-war urban problems and critiques physical redevelopment policies]

Couch (1990) *Urban Renewal: Theory and Practice* [Chapter 1 covers post-war approaches to the reconstruction of urban areas]

Cullingworth and Nadin (2006) *Town and Country Planning in the UK* [see Chapter 2 for detail on the emergence and evolution of town and country planning]

Greed (2000) *Introducing Planning* [comprehensive background on the scope and nature of British town planning]

Pacione (2005) *Urban Geography: A Global Perspective* [Chapter 8 looks at the emergence and development of urban planning and policy, and Chapter 9 focuses on New Towns in the UK as well as in Europe, the Third World and the USA]

Rydin (1998) *Urban and Environmental Planning in the UK* [Chapters 1 and 2 feature the evolution of planning policies and post-war planning of the 1950s–1970s]

Ward (2004) *Planning and Urban Change* [Chapter 4 discusses the emergence of the post-war planning agenda, and Chapters 5 and 6 look at the planning system and policies from 1952–1974]

KEY WEBSITES ON THE EARLY YEARS OF URBAN REGENERATION

From Here to Modernity: www.open2.net/modernity [traces the history of modernism from the 1920s in terms of the built environment]

Online Planning Resources: http://planning.rudi.net [extensive and up-to-date bibliographies on numerous topics relating to planning and regeneration, many applicable to the early years of urban regeneration]

3 Entrepreneurial regeneration in the 1980s

INTRODUCTION

This chapter looks at the distinct approaches to urban regeneration undertaken by Margaret Thatcher's Conservative governments which reigned from 1979 to 1990. These were characterised by an 'entrepreneurial' ethos consisting of neo-liberal philosophies of public–private partnerships, privatisation, deregulation, liberalisation and centralisation. The chapter will outline the philosophies of the 'New Right', which underscored approaches to urban regeneration, and move on to evaluate key policies espousing these values. Examples from across the UK will be illustrated including arguably the most famous piece of urban regeneration in Europe, London's Docklands, which epitomised 1980s-style entrepreneurial regeneration.

THE ROLLING BACK OF THE STATE AND THE RISE OF MARKET-LED URBAN REGENERATION

As illustrated in Chapter 2, urban policy and regeneration initiatives that immediately followed the Second World War were associated with physical redevelopment. These were followed by area-based social welfare projects associated with the inner city problems in the late 1960s. The story continued in 1979 when Thatcher and the Conservatives swept to power after five years in opposition. The period, which has been identified as lasting from 1979 to 1991 (when a new set of urban policies began to be unveiled, a year after Thatcher's departure), was characterised by 'urban entrepreneurialism', which placed greater emphasis on the role of the private sector in urban policy, also termed privatism (see Barnes and Campbell 1988; Barnekov et al. 1989; Deakin and Edwards 1993; Bailey et al. 1995). This era was also characterised by the creation of business elites and growth coalitions (see Hall 2001, pp. 81–85). Growth coalitions involved both public and private sectors in partnership to mutual interest to promote and implement economic development strategies in cities (see Harding 1991; Pacione 2005, Chapter 7). From the early 1980s, the government based its range of urban policy initiatives on the belief that competitive and market economies could deliver equitable and efficient solutions to urban problems (Nevin et al. 1997). 'Privatised' partnerships and business representation on a range of national and local institutions characterised this period, as is clear when looking at key policies (Bailey et al. 1995).

The major shifts in urban policy in the early 1980s related primarily to varying government initiatives, but also to the tensions of priorities between regions, cities and more general issues such as unemployment and the need to promote economic growth, which were being recognised in the mid-to-late 1970s (Herbert 2000). Major social and economic issues had come together to create a sense of urgent need for change (see Table 2.3), and several other significant events during the period between the late 1970s and early 1990s also served to shape policy responses and approaches to urban regeneration (Table 3.1).

The centralisation of power represented the internal structures of the state, with a shift in local government towards local governance with the private sector taking a greater role in urban regeneration. Central government embarked on a confrontational approach to local government with rate capping, cutbacks, abolition and the implementation of quangos, which bypassed local authorities, all symbolising centralisation tendencies. Major agencies such as Urban Development Corporations (UDCs) were established with nominated members from a business background to drive change. Urban entrepreneurialism was promoted, which involved leaner and flatter managerial structures, generic roles, team working and flexibility.

The entrepreneurialism period of the 1980s represented a neo-liberal philosophy involving deregulation, liberalisation and privatisation. Social needs were largely subordinated to the needs of business. During this period of urban regeneration policy, emphasis was given to property-led initiatives and the creation of an entrepreneurial culture (see Hall and Hubbard 1996; 1998). Urban entrepreneurialism is argued to have succeeded urban managerialism during this period as the main form of governance of cities globally (see Chapter 6; Wood 1988; Harvey 1989; Short and Kim 1999).

It has been argued that the advent of the Thatcher government in 1979 'witnessed the fracture of three main pillars upon which post-Second World War social democratic policies were constructed – Fordism, Welfarism and Keynesianism' (Gaffikin and Warf 1993, p. 71). The reorientation of urban policy by the New Right Conservative government was part of a wider agenda to restructure the UK economically, socially, spatially and ideologically around a new consensus of the free market, individualism and a clear rejection of the post-war Keynesian welfare state (see Pacione 2005, p. 178). The New Right philosophy argued that the market was the most efficient means of ensuring the production and distribution of goods and services and challenged the post-war consensus (see Thornley 1991). The thrust of state policy shifted from welfare to enterprise where 'the aim has been to

TABLE 3.1 MAJOR EVENTS AND EFFECTS, 1979–1991

Year	*Event*	*Effect*
1980	Privatisation begins	Adds to manufacturing slump and unemployment
1981	Brixton riot	Urban areas had to be noticed and their problems addressed
1982	Unemployment and inflation rise	Unemployment and inflation double from 1979 figures
1984	Miners' strike	Thatcher sees off the unions
1989	Interest rate hikes and housing market crashes	Many property-led regeneration schemes suffer
1989	Poll tax introduced	The beginning of the end for Thatcher

reverse the post-war drift towards collectivism and creeping corporatism, to redefine the role and extent of state intervention in the economy, to curb the power of organised labour, and to release the natural, self-generative power of competitive market forces in order to revive private capitalism, economic growth and accumulation' (Martin 1988, p. 221).

Keynesian economics, which had dominated government thinking in the post-war period, argued that economic depression could be avoided if government was prepared to increase aggregate demand by deliberately raising public spending above the level of taxation revenue. During the 1980s the Keynesian commitment to the macro-economic goal of full employment was replaced by the objective of controlling inflation by means of restrictive monetary measures and supply-side flexibilisation (Pacione 2005, p. 78). Thatcherism became 'a doctrine for modernising the UK's economy by exposing its industries, its cities and its people to the rigours of international competition in the belief that this would promote the shift of resources out of inefficient "lame duck" traditional industries and processes into new, more flexible and competitive high-technology sectors, production methods and work practices' (Pacione 2005, p. 178). The principal mechanisms for achieving this transformation centred on tax cuts and deficit spending, deregulation and privatisation, all of which had geographically uneven impacts. At the urban level these three macro-economic strategies were combined most strikingly in the concepts of the Enterprise Zone and the Urban Development Corporation which this chapter later discusses.

As part of the broad political and economic agenda, urban policy was also used to restructure central government-local government relations. Pacione (2005, p. 178) showed that five processes characterised these changes:

1. Displacement involving the transfer of powers to non-elected agencies (such as UDCs), thereby bypassing the perceived bureaucracy and obstructiveness of local authorities.
2. Deregulation, involving a reduction in local authorities' planning controls to encourage property-led regeneration (EZs).
3. The encouragement of partnerships between central government and the private sector.
4. Privatisation, incorporating the 'contracting out' of selected local government services, housing tenure diversification, and provision for schools to 'opt out' of local education authority control.
5. Centralisation of powers through a range of quangos (quasi-autonomous non-governmental organisations) now termed NDPBs (non-department public bodies).

Each of these five changes had significant impacts on the formulation and implementation of urban regeneration policies.

TOP-DOWN URBAN REGENERATION POLICY AND PUBLIC–PRIVATE PARTNERSHIPS

Consistent with the rolling back of the carpet of the state and the rise of market-led, entrepreneurial approaches to urban regeneration, a top-down rather than bottom-up approach was pursued by central government. As illustrated in Chapter 2, urban interventions in the UK until the 1960s usually addressed solutions in physical terms. The 'rediscovery of poverty' in the UK's cities signalled an urban crisis and led to a refocusing of urban policy by the Urban White Paper (1977). The election of the Thatcher government represented a watershed in urban policy. Successive Conservative government involvement in urban regeneration shifted the policy emphasis from the public to the private sector. The main role for the public sector was to attract and accommodate the requirements of private investors without unduly influencing their development decisions. This perspective underlay the range of new initiatives, introduced by the 1980 Local Government Planning and Land Act, the most significant of which for urban policy were the Enterprise Zones and Urban Development Corporations.

Therefore the entrepreneurial thrust of urban policy during the 1980s was underwritten by co-operation between the government and the private sector. This approach was based on the US experience where public–private partnerships date back to the end of the Second World War (see Moore and Richardson 1986; Law 1988; Barnekov et al. 1989). In the US, following the Second World War, city governments faced with growing blight in downtown areas and attracted by federal funding from the urban renewal programme, joined forces with private developers in 'quasi-public' redevelopment corporations, which were able to avoid conventional procedures for municipal policy-making. By the 1960s these business–government partnerships had produced a range of downtown redevelopment projects in cities such as Pittsburgh, Baltimore (Levine 1987) and Minneapolis (see Pacione 2005, Chapter 16; Hall 2006, Chapter 6). In the 1970s and 1980s, driven by problems of continued de-industrialisation and economic decline, city governments moved beyond single-project collaboration with developers. In the context of heightened inter-city competition for private investment, city governments became entrepreneurial, providing extensive subsidies and incentives to attract developers, and often becoming co-developers in more risky development projects. Public–private partnerships became the cornerstone of economic development strategies of virtually all US cities and these strategies centred on the creation of a good climate for business (Short and Kim 1999).

This trend is exemplified by the 'Rouse-ification' of downtowns across the US with the production of 'festival marketplaces' in places such as New York (South Street Seaport) and Baltimore (Harborplace). The Rouse Company became the leading downtown developer in the country. There was an uneven distribution of these developments, with a strong contrast between the spectacularly redeveloped waterfront and festival areas, and the social and economic problems and poverty rate in non-white neighbourhoods (Pacione 2005, pp. 337–338; Hamnett 2005). Importing the US model, public–private partnerships came to be another key characteristic of Thatcherite urban regeneration policy during the 1980s (Harding 1990).

PROPERTY-LED REGENERATION

Linked with the trend towards top-down government and public–private partnerships was the focus on property-led regeneration. During the 1980s, property-based development played a prominent role in urban regeneration projects undertaken by public–private partnerships. Property development refers to the assembly of finance, land, building materials and labour to produce or improve buildings for occupation and investment purposes (Ambrose 1994). Post-Second World War governments have sometimes used property as an instrument of macro-economic management, boosting or dampening aggregate demand by pumping public resources into, or withdrawing them from, house-building, roads, railways, schools and other physical infrastructure (Pacione 2005). Property has been used extensively since the 1930s to promote regional development, mainly through public provision of advance factories, industrial estates and associated infrastructure (Healey and Ilbery 1990). During the 1980s property-led regeneration assumed a central place in urban policy, with the key role of the private sector demonstrated most visibly in flagship projects such as the redevelopment of Canary Wharf in London's Docklands (Fainstein 1994; Malone 1996; Pacione 2005).

However, property development alone is an insufficient basis for urban economic regeneration. Turok (1992) critiqued 1980s-style property-led regeneration and assessed whether the approach symbolised a 'panacea' or 'placebo'. Although property development and rehabilitation can improve a residential environment and construction projects can provide scope for employment, a property-led approach fails to consider the development of human resources, such as education and training, which have long-term effects on people's incomes and employment prospects; the underlying competitiveness of local production; and investment in infrastructure, such as transport and communications.

The value of property-led redevelopment is dependent on a locality's situation (Healey et al. 1992). Turok (1992) argued that a strong role for property-based measures may be appropriate where there are problems associated with land or building conditions or where shortages of floorspace restrict inward investment and indigenous growth. Uncontrolled property development may drive local property prices up, encourage land speculation and displace existing economic activities and communities unable to afford higher rents (Brownill 1990; see Chapter 11). The market's preference for short-term rather than long-term investments can divert finance capital into property and away from activities such as manufacturing and wider infrastructure development (Turok 1992).

Nowhere has the impact of property-led redevelopment been more visible than in the urban landscapes of London and New York (Pacione 2005, Chapter 16). In these global cities in the 1980s in particular, the growth of the financial services industry and related expansion of office space interacted to generate a major restructuring of urban space (see Merrifield 1993; Fainstein 1994; Church and Frost 1995). From the mid-1980s, London witnessed the largest office-building boom in its history as a result of the Conservative government opening up a once highly regulated property development arena for speculative ventures. This so-called 'big bang' was followed by an influx of foreign banks and other traders (Hamnett 2003a). Much of the new office development appeared in the traditional heart of the City known as the square mile, which sought to protect itself from a leakage of financial-sector activity to European rivals, as well as to the emerging London Docklands redevelopment with its flagship development of Canary Wharf, which happened to coincide with the office boom. The regeneration of London's Docklands has been part of a longer-term strategy to redevelop the Thames Gateway sub-region that continues in the late 2000s, partly in preparation for the 2012 London Olympics (see Chapters 5 and 12).

1980s URBAN POLICIES

The philosophies of Thatcher's Conservative governments of the 1980s are clearly reflected in the key initiatives that epitomised this policy period. Principal policies with their start dates are shown in Table 3.2, and were among a cast of dozens announced and established in the 1980s (see Pacione 2005, pp. 331–333). These characterised the entrepreneurial property-led development ethos of this urban policy period and central government's belief in the power of the private sector to carry out urban regeneration.

The policy approaches pursued were in contrast to those instigated in the previous policy periods explored in Chapter 2. Following the neglect of public housing during the inter-war depression years and the lack of housing investment since the 1920s, the publication of the Barlow Commission Report in 1940, and the destruction of many inner urban areas during the Second World War, inner cities emerged as a major policy issue as outlined in Chapters 1 and 2. In the post-war period, a series of policies and housing acts were concerned with the development of council housing estates, minimum standards and building of the New Towns associated with the 1947 Town and Country Planning Act. Policies were directed to the residential side of urban life during this period of urban policy.

From then until the 1970s,

> [U]rban policy under both Labour and Conservative administrations rested on two shared assumptions. Firstly, policy was designed as much to provide social and welfare

TABLE 3.2 PRINCIPAL CENTRAL GOVERNMENT URBAN POLICIES, 1979–1988

Policy	Year	Description
Enterprise Zones	(1981)	11 initially designated
Urban Development Corporations	(1981)	London's Docklands and Merseyside
Merseyside Task Force	(1981)	Toxteth
Urban Development Grant Programme	(1981)	Minimum public sector money to encourage private sector investment
Inner City Enterprise	(1981)	Property development company – inner city projects with private sector funding
Enterprise Zones	(1982)	Further designated zones, reaching a total of 25 by 1984
Derelict Land Grant	(1983)	Payable for reclaiming or improving land or for bringing land into productive use
Free Port Zones	(1984)	Six designated at ports and airports
City Action Teams	(1985)	Five agencies, spanning different government departments to aid inner city regeneration
Estate Action (Urban Renewal Unit)	(1985)	Funding directed at run-down local authority estates
Urban Development Corporations	(1987)	A further nine designated
Urban Regeneration Grants	(1987)	Central government finance for urban regeneration
Action for Cities	(1988)	Included new UDC for Sheffield, extension of Merseyside UDC, two new City Action Teams for Leeds and Nottingham, creation of Land Register, City Grant programme for private sector development

Source: **Healey and Ilbery (1990, p. 166); Pacione (2005, pp. 176–177).**

support services to the victims of economic change in the inner cities as to create wealth in those areas. Second, since disinvestment by the private sector was seen as the cause of many inner city problems, the public sector, including local authorities, was regarded as the natural mechanism to promote urban reconstruction (Parkinson and Evans 1990, p. 65).

Following the election of the Conservative government in 1979, this consensus was shattered:

> In the 1980s, however, the Conservative government has increasingly defined the public sector as the cause of inner cities' problems and the private sector as the solution . . . Indeed, 'regeneration' is now the government's goal, as wealth creation has replaced the distribution of welfare as the aim of urban policy. This privatisation has led to increasing emphasis upon efficiency as opposed to equity in government programmes. Equally important, the government has decreed that local authorities are incapable of leading the

recovery of their cities and is gradually stripping them of a wide range of their powers in finance, housing, education, social services, planning and economic development. (Parkinson and Evans 1990, pp. 65–66)

In essence, wealth creation replaced the distribution of welfare as the aim of urban policy and efficiency was favoured as opposed to equity.

Since the late 1970s and through the 1980s, a complex set of measures was introduced to deal with locally based issues, most notably in inner city areas. Healey and Ilbery (1990) and Pacione (2005) offered insights into the range and scale of legislation and policies introduced. Synthesising the research evidence, this chapter moves on to exploring the two flagship policies of the period: Enterprise Zones and Urban Development Corporations.

ENTERPRISE ZONES

One of the key entrepreneurial Thatcherite inner city urban policies was the Enterprise Zone policy. This policy can be viewed in the wider context of the transition to entrepreneurialism in urban policy, the emergence of public–private partnerships and property-led regeneration. Norcliffe and Hoare (1982) traced the genealogy of Enterprise Zones from 1969, when it was proposed that three large semi-rural regions in the East Midlands, Essex and the Solent be freed from planning controls and left to grow with market forces. In 1977, the geographer Peter Hall proposed the creation of Freeports as a refinement of the 1969 idea, based on the rapidly expanding economies of the Far East (see Banham et al. 1969). The argument was that as attempts to revive inner city areas had failed, a radical 'free-market' solution should be found, where planning, employment, welfare, pollution, health and safety, and taxation legislation should be significantly reduced. This aimed to create an entrepreneurial climate of enterprise, in which small businesses could thrive and which would attract overseas entrepreneurs.

In 1978, the 'Saving Our Cities' Conservative pamphlet advocated a Freeport approach, prior to the party being elected to office in 1979. Enterprise Zones were approved in the 1980 Local Government Planning and Land Act, but in the end were a watered-down version of Peter Hall's more radical ideas. The changes in policy approach by the government sought to bring fresh ideas and a new dynamism to declining urban localities. Comparable urban policies were established within a similar context in the US at this time (see Erikson and Friedman 1990a; 1990b).

The locational rationale for the designation of Enterprise Zones was that there should be a reasonable geographical spread among regions with different economic problems; that they should be located in areas with problems of marked economic decline and physical decay, with a shift in emphasis from regions to areas; that within areas, they were located in sites with both need and potential for redevelopment; and that local authorities must be willing to be involved in the initiative. Enterprise Zone policy was introduced in 1981 and early designations closely reflected the geography of de-industrialisation, with zones being usually concentrated in the inner cities of large urban areas with declining manufacturing economies (Butler 1982; Anderson 1983). Six sites were originally proposed for Enterprise Zones although by 1981 the number was expanded to 10 with no further additions envisaged at this time. The early zones included the Isle of Dogs in inner London, Gateshead in Newcastle-upon-Tyne and Clydebank in Scotland. However, in 1983/1984 another 14 were designated including towns such as Rotherham and Scunthorpe. In total, 25 had been designated by 1984 (Healey and Ilbery 1990). In total, 38 Enterprise Zones were designated and the programme was long lasting, eventually ending in 2006 (ODPM 2003d). Those that remained in existence in the mid-2000s were located in the Dearne Valley (South Yorkshire), East Durham, the East Midlands, Tyne Riverside and Lanarkshire in Scotland (ODPM 2003d).

Enterprise Zones were relatively small areas of land in which special incentives were provided to encourage firms to locate within them. Key were financial incentives, relaxed planning restrictions, and reduced government interference. The incentives were in place for 10 years following designation (Healey and Ilbery 1990, pp. 166–168). The policy instruments were:

1. 100 per cent tax allowances for capital expenditure on constructing, improving or extending commercial or industrial buildings;
2. exemption from Business Rates for industrial and commercial premises;
3. simplified planning procedures (planning matters dealt with in 14 days) if the development proposal complied with a published scheme, which effectively granted development planning permission;
4. exemption from industrial training levies;
5. faster processing of applications for firms requiring warehousing free of customs duties;
6. a reduction in government requests for statistical information.

Commercial as well as industrial development was encouraged resulting in mixed landscapes of offices, hotels and retailing in addition to factories (Bromley and Morgan 1985; Bromley and Rees 1988). One of the few dynamic sectors of the UK economy during the early 1980s was the small firms sector in fields such as computing and hi-tech manufacturing. It was these types of firms that Enterprise Zones were designed to attract, with the intention of establishing regional growth multipliers. These firms offered the potential for economic growth coupled with the diversification of the local economy (Healey and Ilbery 1990).

The success of Enterprise Zone policy has been the subject of some debate (see McDonald and Howick 1981; Massey 1982; Norcliffe and Hoare 1982; Anderson 1983; Hall 1984; Lloyd 1984; Bromley and Morgan 1985; Healey and Ilbery 1990). Despite some success, which varied somewhat between zones, Enterprise Zone policy also received considerable criticism.

A number of positive aspects have been recognised from various evaluations, which can act as transferable lessons for future urban regeneration (see Roger Tym and Partners 1982; DoE 1995c; ODPM 2003d). Enterprise Zones certainly created new jobs, an estimated 8000 between 1981 and 1984 in the original 11 Enterprise Zones, many of which were in new companies. 700 firms (start-ups and transfers) were established during this period. In the Swansea Enterprise Zone, early public sector activity led to an increase in employment in the zone and by 1990, 384 new firms had been created in the zone providing around 6000 additional jobs, of which 69 per cent were in the service sector (Bromley 1997). Physical regeneration of derelict land and buildings occurred and new investment was attracted to declining areas (Erikson and Syms 1986). The main successes in Swansea's Enterprise Zone have been the physical regeneration of the formerly blighted Lower Swansea Valley (Bromley and Humphrys 1979; Evans 1981), the social value of the redeveloped areas of public space, and the economic regeneration with the creation of new retail, distribution and other service industries (Bromley and Thomas 1988; Gasson 1992; Thomas and Bromley 1987; Thomas et al. 2004; Tallon et al. 2005).

However, the property-led, economic regeneration and private sector focus espoused by Enterprise Zones resulted in some negative aspects. Two economic concepts that help explain why Enterprise Zones did not achieve the level of success expected are those of displacement, which often saw companies relocating to take advantage of the Enterprise Zone but creating no extra jobs, and deadweight, which refers to investment that would have happened in any case without the Enterprise Zone incentives. 'Boundary' or 'shadow' effects were created around Enterprise Zones with negative effects being experienced by surrounding areas as they automatically became less attractive locations for investment. Early studies of the Swansea Enterprise Zone found that in its first five years of operation, most of the new employment resulted from relocation and closures elsewhere in the city-region and there was little extra job generation. In fact, 80 per cent of relocated firms had moved from elsewhere in Swansea (Bromley and Rees 1988). In the Clydeside Enterprise Zone, 66 per cent of the 1800 jobs created by 1984 came from short-distance transfers, with only 29 per cent from new start-ups and 5 per cent from firms already in the Enterprise Zone before designation (Healey and Ilbery 1990, p. 168). Between 1981 and 1983, 40 per cent of firms relocated from other areas to the original 11 Enterprise Zones, of which 86 per cent had moved from within the same county. Enterprise Zones failed to attract dynamic, innovative, small sunrise firms that generate growth; they simply attracted the traditional enterprises of their respective regions (Healey and Ilbery 1990; Hall 2001). Enterprise Zones therefore failed to diversify their local economies significantly as the types of firms attracted tended to mirror those already present in the surrounding area. In Swansea, for example, the majority of new employment was in the service sector, mainly in retail and distribution. These jobs were principally in smaller enterprises and not replacing the jobs lost in larger manufacturing enterprises or matching the skills of those unemployed in the area (Bromley and Morgan 1985; Bromley and Rees 1988). This corresponds with the experience of other Enterprise Zones across the UK such as in Manchester (Williams 1996). Additionally, the significant retailing function of the peripheral Swansea Enterprise Zone added to the competitive threat being experienced by the city centre (Sparks 1987; Thomas et al. 2004; Tallon et al. 2005).

Enterprise Zones did not generate the number of jobs anticipated, particularly in manufacturing, and their impacts were certainly very small compared to national unemployment levels, particularly those experienced during the 1980s. They were capital- rather than labour-intensive, providing part-time, low skilled, low wage employment. Instead the scheme favoured private sector entrepreneurs. Enterprise Zones were very expensive in job creation. For example, over £130 million was spent on Enterprise Zones between 1981 and 1983, which worked out at £17,000 per job (Healey and Ilbery 1990). A study of 23 Enterprise Zones by the National Audit Office noted that the zones had cost the Exchequer £431 million with only 10,500 permanent jobs and 2500 temporary jobs being created, meaning that the real cost per job was estimated at nearer £30,000. Enterprise Zones reduced the power of local authorities and increased central–local government conflict, representing another political space in which central government could attack local government and local services. Some Enterprise Zones were considered outright failures such as Belfast's Enterprise Zone, which failed to attract 100 jobs by 1984 (Hall 1984).

The combination of these negative issues led to the government announcing in 1987 that no further Enterprise Zones would be created, except in exceptional circumstances. Clearly, there were positive effects of physical regeneration and some new business and job creation. However, negative features have emerged from case study research including the lack of new jobs created, especially in the manufacturing sector, and the localised positive effects of the policy within the zone. Characteristic of the Thatcher era of urban regeneration, increased political interference from central government at the local level characterised Enterprise Zone policy. The legacy of the Enterprise Zone designations today continues to have significant impacts on cities that experienced this form of urban policy (ODPM 2003d).

URBAN DEVELOPMENT CORPORATIONS (UDCs)

Among the plethora of 1980s urban policies, UDCs were described as 'The initiative which most dramatically articulates the government's vision – the jewel in the crown of Mrs. Thatcher's urban strategy – is the creation of the Urban Development Corporations'

(Parkinson and Evans 1990, p. 66). This was the flagship of the Conservative government's urban regeneration programme in the 1980s and 1990s in terms of the amounts of money invested, the political and media attention received, and the extent to which they embodied the ideologies of the New Right (Anderson 1990).

Prestwich and Taylor (1990, p. 239) wrote:

> At an early stage in the life of the Conservative government of 1979 a new radical approach to town planning was revealed by Michael Heseltine . . . he was critical of the hierarchy of planning, planners and plans which was costly to maintain . . . planning had become inflexible and unresponsive to an extent that it was damaging economic development . . . what was required . . . was less planning at local authority level . . . the philosophy soon became reality with the introduction of Enterprise Zones and the two Dockland Development Corporations.

UDCs originated with the 1980 Local Government Planning and Land Act and were introduced by Michael Heseltine when he had responsibility for the Department of the Environment (as were Enterprise Zones, again to emphasise their distinctiveness from regional economic policy which was the responsibility of the Department of Trade and Industry). Under their terms of reference, the primary objective of a UDC was to secure the regeneration of its area by bringing land and buildings into effective use, encouraging the development of existing and new industry and commerce, creating an attractive environment, and ensuring that housing and social facilities were available to encourage people to live and work in the area (Oc and Tiesdell 1992; Imrie and Thomas 1999b).

UDC policy, like that of the Enterprise Zones, was focused on the local as opposed to regional scale and especially at those cities, scattered across the UK, where there were particularly local urban problems of severe de-industrialisation, economic restructuring, environmental degradation, but also of economic potential. UDCs had a relatively short but highly controversial history.

UDCs were government agencies, or quangos, implanted directly upon designated areas and were responsible for the regeneration of these areas. They were run by appointed boards consisting largely of representatives from the local business community and typically exhibited little representation from the local resident community. They aimed to encourage the private sector back to run-down inner city areas through a market-oriented and property-led approach. In order to achieve this area regeneration objective, UDCs were given substantial powers to acquire, hold, manage, reclaim and dispose of land and other property; carry out building and other operations; enhance the environment; seek to ensure the provision of water, electricity, gas, sewerage and other services; provide a transport infrastructure; carry out any business or undertaking for the purposes of the objective of urban regeneration; and provide financial incentives for the private sector (Imrie and Thomas 1999b).

The UDCs had the power to purchase land by agreement, to 'vest' it from public sector bodies, and/or to compulsory purchase it from private sector landowners (Imrie and Thomas 1999b). Funding for UDCs came from two sources: first, central government, and, second, by the proceeds from the disposal of development land. A total of 13 UDCs were designated during the 1980s and early 1990s (see Figure 3.1; Table 3.3) and the designations, like Enterprise Zones, were intended to last for relatively short periods of between five and 15 years.

The first two were designated in 1981 in London and Liverpool/Merseyside. Five more were designated in 1987 in England in the Black Country, Trafford Park (Greater Manchester), Teesside, Tyne and Wear and, in Wales, at Cardiff Bay. In 1988/1989 another four were designated in Bristol, Leeds, Central Manchester and Sheffield, while two more were designated in 1992/1993 in Birmingham and Plymouth. By 1997, Bristol, Leeds, Central Manchester and Sheffield had all wound up, and Cardiff Bay finished in 2000 (see Table 3.3).

Figure 3.1 Urban Development Corporations
Imrie and Thomas (1999b)

After being wound up, the intention was that the economic mechanisms they helped to implant would continue the long-term regeneration of the area.

No formal evaluation was undertaken by the DoE (or successive departments) into the success or otherwise of the Urban Development Corporations. This is because, by the early 1990s:

> UDCs are no longer the flavour of the month in central government; that mantle has passed to the new TECs (Training and Enterprise Councils) which, unlike the UDCs, cover the whole country. The National Audit Office and the Public Accounts Committee criticised the UDCs for adding to the complexity of public sector bodies and schemes of finance involved in inner city regeneration. From a Treasury perspective, they appear able to consume limitless capital resources to produce returns, which are at best uncertain (Coulson 1990, p. 301).

Surveys of the progress of the UDCs were undertaken in the late 1980s and early 1990s (see Stoker 1989; Potter 1990). In addition, ten of the corporations, including London Docklands, Manchester, Liverpool and Sheffield, are discussed and evaluated in one or both editions of Imrie and Thomas (1993a; 1999a), the latter version of which offers a slightly less critical re-evaluation. Among an array of pieces, other useful evaluative case studies focus on Merseyside (Wray 1987; Parkinson 1988; Adcock 1994) and Teesside (Sadler 1990).

TABLE 3.3 URBAN DEVELOPMENT CORPORATIONS IN ENGLAND

			Annual gross expenditure (£m)			*Cumulative outputs*	
		Year	*1992/ 1993*	*1995/ 1996*	*Jobs created*	*Land reclaimed (ha)*	*Private investment (£m)*
First generation	London Docklands	1981	293.9	129.9	63,025	709	6,084
	Merseyside	1981	42.1	34	14,458	342	394
Second generation	Trafford Park	1987	61.3	29.8	16,197	142	915
	Black Country	1987	68	43.5	13,357	256	690
	Teesside	1987	34.5	52.8	7,682	356	837
	Tyne and Wear	1987	50.2	44.9	19,649	456	738
Third generation	Central Manchester	1988	20.5	15.5	4,909	33	345
	Leeds	1988	9.6	0	9,066	68	357
	Sheffield	1988	15.9	16.4	11,342	235	553
	Bristol	1989	20.4	8.7	4,250	56	200
Fourth generation	Birmingham Heartlands	1992	5	12.2	1,773	54	107
	Plymouth	1993	0	10.5	8	6	0
Total		12	621.4	398.2	165,716	2,713	11,240

Source: **Cullingworth and Nadin (2006, p. 365).**

Drawing upon this evidence, a brief flavour of a few of the UDCs illustrates some of their successes and failures. The Tyne and Wear UDC (1987–1998) is arguably one of the more positive examples in terms of the transformation of the built environment. Here a 'can-do' attitude resulted in a string of office and industrial parks – including the Newcastle Business Park and the emerging edge-of-city Quayside – as well as other developments, which all owe their existence to the quango, although some of the wider criticisms resonated (see Byrne 1999).

The Cardiff Bay Development Corporation started in 1987 and had a number of objectives, which were to regenerate the areas of south Cardiff and Penarth Docks and establish Cardiff internationally as a 'maritime city'; reunite the city and its waterfront; create an environment in which people want to live, work and play; and ensure a mix of development, reflecting the hopes and aspirations of the local community (see Thomas and Imrie 1993; 1999; Rowley 1994). Over a period of 15 to 20 years, the UDC aimed to act as the catalyst for the transformation of 2700 acres of land into 6000 new homes, 4 million square feet of commerce, 5 million square feet of industry, providing 30,000 new jobs. There has, however, been much controversy about and opposition to the development of Cardiff Bay (see Brookes 1989; Walton 1990; Rowley 1994; Imrie and Thomas 1995), which tend to resonate with the wider criticisms of the policy. Top among these criticisms was the reliance on property-led regeneration and mega-projects, the high-end nature of the housing developed, the lack of community involvement and natural environmental concerns (Thomas and Imrie 1993; 1999; Rowley 1994). Cardiff Bay continues to develop and by the late 2000s housed buildings including the National Assembly for Wales and the Wales Millennium Centre (see Punter 2007; Hooper and Punter 2007; www.cardiffbay.co.uk).

CASE STUDY: BRISTOL UDC

Bristol UDC represented an example of a UDC that exhibited the ethos of 1980s-style urban regeneration and demonstrated many of the controversial aspects of the entrepreneurial, market-based approach. Bristol's UDC was designated on 7 December 1987 by the Secretary of State for Environment. Despite being in one of the fastest growing regions of the UK, which enjoyed growth in sectors like insurance, banking and finance, medical services, and hotels and catering, a number of economic problems had begun to emerge in Bristol during the late 1980s (Boddy et al. 1986; Bassett 2001; Tallon 2007a). The main problems revolved around job losses in the manufacturing sector including paper and packaging, food, drink and tobacco, and defence industries. These had previously been major elements of the regional economy (see Oatley 1993; Oatley and May 1999; Tallon 2007a).

In 1986, the DoE recognised four problems that the Bristol UDC was later to tackle (Oatley 1993): growing polarisation in the labour market, spatial polarisation in the city and the development of areas of high deprivation, persistent problems of social order, and tensions between a Labour-controlled local authority and the Conservative central government. Specific barriers to urban regeneration that the UDC was charged with addressing included problems with derelict land and premises, an ageing and inadequate road system, land contamination, and fragmented patterns of land ownership.

The designation of the UDC drew immediate criticism from the local authority, which was concerned about five issues (Oatley 1993; Hall 2001): outside intrusion into the key areas of planning and development; the use of public funding to subsidise private development; the lack of the UDC's local accountability; potential clashes with existing local government policies of urban regeneration; and the definition of regeneration within the designated area – the local authority felt that the UDC was more concerned with the regeneration of the local property market than with the social needs of the area.

The result of this combination of criticisms was that the City Council challenged the urban problems that the DoE had originally highlighted and argued that there was no justification for the UDC. This challenge forced a House of Lords inquiry and while finding that the City Council had cause for complaint, it did eventually endorse the designation (Oatley 1993; Oatley and May 1999). The inquiry made a number of recommendations: four areas within the initial designation should be excluded from the UDC; the UDC should contain objectives linked to social as well as physical and economic regeneration; it should not disrupt or displace existing employment; and that the Bristol UDC should not replicate the mistakes made by the London Docklands Development Corporation (see Brownill 1990; 1999).

Despite these recommendations, Bristol Development Corporation stressed the importance of the market in generating new economic activity in their plans, consistent with 1980s-style urban regeneration. The strategy was based upon the transformation of Bristol's image through prestigious architectural projects and marketing campaigns. The proposals included the displacement of 300 jobs from the area while there was also a poorly defined relocation policy (Hall 2001, pp. 74–75). The proposals favoured physical rather than social regeneration, and there were no guaranteed mechanisms outlined to direct employment benefits towards local people (Oatley and May 1999). These proposals led to fears from the local authority, local businesses and local people over disruption, displacement and contravention of local plans. A polarisation of opinion emerged between local interests who wished to maintain and build upon the local economic structure and the UDC who believed that restructuring the local economy was the key to regeneration.

The fears of the local interests were not allayed by the UDC's processes of consultation. Consultation with the local authority and communities was unprepared and informal, and proved to be more an exercise in persuasion than a mechanism to facilitate genuine participation in the development process (see Chapter 8). The consultation resulted in no substantial response from the UDC over the fears raised by the local people or the local authority. The UDC decided that no changes to their proposals were needed despite a number of specific concerns being raised. The UDC seemed more willing to consult with local businesses and land and property interests than with the local people and local authority. These consultation sessions were aimed more at winning over support than involving local interests in development (Oatley 1993; Oatley and May 1999).

By 1992, little had been achieved, many sites remained derelict, and the process had become bogged down in a number of costly and lengthy public inquiries. The recession of the late 1980s particularly affected the property market, demonstrating the limitations of relying on market forces and private capital to regenerate local economies. The lack of alternative strategies precluded any effective response to the prevailing negative economic conditions (Oatley and May 1999). So Bristol's UDC is regarded as one of the least successful, one that disappeared without managing to agree a development plan for its flagship city centre site, Quay Point, let alone attract investors and occupiers to the area (Plate 3.1). Bristol City Council and the Bristol UDC were in conflict from the outset. While the Bristol UDC dithered, the market, then less restrained by planning constraints, chose easier ground outside the city as witnessed in the steady rise of north Bristol as an autonomous office location, and the construction of a regional shopping centre, The Mall at Cribbs Causeway,

Plate 3.1 Temple Quay office quarter developed after the failed Urban Development Corporation
Photograph by author.

which opened in 1998 (Tallon 2007a; see Chapter 10). Overall, the Bristol UDC achieved very little and caused a great amount of hostility and damage.

Drawing upon the above examples and research on the UDCs more generally, a number of positive results can be drawn from the 1980s/1990s UDC experience. There is little doubt that a single-purpose body concentrating on a narrowly defined area, unencumbered by the diverse range of local authority responsibilities, and negating local democratic accountability, can achieve things very quickly by virtue of single-mindedness. UDC areas benefited from investment from central government and levered in European as well as private sector funds, resulting in new employment and environmental improvements. Such schemes helped in changing public attitudes and perceptions of UDC areas which have been largely positive. It is argued by their proponents that urban regeneration stimulates further private investment in the areas, which arguably would not otherwise have been attracted to the area.

However, UDCs have been subject to widespread and deep criticism, particularly from academics. Paradoxically, a single-purpose and single-minded organisation may also limit urban regeneration. In accordance with the government's philosophy that wealth creation was the goal and trickle-down would distribute that wealth to those in need, most UDC policies, certainly in their earlier years, concentrated purely on physical regeneration with little regard to human social provision or to the development of human capital, including low-income housing, community facilities, and education and retraining programmes. Despite massive expenditure, very few new jobs were actually generated. Most economic activity was centred on property speculation, and inward investment attracted was small compared with the public money invested in the UDCs. Significant amounts of money were lost in land deals (Imrie and Thomas 1999b, pp. 18–19). Six UDCs incurred liabilities at a time of falling land values, while committed to capital projects with few capital receipts to offset their costs. For example, Imrie and Thomas (1999b, p. 19) detailed that Black Country Development Corporation showed a loss on property deals between its inception and 1992 of £26.5 million, and likewise Bristol lost £12.9 million and Trafford Park lost £10 million.

Perhaps most controversially, UDCs were appointed and not democratically elected bodies. In a number of cases such as in London Docklands, Liverpool and Bristol, there were serious breakdowns in communication between the UDCs and local authorities (Brownill 1999; Meegan 1999; Oatley and May 1999). To achieve a more broadly based regeneration, UDCs had to work more closely with local authorities. The two early UDCs certainly found this difficult due to the prevailing political ideology. However, many of the arguably more successful second generation UDCs worked much more closely with local authorities and placed urban regeneration within an overall strategic economic planning framework of the local authority areas in which they were situated. More recent news is that a new generation of UDCs was announced in 2003 (see Chapter 5).

CASE STUDY: ENTREPRENEURIAL MEGA-REGENERATION IN LONDON'S DOCKLANDS

The most famous example of urban entrepreneurialism in action in the UK in the 1980s, which was initially driven by an Enterprise Zone and a UDC, was the mega-regeneration of London's Docklands. This provides one of the most interesting case studies of recent urban regeneration and a microcosm of the issues, debates and conflicts surrounding it (Brownill 1990; 1993; 1999; Ogden 1992; Foster 1999). The prevailing context of Thatcherite neoliberal philosophy illustrated earlier sets the backdrop for this urban regeneration.

Historically a marshland area, London's Docklands is centred on Isle of Dogs and stretches around five miles from the City and Southwark in the west to Newham and Greenwich in the east. From the sixteenth century onwards the port of London was the key to the city's wealth. The 'legal quays' between London Bridge and the Tower of London, would be crowded with 1400 vessels that had to wait for up to six weeks to be unloaded. To relieve such congestion, which worsened with increased trade from the empire, from 1802 onwards the largest dock system in the world underwent construction. However, from the mid-nineteenth century, competition from the railways began to erode the river traffic, and with the subsequent development of container ships and the emergence of the port at Tilbury in the 1960s, the city docks began to close. During the 1960s and 1970s the docks and surrounding area went into steep decline and in 1981 the last, Royal Docks, closed. Much of the area was derelict and the employment market collapsed (Ogden 1992; see www.royaldockstrust.org.uk).

For almost two decades the quaysides and surrounding areas were a wasteland with high unemployment and out-migration, until the London Docklands Development Corporation (LDDC) was set up as one of the two first generation UDCs to be designated in 1981. At this time the LDDC planned a resident population of over 100,000 and a working population twice that by the end of the millennium. The LDDC's area comprised 2100 ha (5200 acres), taking in portions of the east London boroughs of Tower Hamlets, Newham and Southwark, which comprised a population of around 40,000 in 1981 (Pacione 2005, p. 334).

Oc and Tiesdell (1992) illustrated the LDDC's powers:

- landowner: compulsory purchase powers and acquisition through vesting orders;
- planning control authority: but not the statutory plan-making authority;
- initial developer: making sites viable for subsequent development, mainly private sector;
- manager: of the Isle of Dogs Enterprise Zone, established in 1982.

The LDDC was not an education, health, housing or social services authority. Its objectives were to redevelop its area by investment in reclamation and infrastructure, together with business and community support for residential and commercial development. By co-operating with other authorities, the aim was to make the area an increasingly pleasant and rewarding place in which to live, work and play. From its inception until its eventual wind-up date 17 years later, London's Docklands proved to be a highly controversial example of urban regeneration, polarising opinion between those who pointed to a blueprint for successful free-market inner city regeneration comprising commercial, office and housing development, and those who viewed it as the architectural embodiment of Thatcherism including fortified capitalist enclaves. The regeneration of London Docklands continues to provoke such extreme reactions, which can be briefly weighed up by looking at the property-led regeneration of Canary Wharf, place marketing and image transformation, flagship developments, gentrification, and exclusion.

As Pacione (2001, p. 321) illustrated, in terms of the rise of property-led regeneration, the proposals for a large-scale office development on Canary Wharf emerged from a perceived need for London to continue to attract its share of internationally mobile finance capital during the second half of the 1980s. In contrast with responses that involved extensions to the existing limits of the City, Canary Wharf was located three to four miles to the east with advantages of location in an Enterprise Zone within the LDDC area. Strong central government support ensured the initiation of the project in 1985.

In 1987, the Canadian developer Olympia and York took over from the original consortium and was committed to developing at least 4.6 million square feet of floorspace by 1992, followed by an additional 7 million square feet. By 1992, 4 million square feet, including the tallest office building in Europe, was completed, but only 57 per cent of space had been let and only 14 per cent was occupied (Pacione 2001, p. 321). At the same time Olympia and York collapsed due to debts of over $11 billion, and the project was placed in administration. Pacione (2001) argued that this resulted in the world's largest real estate failure and demonstrated the limits of property development as a regeneration strategy (see Imrie and Thomas 1993b). Additionally, the slump in the property market and recession in the early 1990s slowed down the urban regeneration process in London Docklands. These events effectively symbolised the end of this period of urban development (Hall 2006). Further regeneration in London Docklands was stimulated following much-needed public sector investment in the transport infrastructure. Renewed private sector investor confidence in the post-recession period subsequently helped the Canary Wharf project regain solvency by the late 1990s (Pacione 2001, p. 321).

The physical regeneration of London's Docklands was underpinned by a series of concerted efforts by both the LDDC and private developers to market and promote a number of complementary positive images for the area (Crilley 1992a; 1993). Docklands was promoted as a place to invest; a place for businesses to relocate; a place for the affluent to work, live and play; and a major tourist destination. The efforts to transform the image of the area were wide-ranging and encompassed high-profile media advertising campaigns in the 1980s and 1990s, the careful development of the built environment and a number of arts-related initiatives (see Hall 2001, p. 138).

Although the legacy of the industrial past is an important feature of the urban landscape of London Docklands, the most prominent landscape element to emerge in 1980s urban regeneration projects was the spectacular or flagship development (Hall 2001, pp. 96–99). These are eye-catching, decorative, spectacular, innovative, and post-modern in architectural style (Tables 1.2 and 12.8; Plate 3.2).

Spectacular and flagship developments are epitomised by Canada Tower at Canary Wharf, and similar international developments are visible at Battery Park City in New York and La

Plate 3.2 Spectacular developments on the Isle of Dogs, London Docklands
Photograph by author.

Défense in Paris. Bianchini et al. (1992), Crilley (1993) and Hubbard (1996a) among others have argued that iconic flagship developments such as in London's Docklands can transform the image, identity and fortunes of a city.

Gentrification was a process promoted by the urban regeneration of London Docklands, and represented one of the more contentious outcomes of central and local government urban regeneration policies which operated in London Docklands during the 1980s (Brownill 1990; 1999). The process which is explored in Chapter 11, was stimulated by the refurbishment of old industrial space to provide studio space for artists and cultural industries, as a way of economically revitalising and diversifying inner-city areas. Subsequently wealthier middle-class groups entered the attractive and regenerated space. Impacts on existing working-class residents included social disharmony as new groups enter community space, and the displacement of working-class groups as house prices rise in the wake of gentrification. Impacts on local communities have been striking in the Urban Development Corporation-led gentrification of London's Docklands (Brownill 1990; 1993; 1999; Smith 1991; Rose 1992).

Waterside and luxury housing developments in London Docklands have been replicated around the UK over the last two decades and aimed at luring wealthy professionals to raise the social and economic trajectory of the locality (Hall 2001, pp. 59–60). This accommodation does little to solve housing crises among local poor populations who are unable to take advantage of these developments as illustrated by Hall (2006, pp. 144–145). First, new housing developed in the late 1980s housing boom was far too expensive for the majority of local households whose combined annual incomes were less than £10,000 in a market where two-bedroom flats were selling for £200,000 and over 80 per cent of the housing was for owner-occupation (Hall 2001, p. 159). Second, the types of properties provided in these developments were aimed primarily at double-income professional couples with no children in their 20s and 30s. Consequently this accommodation primarily consisted of small flats. The accommodation needs of the majority of local family and elderly households were usually very different from those of the young professional, childless couples.

The UDC effort focused on private sector housing construction, while Hall and Ogden (1992) reported that between 1981 and 1984 the waiting list for council housing in the borough of Newham rose from 2650 to 9112. In the London Docklands area as a whole, during the 1980s the proportion of owner-occupied housing rose from 5 per cent to 36 per cent, while local authority housing tenures fell from 83 per cent to 44 per cent (Crilley 1992b).

The conflict between economic and social goals was particularly fierce in the London Docklands, where the strategic plans prepared by the local authorities employed a 'needs-based' approach which emphasised the need to stem existing job losses, to attract new jobs that would match residents' skills, to use the vacant land to address the acute housing problems of east London, and to improve the general environment (Ogden 1992). Central government favoured a demand-led approach with the emphasis on creating a new economy attractive to firms and prospective residents from outside the area consistent with the Thatcherite agenda. In practice, the LDDC switched the planning emphasis from attempting to provide manufacturing jobs, towards office and warehouse schemes and retail complexes, 'turning the East End into the West End' (Brownill 1999).

A final arena of controversy driven by the 1980s UDC-led regeneration of London Docklands related to excluded cultures and local protest (see Foster 1999). Rose (1992) pointed out that opposition to the cultures promoted by urban regeneration in London Docklands were widespread and emanated from community groups such as The Docklands Forum. These were related to the types of housing and jobs created in the London Docklands and the exclusion of working-class communities from urban regeneration decision-making processes. Locals have regularly voiced fears about being physically displaced by extensive developments or socially displaced by wealthier incoming populations associated with the gentrification process (Brownill 1990; Lees et al. 2008).

The urban regeneration of London's Docklands and the wider Thames Gateway has continued apace through the 1990s and 2000s, in some cases responding to the concerns emerging from the 1980s experience, but in other ways continuing and reinforcing processes of property-led regeneration, gentrification and exclusion (see Chapters 5 and 11).

CENTRAL GOVERNMENT HOUSING REGENERATION POLICIES IN THE 1980s

The focus of housing regeneration from 1979 was on public sector estates of cities. In 1979, the DoE established the Priority Estates Project to look at ways of improving council estates (Cullingworth and Nadin 2006). As noted in Chapter 2, council estates had become socially, economically and physically problematic. In recognition of the mounting evidence of the problem, central government attempted to direct measures at public sector housing estates through the creation of the Urban Renewal Unit in 1985, later renamed the Estate Action programme. This comprised finance directed at the physical, social and economic improvement of run-down council estates, with measures including transfers of ownership and management to tenants, sales of estates to private developers, and sales of empty properties (see Lambert and Malpass 1998; Cullingworth and Nadin 2006, pp. 355–356). Lambert and Malpass (1998) argued that Estate Action was in part a result of the narrowing of the socio-economic profile of council tenants, related to the residualisation of the sector and growing segregation within the sector.

Housing regeneration policies such as Estate Action began the trend towards the broader urban regeneration approach to estate renewal, rather than housing improvement. However, Cullingworth and Nadin (2006) noted that although £2 billion was spent over an eight-year period, the results of many Estate Action projects were disappointing in terms of their aims. A key difficulty was achieving social and economic improvement through mainly physical measures. However, the policy was an early example of the trend towards competitive bidding for urban regeneration monies, and in addition reflected the emergence of mixed tenure estates as a government policy objective (see Chapters 4 and 11). The final Estate Action schemes were approved in 1995, and in total the scheme cost almost £3 billion, spent on 317 schemes involving the improvement of 490,000 dwellings and the transfer of 93,000 dwellings (Kintrea and Morgan 2005; Cullingworth and Nadin 2006).

Similarly, Housing Action Trusts (HATs) were announced by the government in 1987 to tackle the management and renewal of badly run-down housing estates. These were seen as the housing equivalent of UDCs in that they were quangos with responsibility for improving the physical, social and economic conditions of estates (see Lambert and Malpass 1998, pp. 101–102; Cullingworth and Nadin 2006, pp. 356–357).

THE LEGACY OF ENTREPRENEURIALISM: CRITIQUING 1980s-STYLE REGENERATION

A critique of urban policy at the end of the 1980s revealed a number of problems facing urban regeneration in the context of ongoing economic, social and racial tensions in many inner cities. In 1989, the Audit Commission published a report, *Urban Regeneration and Economic Development* (Audit Commission 1989), which provided a review of urban policy since 1979. The report was critical of a number of aspects of Thatcher's 1980s-style urban policy (Hall 2001, pp. 72–73). Several principal commentators were also painting pessimistic pictures of the state and the effectiveness of urban policy. Stewart (1987) described inner city policy as a charade and Robson (1994) concluded that the impacts of urban policies in the 1980s were at best modest. Similarly, Robson (1988), Parkinson (1989) and Robson et al. (1994) argued that the weaknesses of urban policy included a lack of national co-ordination, demonstrated by the withdrawal of rate support funds with the effect of counteracting urban aid policies; conflict between central and local government exemplified by nominated bodies usurping the role of elected representatives; and a serious imbalance between economic wealth creation objectives and any social considerations, linked with the enterprise culture. Robson et al.'s (1994) report *Assessing the Impact of Urban Policy* comprised a detailed evaluation of the Conservative's Action for Cities package which included the Inner City Programme, and heralded the shift to the more partnership-based approach to urban regeneration which emerged in the 1990s (see Chapter 4). This research found that urban policy had a measurable positive impact on the economic regeneration of targeted areas, and that the most effective regeneration schemes emerged where there was a local consensus enabling effective partnership. However, major problems with urban policy presented by Robson et al. (1994) included the confusing multiplicity of urban grant programmes and their excessive bureaucracy, the inadequate co-ordination of policy within and between central and local government, and the lack of attention to social problems. By the end of the 1980s and the start of the 1990s, major social and economic problems continued to affect society, economy and the built environment (Table 3.1).

Oatley (1995, pp. 262–265) neatly summarised the principal criticisms of urban regeneration policy apparent in the late 1980s in five categories:

1. The definition of the urban problem and the scale of the response

As discussed in Chapter 1, there have been various explanations of the urban problem since the late 1960s, which have been rooted in different theoretical perspectives on the problem. These have resulted in conflicts over what urban policy aims should be, and how best to achieve them. Successive governments have approached the problem in area-based geographical terms as a problem of place (Oatley 1995). This approach was criticised, as urban policy was based on the idea that concentrations of problems in particular parts of cities can be combated by small area-based initiatives. Oatley (1995) and Stewart (1990) argued that this was an erroneous conception as the focus is on symptoms rather than causes, which are to be found in the processes leading to de-industrialisation, unemployment, poverty, social exclusion and growing inequalities (see Chapters 1 and 5). It was argued that urban policy in the 1990s should focus on understandings and explanations as well as the effects of economic and social changes (Stewart 1990).

The scale and nature of government responses in the 1980s were also criticised relative to the problems being faced by cities (Oatley 1995, p. 263), and were minuscule when compared with mainstream government spending. In fact, while new urban policies were being implemented, funds were simultaneously being cut from other city budgets. For example, Herbert (2000, p. 209) reported that the DoE housing budget fell from £4.5 to £1.9 billion between 1981 and 1987; seven inner-city partnerships lost almost £850 million in revenue support finds; and expenditure on social and community programmes also fell significantly.

2. The fragmentation of policy and the lack of co-ordination

The most serious criticism of government urban policy in the 1980s, as with previous and successive periods, was its lack of co-ordination and confusing nature (Audit Commission 1989; Oatley 1995; Johnstone and Whitehead 2004b). There were a plethora of government departments responsible for urban policy and it had developed into 'a patchwork quilt of complexity and idiosyncrasy with few resources to match the scale of the attendant problems' (Audit Commission 1989, p. 4). Robson (1994) highlighted the example of Manchester's plethora of urban policies, which comprised six Urban Programmes, two Enterprise

Zones, two Task Forces, two Urban Development Corporations, two Safer City projects and a City Action Team.

3. The lack of a long-term strategic approach

Related to the lack of co-ordination and coherence of urban policy in the 1980s was the view that it also lacked long-term strategic objectives. Policies were seen as pragmatic and programmatic rather than demonstrating strategic thinking relating to changing cities and city-regions (Stewart 1987; 1990). Policies were also short-term, relative to the long-standing nature and depth of the problems being addressed.

4. The over-reliance on property-led regeneration

The mode of urban regeneration in the 1980s relied heavily on policy instruments that aimed to attract the private sector to invest in property in run-down urban areas on the assumption that this would result in local economic regeneration and that the benefits would trickle down to deprived communities (Healey et al. 1992). However, research revealed that a property-led approach does not guarantee this (Healey 1991; Turok 1992; Imrie and Thomas 1993a; 1993b; 1999a; Loftman and Nevin 1995). Although there was some relative improvement in unemployment indicators across the 57 targeted Urban Programme authorities, conditions in the most deprived areas had not improved as the old industrial conurbations continued to dominate the ranks of the most deprived places (Herbert 2000).

Property-led regeneration 'is no panacea for economic regeneration and is deficient as the main focus of urban policy' (Turok 1992, p. 376). The approach overlooks human resource issues such as education and training, the underlying competitiveness of production, and investment in essential infrastructure. Many property-led schemes also suffered from the collapse in the property market in the late 1980s. Finally, Imrie and Thomas (1992; 1993b) and Brownill (1990) among others argued that public money in effect subsidises private property development and neglects local need, and that property-led regeneration has its limits.

5. Problems of governance, managerialism and bureaucracy

Criticisms surrounded the political and administrative context within which urban regeneration policy operated in the 1980s. Experimentation and partnership between central and local government were characteristic of the 1970s (see Chapter 2). However, during the 1980s this central–local relationship changed into a hierarchical system of central control with a related absence of risk and experimentation (Oatley 1995). Conflict between central and local government over urban policy featured highly in policies such as Enterprise Zones and UDCs. The perception of local authorities was that they had become excluded from the urban policy arena, and there was a failure in operation of many partnership schemes linking the public and private sectors.

In response to this set of severe criticisms, the Audit Commission (1989) concluded that urban policy in the 1990s should attempt to overcome these problems. The report argued that policy should be more straightforward and less complex, and that local authorities should take a more active role in policy formation and instigation (Jacobs 1992; Hall 2001). It was also argued that 1990s urban policy should place the problem of social inequality and related unemployment, poverty and social segregation at its centre, and not just focus on economic development (Oatley 1995).

THE TRANSITION TO ENTREPRENEURIALISM: SUMMARY

During the period 1979 to 1991, urban regeneration policy was distinctive from that which had been characteristic of the period 1945 to 1979; the focus shifted from social welfare projects to private sector and property-led approaches.

The problems at the beginning of the Thatcher years in government were seen as too much state intervention, individual and group dependency on the state, and the restriction of the free market.

In response to the problems as constructed by Thatcher's government, 1980s urban regeneration tended to be more about profit, property and market-led regeneration. There was a focus on public–private partnerships, an emphasis on better and corporate management, and a marginalisation of local government and community in the regeneration process.

The flagship policies, among a cast of dozens, were those of Enterprise Zones and Urban Development Corporations, which were essentially property-led regeneration projects.

The goal of urban policy was to roll back the state and to free market forces, core wider philosophies of Thatcherism at this time. The approach to countering the urban problem combined social pathology and structural-economic approaches.

Problems with urban policy identified at the end of the period centred on the scale of the response to urban problems, the lack of a co-ordinated policy approach, the lack of a long-term strategy, the limits of property-led regeneration and problems of governance.

STUDY QUESTIONS

1. In what ways were the philosophies of the New Right embodied in 1980s urban policies?
2. Critically evaluate the approaches to urban regeneration undertaken by the Conservative governments of the 1980s.
3. Assess the role of public–private partnerships as an apparatus for urban regeneration.
4. Using an example of a UK city, examine the benefits and limitations of property-focused urban regeneration.
5. With reference to a case study example, evaluate the impact of an Enterprise Zone or Urban Development Corporation on local urban regeneration.
6. Critically assess the role of urban marketing within entrepreneurial regeneration strategies.

KEY READINGS ON ENTREPRENEURIAL REGENERATION IN THE 1980s

Atkinson and Moon (1994a) *Urban Policy in Britain: The City, the State and the Market* [The key text on 1980s urban policy and its interaction with cities and the public and private sectors]

Hambleton and Thomas (eds.) (1995) *Urban Policy Evaluation: Challenge and Change* [an evaluation of urban policy including 1980s approaches]

Healey, Davoudi, Tavsanoglu, O'Toole and Usher (eds.) (1992) *Rebuilding the City: Property-Led Urban Regeneration* [collection of critiques of the 1980s model of property-led urban regeneration]

Imrie and Thomas (eds.) (1993a) *British Urban Policy and the Urban Development Corporations* [first version of an account of eight UDCs within a wider socio-political context of 1980s urban policy]

Imrie and Thomas (eds.) (1999a) *British Urban Policy: An Evaluation of the Urban Development Corporations* [second edition, which presents some updates of UDC case studies alongside new examples within a more general analysis of urban policy]

KEY WEBSITES ON ENTREPRENEURIAL REGENERATION IN THE 1980s

Online Planning Resources: http://planning.rudi.net [search for key 1980s urban regeneration policies]

4

Competition and community in urban policy in the 1990s

INTRODUCTION

This chapter examines the period of central government urban policy that can be traced from 1991, a year after John Major's Conservative government took office after the ousting of Margaret Thatcher as Prime Minister in November 1990, and the year in which a distinct shift in urban policy became apparent. This period, which merged into the subsequent New Labour term of office, attempted to respond to the criticisms of the 1980s model of urban regeneration. Following the pattern of the previous chapter, this one will critically evaluate the Conservatives' urban policies up until 1997, and explore their underlying philosophies, drawing upon examples of key policies and particular projects in the UK's cities.

A brief overview of 1980s-style urban regeneration in the UK, which was the focus of Chapter 3, sets the context for the changes that were to follow from the early 1990s. There were a number of factors driving the need for change from the 1980s model of urban regeneration. During the 1980s, the dominant view was that urban areas had been left behind by a rising tide of the economy resulting in 'islands of decline' in a 'sea of prosperity' (Atkinson 2003). The distinctive policy element of this period was an experiment, that of encouraging the private sector to operate within specific run-down areas in order to redevelop them (Healey et al. 1992). This approach was underwritten by grants, subsidies, tax relief, relaxation of planning controls and so on, designed to attract wealth creators back to cities. It was believed that by allowing investors to make profits, they would then rebuild cities and create jobs and wealth, which would 'trickle down' to those in need. Local government was increasingly marginalised and the property development industry took on a key role in 'property-led regeneration' (Turok 1992). However, little thought was given to how the large number of projects related to one another and so projects ran side by side in a largely unco-ordinated fashion (Audit Commission 1989). Robinson and Shaw (1994) argued that urban policy in the early 1990s was still 'in search of the big idea', and sought to deal with the failings of 1980s urban policy. Additionally, the UK continued to suffer severe social and economic problems, which required policy responses (Table 4.1).

TABLE 4.1 MAJOR EVENTS AND EFFECTS, 1991–1997

Year	Event	Effect
1991–1993	Economic recession	Urban problems intensify
1993	Review of urban policy – *Policy for the Inner Cities* published	Emergence of competitive bidding
1997	Election of New Labour	Re-appraisal of urban policy

THE PHILOSOPHY OF 1990s REGENERATION

The first few years of the 1990s saw a gradual reconfiguration of policy. Urban problems retained elements from the previous period 1979–1991, but there had been growing dissatisfaction with property-led regeneration that characterised the period (Imrie and Thomas 1993b). Essentially, the benefits were not seen as being worth the high costs. There was also an increasing acknowledgement that the communities within the areas being regenerated were not experiencing any significant benefits, particularly socially excluded communities. The key problems for the government were how to ensure excluded communities benefited from urban regeneration, and how to address the incoherent 'patchwork quilt' of area-based initiatives and levels of governance involved in urban policy in the 1980s (Audit Commission 1989).

The emphasis of policy shifted somewhat in the 1990s with the emergence of multi-sectoral partnerships and competitive bidding (see Bailey 1993; Bailey et al. 1995; Oatley 1998a). Policies moved away from two-way public–private partnerships based on property-led regeneration, central government direction and 'trickle down' of investment (Law 1988; Harding 1990). Instead policies of this period were characterised by three-way multi-sectoral partnerships between the public, private, and community organisations and voluntary sector. Indeed the rediscovery of community has been declared as the main success story of urban policy in the 1990s (Davoudi 1995; Hambleton 1995). Attempts were made to ensure that the needs of those who experienced the worst effects of urban decline were integrated into urban initiatives. Local government was also brought back into the urban regeneration fold, although still largely dictated by the centre. A key aim was to change the way in which local government operated, making it less bureaucratic and more responsive both to the market and to the needs of its local community (Davoudi 1995, pp. 333–334).

These shifts heralded a new approach to resource allocation, namely competitive bidding for urban regeneration monies (Oatley 1998a; 1998b; 1998c). As demonstrated in Chapters 2 and 3, the majority of urban regeneration funding which stems from central government was allocated on the basis of the demonstration of need, which was usually measured in terms of the scale, intensity and extent of social and economic deprivation in localities (Atkinson and Moon 1994a). However, during the early 1990s, in the UK and elsewhere, the allocation of funding moved to a competitive bidding process (Oatley 1995b). Funding for central government urban policies was distributed according to the quality of the bids for the money and the economic opportunities available as judged by central government, rather than on the demonstration of need as such (Oatley 1998b, p. 9). Central government attempted to use urban regeneration funds to stimulate innovation within deprived areas rather than to simply alleviate need (Stanton 1996). Oatley (1998b, p. 9) stated that, 'The government and those who supported competition argued that it encouraged greater value for money and had a galvanising effect, motivating people to be innovative in developing proposals and encouraging more corporate and strategic approaches in regeneration activity.'

A major difference between these programmes and those of the 1980s was that they were aimed at creating more sustainable forms of regeneration (Hall 2006). First, they aimed to be catalysts, which created new structures and practices that might be replicated elsewhere. Second, the institutional capacity formed by the programmes, the structures, the partnerships and arrangements that were generated, was as important as the actual material outcomes of the policies (see Oatley 1998a; 1998b; Chapter 7). It was this institutional capacity, once created, that was deemed likely to sustain regeneration into the future. Despite these social and governmental principles behind 1990s regeneration policies, they also contained a number of economic and environmental aims.

The two flagship urban regeneration policies of the era were City Challenge, announced in May 1991, and the Single Regeneration Budget (SRB), launched in November 1993. These policies aimed to incorporate local residents, especially in 'disadvantaged' neighbourhoods, into the decision-making processes that affected their local area, or to empower them.

CITY CHALLENGE

City Challenge began in May 1991 and was hailed as the most promising urban regeneration initiative scheme attempted by central government (Russell et al. 1996). It was a significant innovation (Parkinson 1993; Oatley and Lambert 1995; 1998). Although a rather short-lived initiative, it gave a central role to local government and local communities. It was also notable for introducing the idea of competition as a method of allocating urban resources. Government portrayed competition as the most efficient method of allocating resources and developing innovative regeneration schemes, and the best way to harness local talent and initiative (Atkinson and Moon 1994a; 1994b; Atkinson 2003). Atkinson and Moon (1994b) and Atkinson (1998) noted several key elements to the City Challenge programme that indicated the broad shift in policy away from the primarily private sector and central government policies of the period 1979 to 1991, and reflected the philosophies of the period more generally. The policy emerged out of the problems associated with property-led regeneration.

Atkinson (1998) argued that City Challenge in many ways represented a co-joining of a Victorian paternalism with the concepts of competition, supported by the quote by Michael Heseltine when announcing the programme:

> [W]hen I speak of the need for a sense of partnership in our modern cities, it is today's equivalent of that Victorian sense of competition linked with social obligation. Success and responsibility go hand in hand today just as surely as they did 100 years ago . . . men and women will compete with one another and give their best. People will set the pace and exercise their discretion. And by enriching themselves, their lives and their communities, they enrich society as a whole.

In the eyes of government, several features had to be demonstrated in a good bid for City Challenge money. Urban regeneration projects funded by the programme had to be able to lever in extra private sector investment, add value to existing public–private investment, stimulate wealth and an enterprise culture, widen social provision, benefit the local community, improve quality of life and create long-lasting multi-level partnerships (Oatley and Lambert 1998). Bids for City Challenge needed to be seen in the context of a combination of both the potential for economic development in the areas receiving the urban regeneration money, and a focus on areas of high deprivation.

In terms of the programme's operation, two rounds of bidding for regeneration funds were held by central government (Table 4.2). In Round 1 (1991) there were only 11 local

TABLE 4.2 CITY CHALLENGE PROJECTS

	Winning local authorities	*Rejected local authorities*	*Local authorities not invited to bid*
Round 1 1991	Dearne Valley (Doncaster, Barnsley and Rotherham) Bradford Lewisham Liverpool Manchester Middlesbrough Newcastle Nottingham Tower Hamlets Wirral Wolverhampton	Birmingham Bristol Salford Sheffield	Coventry Newham Danswell St. Helens Stockton Sunderland
	Winning local authorities	*Rejected local authorities*	*Non-bidding local authorities*
Round 2 1992	Barnsley Birmingham Blackburn Bolton Brent Derby Hackney Hartlepool Kensington and Chelsea Kirklees Lambeth Leicester Newham North Tyneside Sandwell Sefton Stockton Sunderland Walsall Wigan	Bradford Bristol Burnley Coventry Doncaster Dudley Gateshead Greenwich Halton Hammersmith and Fulham Haringey Hull Islington Knowsley Langbaurgh Leeds Liverpool Middlesbrough Newcastle Nottingham Oldham Plymouth Preston Rochdale Rotherham St. Helens Salford Sheffield South Tyneside Southwark Tower Hamlets Wandsworth Wolverhampton The Wrekin	Lewisham Manchester Wirral

Source: **Oatley and Lambert (1998, p. 112).**

authority winners from a total of 15 invited to bid, all Labour controlled (Oatley and Lambert 1998, p. 112). These projects started in 1992 and each received £7.5 million a year for a five-year period (£52.5 million in total). Round 2 (1992) saw 20 local authority beneficiaries from 57 invited authorities with Urban Programme status, each receiving £7.5 million a year for five years (Davoudi 1995). This amounted to £1.15 billion over the eight-year life of the project, which on the face of it appeared a significant amount of urban regeneration funding. However, this represented a different approach to spending regeneration funds, rather than an allocation of new funds as such (Cullingworth and Nadin 2006, p. 366).

The emphasis on land and property remained, thus demonstrating continuity with 1980s urban regeneration. However, City Challenge attempted to link urban regeneration more closely to the needs of local communities and particularly disadvantaged residents (Cullingworth and Nadin 2006). Projects tended to involve physical renewal, often in city centres. Housing renewal, town centre regeneration and community projects were commonly funded schemes in the successful bid cities. Cullingworth and Nadin (2006) also noted that an integrated approach was encouraged with property development being linked to wide-ranging issues including economic development, housing, training, environmental improvements, and social programmes including crime prevention and equal opportunities. Successful local authorities prepared an action programme detailing projects that were funded through existing programmes with similar procedures for making and considering applications. The private sector was essential, and was expected to play a significant role. City Challenge project winners were mainly inner city locations, although several were edge-of-urban projects such as Dearne Valley in South Yorkshire.

Despite the financing for such projects, the reality was that the programme came at a cost. Several criticisms of City Challenge began to emerge (Davoudi 1995; Davoudi and Healey 1995a; 1995b). A first set of problems related to governance and community involvement, which often amounted to issues relating to consultation (see Bell 1993; Taylor 2003; Chapter 8). Although City Challenge broadened access to local political power through new voices, it tended to perpetuate certain existing inequalities. This is because the bidding timetable was too tight at just six weeks for local authorities to consult effectively with communities and other groups, and to build new partnerships with them. Essentially the way in which the programme was set up meant that local authorities could not engage the community through lack of time and money. Therefore local authorities tended to rely heavily on existing local community and business networks, as evidenced by the experience of Newcastle's West End (Davoudi 1995; Robinson 1997). City Challenge failed to have much effect on the operation of local government. Inherent antagonisms existed between the partners in the three-way partnership due to differing priorities, for example, regarding inner-city employment markets (Davoudi 1995, p. 341). Many of the projects continued to be represented by physical renewal schemes characteristic of 1980s urban regeneration.

A second set of criticisms were linked to the funding of City Challenge. Most of the money was merely top-sliced out of other programmes (Davoudi 1995) rather than representing significant 'new' money. Additional cuts in local authority finance during the early 1990s meant that often even the winners of City Challenge were not better off than before. There were related tensions between choosing areas ripe for economic development versus those that were highly deprived, and progress was affected by the early 1990s economic recession. All winners of City Challenge received the same amount of funding regardless of need (Oatley 1994; 1995a; 1995b; Davoudi 1995). Partnerships in Bristol (see Malpass 1994; Hooton 1996), Salford and Sheffield went through the arduous bidding process twice without winning any urban regeneration money from the programme. Finally, there is evidence in many of the areas to receive money to suggest that 15 years on underlying social problems have not been resolved.

CASE STUDY: BIDDING FOR CITY CHALLENGE MONEY – THE BRISTOL EXAMPLE

The key elements of City Challenge are reflected in the experience of Bristol, namely competition between urban areas, collaboration within localities, and the leading role allocated to local authorities. Bristol's first bid for City Challenge in 1991 emerged from a less ambitious Estate Action bid for the peripheral estates of Hartcliffe and Withywood on the southern edge of the city (Malpass 1994; Bassett 2001). The City Challenge bid was widened beyond the physical improvement of housing and included high visibility flagship projects. Although some local authorities such as North Tyneside (Davoudi and Healey 1995b) were enthusiastic about bidding, Bristol was much more reticent due to a long-standing mistrust and competition between organisations and authorities. Relations with central government had also been damaged due to the UDC experience in the city (Chapter 3).

Although winning City Challenge money would guarantee five years of urban regeneration funding from central government and levering in of private sector money, there were concerns over the strings attached to the funding in terms of the bidding procedure, the involvement of the private and voluntary sectors in regeneration, and the way that City Challenge was to be managed outside the direct control of local authorities (Malpass 1994, p. 307). However, the Labour local authority decided to bid, albeit without a great deal of enthusiasm and without wide debate or consultation. Malpass (1994) argued that the local authority's lack of enthusiasm was partly due to objections to the principle of competitive bidding, and having to conform to the Conservative central government's rules of the game. The local authority wanted to protect its freedom, powers and assets, and was not experienced in urban entrepreneurialism (Malpass 1994). Despite these reservations, the local authority could not resist the opportunity of levering in significant public and private funding to a deprived area.

The bid focused on deprived peripheral estates, and sought to address the social and economic problems linked to the unpopularity of the 1950s and 1960s tower blocks, the lack of employment due to the closure of the main tobacco factory employer in the area, high levels of multiple deprivation and the consequences of its peripheral location at a distance from employment opportunities and services. The Bristol bid was in contrast to most of the others around the UK, which were directed at inner cities and city centres, and was a housing-led project (Malpass 1994).

Bristol's first bid for City Challenge funding failed due to perceived flaws or inadequacies in the submitted proposals (Malpass 1994; Hooton 1996; Bassett 2001; Tallon 2007a). These included a general lack of enthusiasm from the local authority; a lack of partnership between public, private and voluntary and community sectors; a lack of vision in the proposals for the flagship housing project planned for the major derelict tobacco factory; and a failure to demonstrate the wider impacts of the projects on Bristol. These reasons reflect both the extent of central government involvement in the process and the difficulties in putting together local partnerships between conflicting groups in a matter of weeks.

Bristol's second bid for City Challenge funding in 1992 was more securely grounded and organised. The bid was directed at the same geographical area, but was less housing-focused. Although the second bid responded to the criticisms of the first, the second attempt was beset by conflict between the public and private sectors over the nature of the housing regeneration. This second bid also failed partly due to the stances adopted by local politicians and landowners, rather than due to the quality of the bid (Malpass 1994). Thus Bristol City Council in partnership with the private sector and local communities was one of three places, along with Salford and Sheffield, which twice bid and twice failed to secure City Challenge funding.

Despite the criticisms, City Challenge was regarded by many as a fairly successful advance in terms of urban policy. It at least attempted to reintegrate local authorities and communities into the urban regeneration process. De Groot (1992), Russell et al. (1996) and Oatley and Lambert (1995) argued that City Challenge was a success and provided a model for future urban regeneration, particularly regarding the value of the multi-sectoral partnerships that developed and value for money in terms of private investment levered in.

City Challenge was just one programme in a cast of many for urban areas at the time, with many 1980s policies still continuing (Table 4.3). In 1993, the government undertook a major review of urban policy and no third round of City Challenge was offered. Instead, the review produced a paper – *Policy for the Inner Cities* – that in turn sparked the creation of the Single Regeneration Budget.

SINGLE REGENERATION BUDGET (SRB)

The Single Regeneration Budget was launched in April 1994. SRB largely adhered to Conservative convention on the causes of urban decline, which acknowledged the diversity of causes from a distressed labour market, exhausted infrastructure, crime and social disorder problems, and poor health (Atkinson 2003). The entrenched nature of the problems meant that neither the market nor the actions of mainstream government departments was in a position to tackle these problems adequately (CLG 2007d). SRB acknowledged the need for greater co-ordination in the regeneration process in response to continued criticism in this respect. The initiative brought together 20 programmes that were previously administered by five central government departments, all under the overall control of the then Department of the Environment (DoE) (Table 4.3; see Oatley 1998d, p. 147; Brennan et al. 1999, p. 2071). This was an attempt to ensure a streamlining of policy along with added coherence and co-ordination. This central restructuring was complemented by the creation of Government Offices of the Regions (GORs) in each of the English regions composed of the regional offices of the Departments of Trade and Industry, Employment, the Environment and Transport, plus a senior representative from the Home Office (Atkinson 2003).

As attempted with City Challenge, regeneration partnerships at the local level were the main means by which SRB was implemented. They were to be multi-sectoral partnerships involving the public, private, and voluntary and community sectors. SRB aimed to provide a co-ordination mechanism through which the mainstream programmes of education, health, housing, crime and safety could be brought together to help solve the problems of deprived areas in strategic, co-ordinated and cost-effective ways (CLG 2007d). This process has more recently been termed as 'bending' of mainstream activity (see Chapter 5) in terms of both customising policy delivery and channelling more funding into the areas concerned. A core principle of SRB explicitly stated that local people should be engaged in and benefit from regeneration.

The creation of such partnerships was a condition for receipt of funding, which was, like City Challenge, allocated on a competitive basis. It was among the first group of urban policies to allocate funding following a competitive challenge fund model that resulted in a geographical pattern of allocating funds that significantly differed from those of earlier central government urban regeneration policies. In a sense, it represented a spatial 'free-for-all' with funding distributed around England (Hutchinson 1997).

SRB thus argued for the need for a strategic approach to regeneration: a partnership model involving the public, private, and community and voluntary sectors (CLG 2007d); competitive bidding for available funds; and payment by results. These principles represented a continuation in many ways with those of City Challenge.

The overall objective of SRB was to enhance the quality of life of local people in areas of need by reducing the gap between deprived and other areas, and between different groups. Economic objectives were to enhance employment, education and skills of local people,

TABLE 4.3 BUDGETS SUBSUMED WITHIN THE SINGLE REGENERATION BUDGET, 1994/1995

Programme	£ *million*
Urban Development Corporations*	286
Housing Action Trusts*	88
English Partnerships*	181
Estate Action	373
City Challenge	213**
Urban Programme	83
Task Force	16
City Action Teams	1
Section 11 (part)	60
Ethnic Minority Grant/Business Initiative	6
Safer Cities	4
Programme Development Fund	3
TEC Challenge	4
Local Initiative Fund	29
Business Start-up Scheme	70
Education Business Partnerships	2
Compacts	6
Teacher Placement Service	3
Grants for Education Support and Training	5
Regional Enterprise Grants	9
Total	1,442

* Programmes ring-fenced in the SRB.

** Programme supported by £19 million from the Housing Corporation.

Sources: **Herbert (2000, p. 210) and Oatley (1998d, p. 147).**

particularly the young and disadvantaged; and to encourage sustainable economic growth and wealth creation through competition.

Alongside the economic objectives, the social objectives of SRB were to improve housing through good design, management and maintenance; promote initiatives to benefit ethnic minorities; crime prevention and community safety; and improving the quality of life for local people. A significant part of SRB funding was designed to address the multi-faceted problems of neighbourhoods in a holistic fashion (Oatley and Lambert 1998).

In all there were six rounds of bidding between 1994 and 2000. The six rounds resulted in 1027 schemes and £26 billion of funding, £9 billion of which came from the private sector, over the six rounds of funding, which could last for up to seven years (Rhodes et al. 2003; Cullingworth and Nadin 2006, p. 371). Overall, SRB funding accounted for only 22 per cent of total expenditure with the other 78 per cent coming from local authorities, Training and Enterprise Councils (now Learning and Skills Councils), the voluntary and private sectors, and European funding (CLG 2007d).

Urban regeneration schemes that were funded by SRB had the option of being thematic, rather than spatial, which focused on the needs of particular groups (see Bassett et al. 2003). The flexible format of SRB funding allowed regeneration projects to be customised by geography, size, duration and objective (see DTLR 2002 for 10 case studies). Bids could be submitted for virtually any sort of regeneration activity (CLG 2007d), and areas with no tradition of making bids or receiving funds for regeneration were able to apply. This was in contrast to the boundary-driven approach characterised by policies such as the Urban Programme (see Chapter 2) and City Challenge.

SRB was delivered by the GORs over its first four rounds and through the nine Regional Development Agencies established in England after 1999 for the last two. Therefore, SRB developed as an integrated approach to funding urban regeneration from central government, through regional administration and central government decision, to complete devolution to the regions (Cullingworth and Nadin 2006, p. 370). SRB projects were carried out through wide-ranging local partnerships involving local organisations, local businesses, voluntary and community sectors, and local authorities, which arguably was its most important outcome (Fordham et al. 1998). It encouraged best practice and good value for money.

A final evaluation of SRB (CLG 2007d) illustrated key positive outcomes, although a caveat was that these were a small but valuable contribution. The evaluation noted that in SRB areas household incomes and employment levels improved; satisfaction with housing and views of the areas improved; perceptions of safety increased; and overall CLG (2007d, p. 4) noted that 'the most successful activities were improvements to the physical fabric of the area, building the community and enhancing social cohesion'.

While many academic commentators gave the SRB a guarded welcome (Atkinson 2003), concerns emerged over its effectiveness and ability to achieve its objectives (Hall 2001, pp. 76–77). These criticisms can be categorised under three main issues: finances available for urban regeneration; the real extent of community participation in the regeneration projects; and the co-ordination of urban policy.

The most serious criticism surrounded the criteria used for the allocation of funding and the actual amounts of funding available (Brennan et al. 1999). SRB replaced a number of existing budgets including the Urban Programme (Table 4.3; see Chapter 2), which effectively reduced deprivation as a criterion for fund allocation. SRB was also accused of being a means by which central government disguised the fact that the rationalisation of urban regeneration budgets actually reduced the amount of central government funding available, 'making less seem more' (Atkinson 1999b).

> Critics of the Challenge model argued that it was a distraction, used to mask the decline of regeneration resources and mainstream expenditure and a way of rationing scarce resources. Critics argued that it placed the government and the newly restructured regional offices in a powerful position, which undermined the avowed aim of government to encourage local empowerment and ownership of proposals. Furthermore, competition between localities reduced the scope for inter-local co-operation and the new regime of governance tends to weaken local democratic accountability (Oatley 1998b, p. 9).

Some startling allocations have failed to include some severely deprived areas, which received money under previous programmes. For example, while deprived areas of places such as Leicester, Nottingham and Walsall failed to receive SRB funds, more prosperous localities within Bedford, Eastbourne, Hertfordshire and Northampton each benefited from around £17 million of funding (Hall 2001, p. 77). However, other research on funding allocations for the first three rounds of SRB by Brennan et al. (1999) found that it had successfully targeted the most severely deprived local authority districts in England. In this respect, it is argued that SRB performed better than Enterprise Zones, UDCs and City Challenge.

Community participation was largely illusory and there remained very little evidence that deprived communities were benefiting from urban regeneration projects such as those carried out through the SRB programme (Atkinson and Cope 1997). The business perspective prioritised economic and employment aims in SRB bids, as would be expected. Herbert (2000) noted that an analysis of successful bids from the first round of SRB bidding showed that whereas 34 per cent targeted ethnic minorities and poor housing, 95 per cent had employment as the main objective. However, there remained a compelling need to add a community-led dimension, and to recognise the centrality of social integration as a goal and social exclusion as a risk.

Finally, the co-ordination of urban policy at and between levels of governance remained fragmented and confusing, presenting little evidence that co-ordination had actually increased despite the organisational changes introduced through the SRB.

The final evaluation concluded that the successes of SRB could be built upon in future area-based initiatives (see Chapter 5), in particular enticing business back to neighbourhoods and focusing on neighbourhood-scale regeneration (see CLG 2007d; Rhodes et al. 2003). The features and approaches of SRB schemes have been adopted in post-1997 policies and practices, particularly in adopting an integrated strategic approach, mainstreaming regeneration, engaging the community, and operating flexibility and innovatively (Rhodes et al. 2003; CLG 2007d; see Chapter 5).

OTHER COMPETITIVE BIDDING URBAN POLICIES IN THE 1990s

Although City Challenge and Single Regeneration Budget were the two flagship policies to emerge during the early 1990s, a couple of other initiatives merit a mention and reflect the wider nature of urban policy during the period 1991 to 1997.

CITY PRIDE

The City Pride scheme was announced in November 1993 and aimed to shift power from Whitehall to local communities and to make government more responsive to local priorities (Williams 1995a; 1995b). The policy aimed to make localities, and businesses situated within them, competitive within the global arena. City Pride partnerships envisaged a challenge to the civic and business leaders of the cities of London, Manchester (see Williams 1995b) and Birmingham to prepare a prospectus detailing a long-term strategic vision of their city's strategic development over the following decade (see Williams 1998). Hall (2006, p. 61) noted that this was one initiative based on the institutionalisation of place marketing, which was typical of this period.

For example, key urban regeneration themes within Manchester's prospectus included repopulating the city centre (see Chapter 11), diversifying the city's economic base, increasing employment and reducing poverty, reducing dereliction, making the city more competitive (see Chapter 6), and implementing sustainable development and regeneration (see Chapter 9) (Williams 1995b, p. 349). It was concluded at the time that the successful advancement of City Pride depended upon a clearer articulation of the benefits for the whole community, the sustainability of partnership arrangements, and a transfer of leadership roles to cities (Williams 1995b, p. 357), issues that remain relevant in the late 2000s.

The initiative was characteristic of the 1990s era of 'new' urban policy reflecting partnerships between public and private bodies at the local level, with central government's role being to provide the framework to enable this (Paddison 1993). Although there was no specific funding regime associated with City Pride, the aim was to provide a holistic vision making the best use of existing funding and levering in additional resources (Williams 1995a; 1995b).

THE NATIONAL LOTTERY

The National Lottery was launched in November 1994; by 2008 it had generated £21 billion for 'good causes' and had funded over 300,000 projects (www.national-lottery.co.uk). The good causes currently encompass the arts, sports, heritage, health, education, environment, community and charity sectors. Grants are distributed by way of a bidding process through The Big Lottery Fund, The Millennium Commission, The Sports Councils, The Heritage Lottery Fund, The Arts Councils, and The New Opportunities Fund. Clearly, many schemes with urban regeneration ideals at their heart have been funded in part or in their entirety by the National Lottery.

Although not specifically an 'urban policy' as such, the National Lottery has played a significant role in a variety of types of urban regeneration projects in the UK. Griffiths (1998b) argued that the National Lottery's significance to cities does not lie simply in the fact of the scale of funding. The National Lottery shared many of the characteristics of 1990s urban policies as funding in the form of lottery proceeds were also distributed on a competitive basis in respect of the quality of the applications rather than on the demonstration of need, with limited targeting on social disadvantage.

Many projects funded by the National Lottery were completed in the late 1990s for the millennium. Hall (2001) noted that around 10,000 millennium projects were funded by a combination of National Lottery and private sector funding across the areas of heritage, culture, arts, sports, the environment and community regeneration. It is estimated that around £5 billion was spent on millennium projects, around half of this from the National Lottery. Many projects were in the mode of 1980s-style flagship and property-led urban development. These included the infamous Millennium Dome in Greenwich, Birmingham's Millennium Point, Portsmouth Harbour's regeneration, Manchester's Lowry Museum, and the @Bristol complex. However, Hall (2001, p. 167) listed some environmentally focused regeneration projects that benefited, such as the National Cycle Network and the Millennium Forest in Scotland.

Hall (2001, p. 167) further argued that in terms of urban regeneration, millennium projects represent 'the most significant wave or urban reimagination and regeneration since the property-led approaches of the 1980s', and thus throw up a range of similar questions regarding their impacts (see Chapters 10 and 12). Although smaller-scale community-based projects have brought benefits to participating communities, flagship or spectacular projects have created long-term debt at the local level in securing external funding, and have replicated some mistakes of the 1980s. These included competition between similar leisure, culture or education schemes, and the complete failure of some property-led schemes such as the National Centre for Popular Music in Sheffield, which closed after a year. The £780 million cost of the Millennium Dome has also been criticised for not being worth the benefits gained locally.

A final important advance of note in the 1990s was the establishment of English Partnerships in 1993 as a non-departmental public body, as the operating name of the Urban Regeneration Agency and Commission for New Towns. Its main aim was to promote regeneration through the reclamation or redevelopment of land and buildings; the reuse of vacant, derelict and contaminated land; and the provision of floorspace for industry, commercial and leisure activities; and housing. It aimed to provide gap funding between the costs of undertaking development and the end value, which had previously been the role of city grants and derelict land grants of the 1980s urban policy regime (Cullingworth and Nadin 2006, pp. 367–368; see Chapter 3; www.englishpartnerships.co.uk). English Partnerships continued beyond the period 1991–1997, although with some changes to its remit (see Chapter 5).

CRITICISMS OF 1990s 'NEW' URBAN POLICY

Some of the failures of the 1980s model of urban regeneration were addressed by competitive bidding policies in the 1990s, such as City Challenge and SRB. Key responses to these failures were the reintegration of local authorities and the voluntary and community sectors into the urban regeneration process, the attempt to target funding at the most deprived areas, and the apparent streamlining of policy initiatives.

Although there had been a clear change in direction of urban policy and a desire to respond to past criticisms (Stewart 1994; Oatley 1995a), during the latter days of the Conservative government in the late 1990s, research continued to reveal continuities in terms of problems with urban policy, and the 1990s brand introduced further controversies relating to finances available for urban regeneration, community involvement and co-ordination of urban policy (Atkinson 2003).

The emergence of the competitive bidding regime indicated a further shift away from pre-1979 concerns associated with welfare further towards urban competitiveness and economic priorities (see Chapter 6; Deakin and Edwards 1993). The abolition of the Urban Programme symbolised the effective reduction in deprivation and need as criteria for fund allocation (Oatley 1998b, pp. 10–11). Nevin et al. (1997) argued that due to the downgrading of need as a criterion, urban regeneration strategies that brought benefits to disadvantaged groups would be difficult to achieve. The social dimension was still a minor part of the equation. The problem of limited improvement in the most deprived areas was exacerbated in the first half of the 1990s with the 'effective withdrawal of inner-city policies' (Herbert 2000, p. 209).

As Oatley (1995a, p. 265) stated, associated with this shift was a continued reduction in public funds available for urban regeneration. The £1.4 billion available for SRB in 1994/1995 was in fact £300 million less than the former 20 separate programmes the year before, despite the recognised scale and nature of the urban problem. Policies like City Pride involved no new direct funding for urban regeneration. Herbert (2000) provided further evidence by observing that between 1990 and 1996 urban funding was cut by 40 per cent.

A further criticism was that community participation in schemes such as City Challenge and SRB was largely illusory due to the existing power relationships and the ways in which the programmes were set up. There was still very little evidence that deprived communities were benefiting from urban regeneration, despite efforts to respond to this recognised issue (Atkinson 2003).

Urban policy remained a confusing landscape in the mid-1990s. Attempts had been made to mitigate this criticism as the SRB merged 20 existing programmes and rendered the 1980s policy programmes irrelevant. This seemingly brought to fruition the promise of co-ordination and more strategic thinking, as did bringing local authorities back into the urban policy and regeneration process. However, commentators remained unconvinced and Robson (1994) feared that the swing back towards local authorities went too far and that there was little evidence of a strategic vision for urban policy. Herbert (2000) argued that schemes such as City Challenge did not amount to a clear inner city strategy.

Overall, urban policies of the 1990s appeared to have been constrained by the legacy of the policies of the 1980s and existing local government structures. These exerted a disproportionate influence over the direction of policy rather than the opposite, intended effect (Hall 2001). There was little evidence, despite the organisational changes introduced, that the co-ordination of urban policy had actually increased as policy at and between levels of governance remained fragmented and confusing (Atkinson 2003).

SUMMARY

The aims of urban regeneration programmes in the period 1991–1997 were to produce a more self-sustainable form of regeneration, to broaden access to power, and to transform the operations of local government.

City Challenge and Single Regeneration Budget were significant innovations in urban regeneration policy.

City Challenge was the first in a new wave of initiatives based on competitive bidding for urban funding and at the time was termed a 'revolution in urban policy'.

City Challenge acted as a model for restructuring urban policy in 1993, which led to the Single Regeneration Budget. These policies typify the more balanced, integrated approach to urban regeneration that developed since the more narrowly focused, market-oriented, property-led approaches of the 1980s.

It remains perhaps too early to provide a fully rounded evaluation of these policies, but the signs are not particularly encouraging. Despite representing a major shift away from the much-criticised models of urban regeneration in the 1980s, both programmes have also been subject to a number of major criticisms.

Criticisms essentially condensed to the reduction in finances available for urban regeneration and the elevation of competition over need as a funding model, the fallacy of community involvement, and the continued lack of co-ordination of urban policy.

The urban policy agenda moved on and was re-assessed with the formation of the New Labour government in May 1997.

STUDY QUESTIONS

1. Discuss the criticisms of urban policies characterised by competitive bidding.
2. Compare and contrast the Challenge Fund model of urban regeneration with the property-led approaches of the 1980s.
3. What were the problems with urban policy that had been identified by the end of the 1990s, and how could subsequent policies be adapted to counter these concerns?
4. Drawing upon academic and official sources, critically evaluate a City Challenge or SRB project within a UK urban area.

KEY READINGS ON 1990s URBAN POLICY

CLG (Communities and Local Government) (2007) *The Single Regeneration Budget: Final Evaluation – Urban Research Summary Number 25* [brief but informative review of the SRB programme; full version available at www.landecon.cam.ac.uk/research/reuag/uars/projects/urgsrb.html]

Hambleton and Thomas (eds.) (1995) *Urban Policy Evaluation: Challenge and Change* [offers evaluations of early 1990s competitive bidding policies]

Oatley (ed.) (1998a) *Cities, Economic Competition and Urban Policy* [the key edited text on 1990s urban policy, giving a theoretical overview and an appraisal of key initiatives]

KEY WEBSITES ON 1990s URBAN POLICY

Communities and Local Government: www.communities.gov.uk [current government department responsible for urban policy which contains archived information and evaluations of 1990s urban policies]

Department of Land Economy, University of Cambridge: www.landecon.cam.ac.uk [carried out the national evaluation of SRB and has published a number of reports since the mid-1990s]

National Lottery: www.national-lottery.co.uk [distributed finances among projects, many of which are linked to urban regeneration in their nature]

5 New Labour, new urban policy? Urban regeneration since the late 1990s

INTRODUCTION

The final chapter tracing central government-driven urban regeneration policy in the post-war period examines New Labour's brand of urban policy from the late 1990s until the late 2000s. The philosophies underpinning this are reviewed, and comparisons and contrasts with previous approaches are elucidated. Key policies are critiqued and case examples are highlighted. Policies pursued during this period can be broadly divided into two: those policies focusing on the regeneration of neighbourhood and community in areas traditionally dominated by housing, in the context of social exclusion and related urban problems; and those policies based around regenerating declining regions, city centres, and areas of cities formerly dominated by industry and commercial uses rather than housing. The government department responsible for urban regeneration, Communities and Local Government, makes a similar distinction between communities and neighbourhoods (covering areas such as sustainable communities, neighbourhood renewal, social exclusion, community cohesion and civic renewal), and cities and regions (encapsulating city-regions, regional development and regeneration).

The year 1997 marked another dividing line in central government urban policy in the UK. The early period of the 1940s until 1979 was characterised initially by town and country planning policies, and the development of new housing, and latterly by area-based social and community initiatives (Chapter 2). The property-led, entrepreneurial era of urban regeneration can be traced from 1979 until 1991, following which, an era of competitive bidding for urban funds emerged. In May 1997, the Conservatives were removed from power after 18 years, heralding a new approach to economic and social policy in the UK. A hallmark of New Labour's urban policy has been recognition of the interrelationship between the economic and social dimensions of urban policy within the context of the so-called 'urban renaissance'.

New Labour came to power in 1997 with a commitment to the renaissance of the UK's cities. Prior to its election, New Labour had signalled its intention to address urban problems in a more coherent manner and ensure that combating social exclusion would be a key part of urban policy. New Labour had broadly supported the SRB and retained it once in power for three years, albeit with some refocusing (Chapter 4). New Labour appeared to accept the general diagnosis of urban problems offered by previous Conservative governments, when it stated that areas of multiple deprivation in cities had been largely by-passed by national economic success, suggesting that the causes of urban decline lay in the areas themselves rather than in the wider social context (Johnstone and Whitehead 2004b). What was new was the emphasis on social exclusion (see Chapter 1), the revival of citizenship, democratic renewal and the participation of communities to spearhead urban change (see Purdue et al. 2000; Imrie and Raco 2003a).

This policy period can first be analysed by exploring the characteristics of urban policy by taking the 1977 and 2000 Urban White Papers as historical markers to assess continuity and change in urban policy over the two and more decades. A following section will look at the role of neighbourhood and community in New Labour's urban regeneration policy, by examining the Social Exclusion Unit, the New Deal for Communities (NDCs) and the Neighbourhood Renewal Unit as key developments during the last ten years. The final section will outline, discuss and evaluate other key delivery vehicles associated with urban regeneration and sustainable communities initiated by New Labour since 1997.

THE URBAN TASK FORCE AND THE URBAN WHITE PAPER

In 1998, the architect Richard Rogers was appointed to chair the Urban Task Force, which sought to identify the causes of decline in urban areas and to recommend practical solutions that would lead to the sustainable regeneration of these areas. Its influential final report *Towards an Urban Renaissance* was published in 1999 (Urban Task Force 1999), which contained over 100 recommendations and popularised the concept of 'urban renaissance'. It encouraged design excellence, brownfield development and higher densities. Drawing upon many of the Urban Task Force's recommendations, an Urban White Paper was published (DETR 2000a), the first for 23 years, which set out the strategy for achieving an urban renaissance. This was accompanied by £1 billion of tax measures to increase investment in urban areas (Colomb 2007).

Research in the early 2000s sought to analyse continuities and change in urban policy by examining the 1977 and 2000 Urban White Papers as indicators of 'new horizons' and 'old barriers' in urban policy (Johnstone and Whitehead 2004b). It is important to recognise the changes in the nature of urban problems and the political responses to them and also to understand the persistent structural problems, which tend to recreate urban poverty and social injustice, representing the old barriers.

Although not all urban policy is geared towards issues of social injustice, many early urban policy initiatives such as the Community Development Projects and more recent initiatives such as the Neighbourhood Renewal Strategies have been focused on or inspired by issues of social injustice. These can be expressed in terms of social class and poverty or racial discrimination and lack of opportunities. Urban policy has provided consistent political focus on the most marginalised groups: the 'others' of affluent society. Urban policy has also provided a key insight into issues of territorial justice (the geographical dimension) both in relation to how territorial justice is produced and how territorial justice and consequences of uneven development are addressed by the state and key agencies (Johnstone and Whitehead 2004b). Crucially, urban policy, unlike many of the mainstream spending areas such as health and education, has an explicit spatial agenda, whether through the static urban priority areas of the Urban Programme or the spatial free-for-all of the SRB. Urban policy has consistently tracked uneven development and highlighted the links between social and territorial justice.

It is useful to explore continuity and change in urban policy through looking at the two Urban White Papers. One of the most striking differences between them is the way in which they present the urban, as both documents emerged from very different origins. Both Urban White Papers were responses to reports into the state of urban areas in the UK. In 1977, the

White Paper (DoE 1977) was a response to the Inner Area Studies that were carried out in Birmingham, Lambeth and Liverpool (see Chapter 2). In 2000, the Urban Renaissance White Paper (DETR 2000a) was the government response to the Urban Task Force report (Urban Task Force 1999). In this context, where the 1977 White Paper took its lead from Birmingham, Lambeth and Liverpool and the poverty and decline of cities, Birmingham was replaced in 2000 by Barcelona and Liverpool with Amsterdam, as bases for understanding and addressing urban issues (Johnstone and Whitehead 2004b).

It has been argued that this changing focus shows that images of poverty, dependence and decline have been replaced with depictions of urban areas as economic power-houses and vibrant cultural communities. This changing representation of cities from 'spaces of despair' to 'spaces of hope' (Harvey 2000) has implications for how urban policy is being conceived. This is most clearly expressed in the apparent relegation of urban poverty debates and the elevation of urban place marketing and 'boosterism' in the urban renaissance programme.

In relation to social justice, it is interesting to note similarities in both Urban White Papers regarding the causes of urban problems and urban social injustices. They both stress strongly the structural economic causes of urban decline, while emphasising in different ways the related problems of social exclusion, disadvantage and physical decay. Despite these similarities, the emphasis on social justice or injustice is in marked contrast. While it appears the 1977 White Paper was developed in order to address social disadvantage, by 2000 issues of social injustice are buried beneath a gloss of architectural aesthetics and discussions of planning design (see Lees 2003a; 2003b; Hoskins and Tallon 2004). Urban poverty and disadvantage are no longer what urban policy appears to be primarily about. This is perhaps most clearly illustrated by a failure to effectively address issues of race and ethnicity (Johnstone and Whitehead 2004b).

In relation to territorial justice, there are very different understandings of the spatial concern expressed in the two documents. In 1977, territorial injustice is seen to be a problem largely of the inner cities of large metropolitan areas. The 2000 urban renaissance document clearly reveals a wider spatial concern for issues of justice, with urban policy increasingly being developed in the context of rural concerns over green-field development and a wider concern with the city, town and suburb representing a form of 'post-inner city urban policy'. The changing spatial focus appears to be deflecting attention away from evolving patterns of deprivation and poverty within the city, particularly inner cities (Johnstone and Whitehead 2004b).

Overall, the Urban Task Force and subsequent Urban White Paper have been criticised for their excessive focus on design to the detriment of wider economic and social factors (Lees 2003a; 2003b; Johnstone and Whitehead 2004b; Hoskins and Tallon 2004; Colomb 2007). Richard Rogers lamented the government's slow progress in implementing an urban renaissance and an independent update report (Urban Task Force 2005) was similarly critical and renewed calls for a 'strong urban renaissance'. The Urban White Paper contained little that was explicitly new (Atkinson 2003). However, it attempted to create a vision of urban living and a framework in which other initiatives and themes could be brought together.

THE STATE OF THE CITIES

Linked with the urban renaissance agenda since the late 1990s, the government commissioned a number of reports that sought to assess the state of the cities, usually focusing on those in England (DETR 2000b; 2000d; ODPM 2005d; 2006a). The 2006 *State of the English Cities* report was an independent report that provided a comprehensive audit of urban performance in England; a review of the impact of government policies upon cities; insights into how cities are changing; an assessment of the drivers of urban change; a review of lessons learned; an exploration of the opportunities and challenges cities face in an international context; and an assessment of policy implications to enable cities to build upon the progress that has been made in recent years (ODPM 2006a).

An Urban White Paper commitment (DETR 2000a), the main themes identified and researched were demographics, economic competitiveness and performance, liveability, social cohesion, and governance. ODPM (2006a) argued that a state of the cities assessment was important in the context of a growing recognition of the importance of cities in creating sustainable communities. It also argued that a better understanding was needed of the processes and dynamics of change that are reshaping cities.

A key finding of both DETR (2000a) and ODPM (2006a) was that policy integration, partnership and local authority leadership were essential to effective urban regeneration. DETR (2000a) also identified the need to rethink scales of intervention to tie city policies to the wider frameworks of regional and sub-regional strategies (see also Goodwin 2004; Atkinson 2007). Urban Task Force (1999) and ODPM (2006a) argued for urban policy to be based around larger areas along the lines of the 'city-region' (see Boddy and Parkinson 2004).

Weaknesses in urban policy highlighted by ODPM (2005d; 2006a) reflect some wider issues, namely the fragmentation of urban policy, the sheer plethora of initiatives, the short-term nature of policy responses, and resource constraints. However, the *State of the English Cities* (ODPM 2006a) noted advances regarding the recognition of the role and importance of cities in economic competitiveness, an increased awareness of the spatial impacts of sectoral policies, and indications of improvements in the relationship between central and local government (see Chapter 6).

A wider report on the competitiveness of English cities compared with their European competitors revealed that the English 'core cities' of Birmingham, Bristol, Leeds, Liverpool, Manchester, Newcastle, Nottingham and Sheffield (see ODPM 2004b; Comedia 2002), were recovering from a period of decline. However, these cities 'lag behind their European counterparts . . . [in] innovation, workforce qualifications, connectivity, employment rates, social composition and attractiveness to the private sector' (ODPM 2004b, p. 50). Within the broader context of the urban renaissance and the state of English cities, this chapter moves on to looking at, first New Labour's neighbourhoods and communities policies, and, second, the city, regional and regeneration policies established since 1997, now linked with the sustainable communities agenda.

NEW LABOUR'S POLICY FOCUS ON NEIGHBOURHOOD REGENERATION AND SOCIAL EXCLUSION

AREA-BASED INITIATIVES (ABIs)

As was demonstrated in Chapters 2 to 4, over the post-war decades urban policy has attempted numerous area-based initiatives to address deprivation, including the Urban Programme, Urban Development Corporations and City Challenge. The approach had various shortcomings, such as lack of community involvement, an excessive emphasis on economic and property development, and insensitivity to local needs (Healey et al. 1992; Robson et al. 1994). Despite this, in the early years of New Labour's administration, area-based or 'neighbourhood' explanations of deprivation gained a new momentum (Chatterton and Bradley 2000). ABIs announced at this time included SRB (rounds five and six), NDCs, Sure Start, and Employment, Education and Health Action Zones. Chatterton and Bradley (2000) indicated that these varied from earlier attempts in that they were co-ordinated from the Social Exclusion Unit and 18 Policy Action Teams, and represented a more integrated and joined-up approach.

However, in an early critique of New Labour's raft of area-based regeneration policies, Chatterton and Bradley (2000) argued that such policies would be ineffective in addressing deprivation. It was argued that policies such as NDCs and various Action Zones were likely to replicate the failures of previous initiatives, as they continue to simplify the complex processes underlying regeneration (Stewart 1990; Oatley 1995a). Area-based policies arbitrarily draw boundaries around regeneration areas, they continue a focus on place commodification, and they represent accountability structures that rest with statutory bodies rather than the community (Chatterton and Bradley 2000).

SINGLE REGENERATION BUDGET UNDER NEW LABOUR

New Labour's management of Single Regeneration Budget was continued from the Conservative-government-managed version that ran from 1994 (see Chapter 4). The last two rounds of SRB under New Labour attempted to refocus and target the funding more directly to places most in need, with 80 per cent of funding channelled to comprehensive schemes in the 50 most deprived local authority areas, and the other 20 per cent targeted at deprived coalfields, rural areas and coastal towns (Cullingworth and Nadin 2006, pp. 270–271). From April 2002, the Single Regeneration Budget was subsumed into the Regional Development Agencies' Single Programme and this met the resource obligations to SRB bids from Rounds 1 to 6 (CLG 2007d). The last SRB projects reached their end in March 2007 and many projects ceased to continue because of a failure to secure replacement funding. In this respect, SRB has failed to represent a sustainable long-lasting form of urban policy in the case of many projects.

The final review of SRB (CLG 2007d) offered lessons for future urban policy. It was noted that there remained a need for both more customised policy delivery for deprived areas and more funding for deprived areas relative to other areas along the lines of the final two rounds of SRB. How to achieve this targeting remained a question post-SRB. The final evaluation made clear that a priority of SRB was to enhance the involvement of the private sector in local urban regeneration, which had been missing in many deprived urban areas at a local level prior to the 1990s.

NEW LABOUR'S 'THIRD WAY' AND THE FOCUS ON SOCIAL EXCLUSION

At the heart of many of New Labour's early public policies was the attempt to bring together the market and the state, representing what has been termed a 'Third Way' (Giddens 1998; 2000; Tiesdell and Allmendinger 2001; Imrie and Raco 2003b; Johnstone and Whitehead 2004b). This perhaps can be represented as 'an uneasy and problematic marriage of the large-scale anti-poverty programmes of the postwar social democratic state, with the economic imperatives of Thatcherite neo-liberal urban policy' (Johnstone and Whitehead 2004b, p. 9). There was initially a concerted effort by New Labour to maintain controls over public spending on urban policy, initially retaining the SRB. However, the government also sought to target urban regeneration spending on the most deprived urban areas through a raft of area-based social welfare policies (Johnstone and Whitehead 2004b). New Labour has not embraced Keynesian redistributive economic policies characteristic of Labour administrations of the 1960s and 1970s, thus representing continuity with the approaches adopted by previous Conservative administrations.

In a similar vein, Clarence and Painter (1998) called New Labour's approach a collaborative discourse, which can be seen as a shift away from the input-driven systems of the post-war welfare state and the market-oriented approaches that came from the policies of the New Right in the 1970s and 1980s that aimed to control spending. The emphasis of policy shifted towards the in-vogue ideas of 'joined-up thinking', 'cross-cutting issues' and 'citizen-centred' services. While broadly welcoming the holistic approach attempted by City Challenge and the SRB (see Chapter 4), the government noted that policies had generally had a limited impact on tackling social exclusion and empowering local communities, and that urban policy had been short-term when the problems needed long-term action. The government therefore recognised the importance of addressing urban social exclusion and that urban regeneration required long-term action. Social exclusion was given an immediate high priority, signalled by the setting-up of the Social Exclusion Unit (SEU) in 1997 (Chapter 1).

The SEU published a consultation document in 1998 called *Bringing Britain Together: A National Strategy for Neighbourhood Renewal*, and represented an attempt by government to develop a new approach to urban regeneration, building upon the SRB rather than tearing it down. It did, however, note some past failures of regeneration initiatives, namely:

> [T]he absence of effective national policies to deal with the structural causes of decline; a tendency to parachute solutions in from outside, rather than engaging local communities; and too much emphasis on physical renewal instead of better opportunities for local people. Above all, a joined up problem has never been addressed in a joined up way. Problems have fallen through the cracks between Whitehall departments, or between central and local government. And at the neighbourhood level, there has been no one in charge of pulling together all the things that need to go right at the same time (SEU 1998, p. 9).

The aim of New Labour was to create a holistic and strategic national approach within which ABIs such as the SRB would play a key role. In addition, nine Regional Development Agencies (RDAs) were established in 1999 with the aim of developing a regional strategy to address urban (and other) problems.

Following the publication of *Bringing Britain Together*, 18 Policy Action Teams (PATs) were created in 1998 bringing together practitioners, academics and residents from deprived neighbourhoods. These covered themes such as jobs, neighbourhood management, unpopular housing, anti-social behaviour, community self-help, schools and joining it up locally (www.neighbourhood.gov.uk). The PATs made almost 600 recommendations, which fed into the preparation of a consultation document on the National Strategy for Neighbourhood Renewal. Cullingworth and Nadin (2006) argued that despite the government's protestations that this heralded a new approach to policy making, there was little new thinking demonstrated in the proposals. Oatley (2000) pointed out the resemblance to the 1977 Urban White Paper (see Chapter 2), criticised the reliance on ABIs, and concluded that there was little hope for a significant impact on urban poverty.

In the vein of the focus on social exclusion and neighbourhood renewal, New Labour sought to address those neighbourhoods and communities most in need by using a set of Indices of Multiple Deprivation (see Chapter 1). This is a statistical instrument to measure levels of disadvantage in small areas to target urban policy spending. The latest Indices of Multiple Deprivation were introduced in 2007 and combine a number of indicators chosen to cover a range of economic, social and housing issues, into a single deprivation score for each small area in England. This allows each area to be ranked relative to one another according to their level of deprivation by the indices of income; employment; health and disability; education, skills and training; barriers to housing and services; living environment; and crime (www.communities.gov.uk).

THE NEW DEAL FOR COMMUNITIES (NDC)

In policy terms, NDCs were the result of the 1998 Social Exclusion Unit report, and in addition to SRB was the flagship urban regeneration initiative focused on neighbourhood and community renewal. This programme was heralded as New Labour's answer to social exclusion and was designed to tackle multiple deprivation in the poorest areas (SEU 1998; Atkinson 2000b; Foley and Martin 2000; Lawless 2004; 2006; Rallings et al. 2004). The aim was to help some of the most deprived neighbourhoods in the country by giving grants to community-based partnerships for neighbourhood renewal. NDCs eliminated competition, which had been characteristic of the last decade of urban policy. Community partnerships in specific deprived areas across England were eligible, and 17 first-round NDCs were originally approved in 1998/1999 and established in 2000 as 'pathfinders', followed by 22 second-round partnerships in 2001 taking the total to 39 (see Table 5.1). Up to £1.9 billion was to be invested over 10 years (making them longer lasting than SRB projects), with each receiving £40–50 million over their lifetime. Each NDC partnership was to also plan for maintaining neighbourhood initiatives after the funding had ceased. The intention was that the policies rolled out of the Neighbourhood Renewal Unit would apply the lessons learnt through the NDCs.

The key characteristics of the NDCs were community involvement and ownership; joined-up thinking and solutions; action based on evidence about 'what works'; long-term commitment to deliver real change; and communities at the heart in partnership with key agencies. Five inter-related key areas were targeted under NDC policy comprising worklessness, health, education, crime and community safety, and housing and the environment (Plate 5.1).

NEIGHBOURHOOD RENEWAL FUND (NRF)

A less adventurous and less generously funded source of regeneration funding was created by the 2000 Comprehensive Spending Review – the Neighbourhood Renewal Fund (NRF), which began in 2001 (Atkinson 2003, p. 165). This illustrated the government's continued commitment to Neighbourhood Renewal. It was directed at the 88 most deprived local authority areas and aimed to improve the public services of these localities. Initially the NRF was set up to operate for three years with total spending of around £900 million. The 2002 Comprehensive Spending Review extended its life for a further three years and allocated an additional £975 million for the final two years. However, this was criticised for being relatively short-term when compared with seven-year SRB projects and 10-year NDCs.

LOCAL STRATEGIC PARTNERSHIPS (LSPs)

Local Strategic Partnerships (LSPs) were created to ensure strategic and joined-up working at the local level to contribute to neighbourhood regeneration. They were originally associated with the 88 most disadvantaged local authority areas contained within the 2001 National Strategy for Neighbourhood Renewal action plan (SEU 2001; see Oatley 2000), and allocated and oversaw the use of the Neighbourhood Renewal Fund (see Johnson and Osborne 2003; Keith 2004). LSPs are the main policy vehicle for delivering regeneration in England; they are cross-sectoral umbrella partnerships that bring together the public, private, voluntary and community sectors to provide a single overarching local co-ordination framework within which other, more specific partnerships can work. They typically include local government, local health and education authorities, the police and community representation (Smith et al. 2007a, p. 227).

TABLE 5.1 NEW DEAL FOR COMMUNITIES

Round 1 Partnerships (2000)	*Grant approved (£m)*
Birmingham (Kings Norton)	£50,000,000
Bradford (Little Horton)	£49,975,000
Brighton (East Brighton)	£47,200,000
Bristol (Barton Hill)	£49,994,876
Hackney (Shoreditch)	£59,400,000
Hull (Preston Road)	£54,969,000
Leicester (Braunstone)	£49,500,000
Liverpool (Kensington)	£61,912,300
Manchester (Beswick and Openshaw)	£51,725,000
Middlesbrough (West)	£52,126,000
Newcastle (West Gate)	£54,900,000
Newham (West Ham and Plaistow)	£54,565,000
Norwich (North Earlham and Marlpit)	£35,200,000
Nottingham (Radford)	£55,112,000
Sandwell (Greets Green)	£56,000,000
Southwark (Aylesbury Estate)	£56,200,000
Tower Hamlets (Ocean Estate)	£56,600,000
Total:	£895,379,176
Round 2 Partnerships (2001)	*Grant approved (£m)*
Birmingham (Aston)	£54,000,000
Brent (South Kilburn)	£50,060,000
Coventry (Wood End, Henley Green, Manor Farm)	£54,000,000
Derby (Derwent)	£42,000,000
Doncaster (Doncaster Central)	£52,000,000
Hammersmith and Fulham (North Fulham)	£44,268,000
Haringey (Seven Sisters)	£50,121,000
Hartlepool (West Central Hartlepool)	£53,794,000
Islington (Finsbury)	£52,900,000
Knowsley (North Huyton)	£55,800,000
Lambeth (Clapham Park)	£56,000,000
Lewisham (New Cross Gate)	£45,000,000
Luton (Marsh Farm)	£48,835,000
Oldham (Hathershaw and Fitton Hill)	£53,530,000
Plymouth (Devonport)	£48,725,000
Rochdale (Old Heywood)	£52,300,000
Salford (Charlestown and Lower Kersal)	£53,000,000
Sheffield (Burngreave)	£52,000,000
Southampton (Thornhill)	£48,700,000
Sunderland (East End and Hendon)	£53,895,000
Walsall (Bloxwich East and Leamore)	£52,000,000
Wolverhampton (All Saints and Blakenhall)	£53,500,000
Total:	£1,126,131,000
Overall Total (£m):	£2,021,510,176

Source: **www.neighbourhood.gov.uk.**

Plate 5.1 'Community at Heart' New Deal for Communities, Barton Hill, Bristol

LOCAL AREA AGREEMENTS (LAAs)

Local Area Agreements (LAAs) are of growing importance in co-ordinating central and local government expenditure since being piloted in 2004 and following a decision in 2005 to roll them out to all upper-tier local authorities in England by 2007. These are three-year agreements based on a 'Sustainable Communities Strategy' that sets out the priorities of action between central government, represented by the Government Office and a local area represented by the local authority and the LSP. The priorities are shaped to reflect certain central government targets and the local area agreement includes the notion of freedom and flexibility to join up budgets and services to meet local need (Smith et al. 2007a, p. 226). In Scotland, Community Planning Partnerships are the equivalent and perform a slightly different role (see www.communitiesscotland.gov.uk).

MAINSTREAMING

Linked with targeting exceptional pots of money in time-limited programmes, and in the context of joined-up policy-making, New Labour has sought to transform existing public services and universal funding streams so that they tackle particular problems as well as providing a general service in deprived areas (Smith et al. 2007a). This strategic 'mainstreaming' approach to urban regeneration has been a key theme of post-1997 urban policy aimed at neighbourhood regeneration and addressing social exclusion. This approach recognises that area-based programmes such as NDCs are unable on their own to tackle deeply ingrained and complex problems of disadvantage. Disadvantage and exclusion must rather be addressed through existing 'core' or mainstream expenditure on the public services that are already present in these neighbourhoods. Mainstreaming therefore goes beyond merely spending more money in disadvantaged areas. It aims to get public services such as the police, education authorities and health trusts to respond more effectively to local issues and conditions (see Bramley et al. 2005; Smith et al. 2007a). It can be argued that the managerial and performance indicator culture of the New Labour government acts to constrain some of these policies, especially in respect of tailoring to local context (see Smith et al. 2007a).

Other post-Urban White Paper strategies linked with neighbourhood renewal and social exclusion include the Neighbourhood Renewal Unit (NRU), which was created to carry forward the National Strategy for Neighbourhood Renewal (see SEU 2001; Smith et al. 2007a). The Regional Co-ordination Unit (RCU) was to ensure that any central policies with a local impact complement one another and that RDAs and GORs work together.

EMPLOYMENT, TRAINING, ENTERPRISE AND URBAN REGENERATION

New Labour has placed a focus on employment, training and enterprise within neighbourhood renewal and social exclusion policies. Employment outcomes have been one of the key economic considerations within urban policy and economic competitiveness more generally (Chapter 6). However, Cullingworth and Nadin (2006, p. 381) argued that employment policies have never been successfully integrated with physical planning policies, and the same case can arguably be made for urban regeneration policies. This is partly due to organisational separatism, and local authorities have little responsibility for employment and training. Initiatives have been devolved to GORs, which expanded from the early 1990s when 72 Training and Enterprise Councils (TECs) in England and Wales, and 22 Local Enterprise Councils (LECs) in Scotland were established (Cullingworth and Nadin 2006; Bennett 1990).

Drawing on the US experience, these were privatised organisations in line with the government philosophies of the time (see Chapter 3) with objectives to localise and decentralise training policy (Bennett 1990). Cullingworth and Nadin (2006) illustrated the underlying rationale of TECs and LECs, which were to give local businesses a central place in guiding support programmes for employment and training. The aims were to formulate a better response to local employment needs, achieve a more business-like mode of operation, and to increase the leverage of private sector funds for training (Lewis 1992).

Under the New Labour government in 2001, the TECs were merged with the Further Education Funding Council in the Learning and Skills Council (LSC) following an Education White Paper (Cullingworth and Nadin 2006). The LSC had a budget of £8 million in 2003/2004 and was responsible for all post-16 education and training other than higher education, including youth and employment training. The LSC works through a network of 47 regional offices tailoring policy to local circumstances (see www.lsc.gov.uk). This approach can be seen as 'soft' regeneration, and education and training aims are incorporated within area-based policies such as NDCs.

URBAN POLICY IN SCOTLAND, WALES AND NORTHERN IRELAND

Many of New Labour's urban policies such as the Urban White Paper and policies rolled out from the Neighbourhood Renewal Unit have been focused on England. Since the late 1990s, responsibilities for many areas of public policy have been devolved to administrations in Scotland, Wales and Northern Ireland. In terms of urban policy, the approaches taken by these countries have been fairly consistent and similar to the post-1997 New Labour approaches witnessed in England, particularly the focus on social exclusion, deprived communities and neighbourhoods, and the renaissance of run-down urban areas. The Scottish Parliament and the Welsh Assembly have adopted policies not unlike NDCs and the Neighbourhood Renewal Fund seen in England (Cullingworth and Nadin 2006, pp. 383–385).

In Wales, the Welsh Assembly distributed £83 million to its 88 most deprived communities between 2001/2002 and 2004/2005 through its Communities First initiative (National Assembly for Wales 2001). Communities First is the Welsh Assembly's flagship urban policy programme to tackle deprivation in Wales. It aims to promote social justice; create an equitable environment where people feel empowered; promote a culture in which diversity is valued and equality of opportunity is a reality; and involve local people in leading the Communities First process. It is a long-term investment in the most deprived Welsh communities, and is embedded in 142 communities across Wales (www.communitiesfirst.info). In terms of urban regeneration in Wales, the Ministers for Social Justice and Local Government, and Sustainability and Housing had responsibility in 2008 (see www.assemblywales.org).

Similarly, the Scottish Parliament's Minister for Communities has responsibility for housing and regeneration. Urban policies include Communities Scotland, which is an agency focused on regeneration in Scotland (www.communitiesscotland.gov.uk). Albeit devolved responsibilities, the Scottish Government (www.scotland.gov.uk), currently has in place a plethora of area-based regeneration and community renewal policies along the lines of England's urban policies.

The revived Northern Ireland Assembly and Government spread matters relevant to urban regeneration across a number of areas of government transferred from the Westminster Parliament, including regional development and social development (www.niassembly.gov.uk; www.northernireland.gov.uk). The Department for Social Development has the overall remit for urban regeneration, and like other parts of the UK, focuses its activity on neighbourhood renewal, and neighbourhood partnerships in areas of deprivation, particularly in Belfast and Londonderry (see www.dsdni.gov.uk). This again reflects the overall approach of New Labour since the late 1990s, and policy approaches appear to have been largely transferred to the other parts of the UK. Some examples of these are reflected within the underpinning themes within urban regeneration in the UK in Section III.

SUSTAINABLE COMMUNITIES AND URBAN REGENERATION INITIATIVES

Communities and Local Government (CLG) is the government department that currently oversees urban regeneration in the UK. In late 2008, the main regeneration delivery vehicles or programmes were listed as: Coalfield Areas and Communities; English Partnerships; Urban Regeneration Companies; and Single Regeneration Budget (now subsumed into the RDA Single Programme) (www.communities.gov.uk). The government argues that building sustainable communities is at the heart of regeneration, and by working with English Partnerships and other agencies including the Regional Development Agencies, the government aims to develop more land for residential and commercial uses, provide resources to stimulate regional economic growth, create jobs, and provide strategic direction and investment to coalfield communities. Illustrating the sheer plethora of programmes and initiatives to promote regeneration in the UK, English Partnerships operates, or is a participant in, around 20 programmes and initiatives on behalf of CLG (Table 5.2).

In late 2008, priority areas in terms of urban regeneration, which are discussed later in this chapter, were defined by English Partnerships as:

TABLE 5.2 ENGLISH PARTNERSHIPS' URBAN REGENERATION PROGRAMMES, 2008

First-Time Buyers' Initiative
Hospital Sites Programme
Housing Action Trusts
Housing Gap Funding
Housing Market Renewal Pathfinders
Land Sales Programme
Land Stabilisation Programme
The London-Wide Initiative
Millennium Communities Programme
Milton Keynes Partnership Committee
National Brownfield Strategy
National Coalfields Programme
National Land Use Database
New Towns
Regional Housing Boards
Register of Surplus Public Sector Land
Strategic Joint Ventures
Town Centre Regeneration
Urban Development Corporations
Urban Regeneration Companies

Source: **www.englishpartnerships.co.uk.**

- the 20 per cent most deprived wards in the country, as defined by the CLG's Indices of Multiple Deprivation;
- the Coalfields;
- Urban Regeneration Company areas;
- areas of major housing growth in the South-East including the four growth areas;
- Housing Market Renewal areas, including Housing Market Renewal Pathfinders;
- the Northern Growth Corridor and other emerging cross-regional initiatives;
- strategic brownfield sites or hardcore brownfield land in, or adjacent to, any of the above priority areas or in areas of housing pressure or housing abandonment.

Direct intervention through such urban regeneration initiatives is often required to overcome market failures and lack of investment in areas of dereliction. This is because in areas of decline there are cycles of deprivation, a loss of community pride and a stigma attached to the area, which creates a negative perception that puts off potential investment. The built environment is often run-down and of poor quality and public sector intervention is required to revitalise it and create a positive climate for investment and development. Physical regeneration is a key aim, but there are also the inextricably linked social and economic dimensions (see Chapter 1). Policies and funding come from central government, for example, from the Neighbourhood Renewal Fund and NDCs, while at the European Union level policies include, for example, structural funds, European Regional Development Fund, European Social Fund and Objective 1 and 2 programmes (see Cullingworth and Nadin 2006, pp. 381–383).

Public sector intervention is necessary to stimulate private sector investment and there has been a growing recognition of the need for the public and private sectors to work in partnership (Bailey et al. 1995; see Chapter 3). Practical problems concern the role of the local authority, and how to ensure fair competition for developers, which depends on land ownership. Gap funding could be used, particularly in terms of the provision of infrastructure and the need for strategic interventions (Bennett 2005). Central government responses have recognised the need for some form of specialist mechanism or special purpose body to combine the powers, resources and leadership needed to drive through change by way of what are sometimes referred to as urban regeneration 'delivery vehicles'. A number of these urban regeneration initiatives since 1997 are highlighted below. These represent a small selection, but are considered the most important in terms of their funding and impacts, and are reflective of the underlying political philosophies of the recent policy period.

ENGLISH PARTNERSHIPS (EP)

Launched in 1993, English Partnerships (EP) is the government's regeneration agency for England and almost finished with the establishment of the Regional Development Agencies in 1999. However, reports of its demise turned out to be premature; EP returned with a significantly increased role following the launch of the government's 2003 Sustainable Communities Plan as made clear by Table 5.2 (www.englishpartnerships.co.uk; ODPM 2003a). EP's focus is on physical development and regeneration, and it wields land assembly and compulsory purchase powers, which it uses to purchase derelict land and bring it back into active use (www.englishpartnerships.co.uk). EP either develops sites itself or awards gap funding to developers to do so. It also sits on the boards of the Urban Regeneration Companies, which are explored later in the chapter, and provides regeneration advice and consultancy. It now concentrates its action on developing its land assets and portfolio of strategic sites, creating development partnerships, improving the environment through land renewal and development, and finding new sources of funding to match public resources (Cullingworth and Nadin 2006, p. 367). EP sets itself targets in terms of land reclaimed, new housing developed, employment floorspace created and private sector investment attracted, which have each shown a steady upward trend during the 2000s. From 2009, English Partnerships is to merge with the Housing Corporation in England, currently the national affordable homes agency that represents and funds social housing, to form the Homes and Communities Agency to help the delivery of sustainable communities (www.homesandcommunities.co.uk).

MILLENNIUM COMMUNITIES

One aspect of the work of EP has been to showcase 'best practice' through demonstration of innovative initiatives such as the Millennium Villages or Communities, the notable example being at Greenwich (DETR 1999; ODPM 2003a). The programme was launched in 1997 and will involve around 6000 homes in seven locations across England (Table 5.3; Figure 5.1). The aim is to influence the practice of the house building industry and local authorities by showcasing how the design and construction of new communities can incorporate sustainability principles including good transport links, energy efficiency, a mix of housing and employment opportunities, and community involvement (Cullingworth and Nadin 2006; www.englishpartnerships.co.uk).

The Greenwich Millennium Village project started in 1999 and aimed to deliver 1400 new homes in a 'sustainable and socially cohesive environment' by 2003. The 32-acre village aimed to set new standards in energy and water efficiency in line with contemporary sustainability concerns (see Chapter 9). Further illustrating social and environmental sustainability aims, of the 1400 new dwellings, 172 are available for rent, 54 for shared ownership and a further 40 reserved for flexible tenure occupation (www.englishpartnerships.co.uk/gmv.htm). Thus, the village was seen as more than just another scheme to mark the Millennium; it was intended as the blueprint for future house building in the UK and should be seen within the

TABLE 5.3 MILLENNIUM COMMUNITIES PROGRAMME

Greenwich Millennium Village, London
Allerton Bywater Millennium Community, near Leeds
New Islington Millennium Community, Manchester
South Lynn Millennium Community, King's Lynn
Telford Millennium Community
Oakgrove Millennium Community, Milton Keynes
Hastings Millennium Community

Source: **www.englishpartnerships.co.uk.**

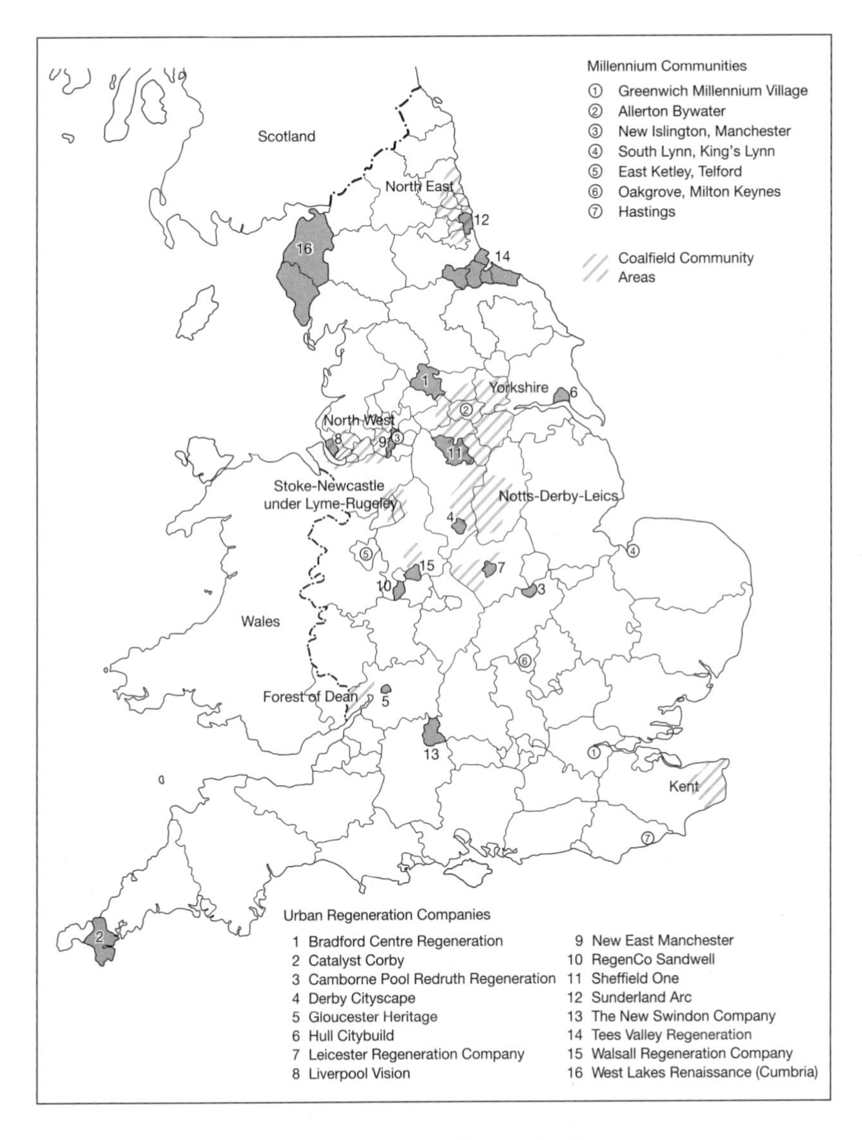

Figure 5.1 Millennium Communities and National Coalfields Regeneration Programme
Source: Cullingworth and Nadin (2006, p. 369).

wider context of the government's aim to restrict greenfield development and encourage the redevelopment of brownfield sites (DETR 2000a; ODPM 2003a). The village is being developed on a previously contaminated site by a private sector consortium and is close to completion. Plans have been made for further Millennium Communities (see Table 5.3; www.fmb.org.uk; Chapter 11).

The programme overall has been criticised for 'seeking trend-breaking results through a fairly conventional large scale top-down commercial development process, such that non-commercial outcomes could be achieved only through the imposition of conditions and use of subsidies' (Cullingworth and Nadin 2006, p. 368). It remains to be seen whether the Millennium Communities will influence mainstream volume house builders' approach to difficult development sites.

The principles have also been incorporated within the more recent announcement by Prime Minister Gordon Brown in 2007 of the intention to develop 10 new 'eco-towns' (see Chapter 9).

ENGLISH COALFIELDS REGENERATION PROGRAMMES

Former coalfield communities across the UK have experienced significant social and environmental problems since the mine closures of the 1980s and 1990s, including chronic health issues, skills shortages, poor educational attainment, poor access to jobs and weak transport links (www.communities.gov.uk). Several central government programmes have been set up as a response to the problems generated by the closures and to the Coalfields Task Force report recommendations (DETR 1998b). These programmes are attempting to address the problems faced by these communities and to improve the quality of life for people living in these areas of the country (CLG 2007b).

The National Coalfields Programme is delivered by English Partnerships and the Regional Development Agencies and is funded to remediate 107 coalfield sites across England (as at late 2008) and to encourage new business and job creation. It has one of the largest portfolios of contaminated land in Europe comprising around 4,550 hectares (www.englishpartnerships.co.uk; Cullingworth and Nadin 2006, p. 368). The National Coalfields Programme is supported by the sustainable communities plan (ODPM 2003a), and has a budget of £386 million. Sites of coalfield communities and former coalfield areas in England cover the North-East (17 sites), North-West (8), Yorkshire (33), West Midlands (6), East Midlands (26), South-East (4) and South-West (6). The National Coalfields Programme is on course to create 42,000 new jobs, 2 million square metres of commercial floorspace, 8000 new homes, and generate over £1 billion of private sector investment and benefits for local communities (www.communities.gov.uk).

In addition, the Coalfields Enterprise Fund is a commercial venture capital fund set up to support the growth of businesses and encourage entrepreneurship in England's former coalfields. The fund aims to attract over £20 million of new investment into these areas, to stimulate new-start companies, to assist established businesses seeking to expand, and to encourage industrial diversification (see www.coalfields-enterprise-fund.co.uk).

Finally, the Coalfields Regeneration Trust is an independent UK-wide grant-making body launched in 1999 as part of the government's response to the Coalfield Task Force Report (DETR 1998b). Funding of over £150 million in England alone has been allocated to the Coalfields Regeneration Trust, which is a key element of the programme to regenerate the coalfield communities of England, Scotland and Wales socially and economically. The Scottish Executive and the Welsh Assembly provide funding for the Trust in Scotland and Wales.

REGIONAL DEVELOPMENT AGENCIES (RDAs)

In 1999, New Labour set up nine bodies to oversee regional development in England, the Regional Development Agencies (RDAs) (Table 5.4; www.englandsrdas.com).

The RDAs were given a largely economic development-related remit. They were charged with creating jobs, reclaiming land, assisting business start-ups and investing mainly in the deprived areas of their regions. Site assembly and compulsory purchase powers were awarded to enable this remit. Essentially, RDAs took over the work of English Partnerships, which has since been adapted to concern itself more with housing and sustainable communities (Cullingworth and Nadin 2006). RDAs inherited English Partnerships' substantial land and property holdings, although these varied considerably. For example, SEEDA had very few, and this has conditioned its operation towards that of a land buyer rather than a land seller.

RDAs are funded by a number of government departments, principally the CLG. Their funding is tied to the achievement of targets in four main areas: numbers of jobs created, performance of businesses the RDA invests in, amount of brownfield land reclaimed, and amount of private investment levered into deprived areas. The early days of RDAs were occupied with the production of master plans for parts of their regions. These were based on extensive consultation as the agencies attempted to develop regeneration solutions that were appropriate to different areas. More recently there has been a feeling that RDAs have consulted and planned enough, and that the time has come to start delivering regeneration objectives. Many local authorities and communities want to see the RDAs providing more leadership relating to this. This may be difficult due to a lack of skills within the RDAs. Regeneration professionals are often in great demand, and the rewards for this sort of work in the private sector can outstrip those offered by the RDAs. Cynics may argue that RDAs were starved of money from the outset, but the reality is that they have a considerable degree of autonomy to pool money from different government offices.

RDAs also work with the private sector in partnerships and some have taken this approach to heart and have operated almost as land and property traders. Yorkshire Forward sold its entire investment property portfolio, valued at £53 million, to a private property management company in 2002. One North East employs UK Land Estates to manage its £120 million property portfolio, with an option to sell it to the company before 2014. At the other end of the scale, SWRDA favours a more traditional approach of in-house management and sale of surplus stocks through auction, although they are similarly reducing their property holdings. The RDA's approach is to hold on to assets until they have reached their regeneration purpose, for example, an industrial estate that is 90 per cent let, and then place them on the market. Somewhere in between are the NWDA, AWM and EMDA, watching closely how ONE and Yorkshire Forward fare. NWDA has outsourced its investment portfolio, but keeps its development portfolio in-house, whereas EMDA favours

TABLE 5.4 REGIONAL DEVELOPMENT AGENCIES IN ENGLAND

Regional Development Agencies
Advantage West Midlands (AWM)
East of England Development Agency (EEDA)
East Midlands Development Agency (EMDA)
London Development Agency (LDA)
North West Development Agency (NWDA)
One North East (ONE)
South East England Development Agency (SEEDA)
South West of England Development Agency (SWRDA)
Yorkshire Forward

Source: **www.englandsrdas.com.**

partnership working, and is aware of the different approaches between the private sector and the RDA.

RDAs deliver regeneration objectives in a very direct manner. The process could follow a pattern of purchase of a site, remediation of the site, award of gap funding grants to assist reclamation, development of the site either as a joint venture with a private company or by the private sector exclusively, and finally disposal of the developed site to an investor (www.englishpartnerships.co.uk).

To demonstrate the impact that an RDA is having on a relatively peripheral region, the SWRDA's role in regional planning and economic development can be illustrated using some facts from 2006/2007. SWRDA created 5549 new jobs, helped 1936 people to enter employment, and assisted 25,348 people with skills development. SWRDA created or attracted 271 new businesses to the region, helped 15,898 firms to improve performance, and assisted 647 enterprises in new collaborations. The RDA also attracted over £255 million of investment to regenerate deprived areas, 59 per cent of which was levered in from the private sector, and regenerated over 86 hectares of brownfield land (www.swrda.co.uk). In 2006/2007, all of England's RDAs met or exceeded their targets, but caution has been expressed over conflict between the objectives of RDAs and central government (see Cullingworth and Nadin 2006, pp. 60–61).

URBAN REGENERATION COMPANIES

In looking to draw lessons from past initiatives, the Urban Task Force (1999) report examined how the conditions for delivering sustained regeneration could be created at the local level. Effective partnership working was seen as vital to this process, and the report recommended the setting up of Urban Regeneration Companies (URCs) as a mechanism for bringing together the key stakeholders to drive forward the regeneration of a particular area (www.urcs-online.co.uk). This recommendation was supported by government and in 1999 three pilot URCs were established in Liverpool, East Manchester and Sheffield. By late 2008, 22 URCs had been set up in England, Wales (Newport Unlimited) and Northern Ireland (Ilex) (Table 5.5). In addition, three Economic Development Companies (EDCs) that were previously URCs have been established, Creative Sheffield (formerly Sheffield One), Hull Forward (formerly Hull Citybuild) and Liverpool Vision. Additional URCs have been created in the West of Scotland (ODPM 2004e; LGA 2005; www.urcs-online.co.uk). There is no government limit on the number of URCs that can be created, although they must fit with the regional economic strategy and local circumstances and be supported by the relevant Regional Development Agency (see ODPM 2004e; 2004f; 2004g).

URCs are private sector-led organisations co-ordinating development and investment in specific run-down urban areas. They are funded by English Partnerships and the Regional Development Agencies. Like UDCs (see Chapter 3 and below), they have finite life spans of around 10 to 15 years. Other similarities include their emphasis on vision, leadership, dynamic style, and the engagement of the private sector to carry out regeneration. However, they do not have planning or land acquisition powers, and are seen more as a co-ordinating body that can kick start development in an area suffering market failure. They must not duplicate existing activity, and should fill a gap in provision by speeding up the delivery of a critical mass of projects. The government sees the primary role of URCs as addressing significant latent development opportunities and bringing about regeneration through developing and implementing a clear and agreed vision for their area. Their main focus should be on physical regeneration and the re-use of brownfield land (ODPM 2004e).

URCs are significant in that they comprise an area of 303,400 hectares and 1.2 million people live and work within their boundaries. They anticipate creating 163,000 jobs over their lifetime and providing an additional 69,000 new housing units. They have the potential

TABLE 5.5 URBAN REGENERATION COMPANIES

Company Name	*Established*
1st East (Lowestoft and Great Yarmouth)	2005
Bradford Centre Regeneration	2002
Central Salford	2004
CPR Regeneration (Camborne, Poole and Redruth)	2002
Derby Cityscape	2003
Gloucester Heritage	2004
Ilex Urban Regeneration Company (Londonderry and Derry City Council area)	2003
Leicester Regeneration Company	2001
New East Manchester	1999
Newport Unlimited	2002
North Northants Development Company	2006
Opportunity Peterborough	2005
ReBlackpool	2005
RegenCo (Sandwell)	2003
Renaissance Southend	2005
Sheffield One	2000
Sunderland arc	2004
Tees Valley Regeneration	2002
The New Swindon Company	2004
Walsall Regeneration Company	2003
West Lakes Renaissance (Furness and West Cumbria)	2003
Wolverhampton Development Company	2007

***Source:* www.urcs-online.co.uk.**

to attract £14.2 billion of private sector investment (www.englishpartnerships.co.uk). One recent example is Gloucester Heritage, which began in 2004. This URC aims to reclaim and develop 100 acres of brownfield land, repair and re-use 82 historic buildings, develop 300,000 square feet of retail floor space, build between 3000 and 3500 new homes, lever in £1 billion of private sector investment over 10 years, and improve infrastructure with a new mainline railway station and inner relief road (www.urcs-online.co.uk). Other examples are illustrative of the activities of URCs in the 2000s. In Liverpool city centre, a Strategic Regeneration Framework was drawn up with help from consultants, and emphasises stakeholder engagement. In East Manchester, economic regeneration and housing development are the focus, partly being achieved through the URC playing a brokering role in pre-application meetings between developers and the planning authority. The Sheffield city centre URC has promoted the growth of the high-technology sector, better access, an improved public realm, and improved leisure, culture and retail facilities. It has also managed to secure European Union funding (www.urcs-online.co.uk).

CASE STUDY: ILEX URC

One of the URCs to be created has been established in Northern Ireland, where urban problems and issues resemble those of the rest of the UK. Ilex URC in north-west Northern Ireland was established in 2003 by the Office of the First Minister and Deputy First Minister and the Department for Social Development in Northern Ireland. Ilex URC was established to plan, develop, sustain and promote the physical, economic and social regeneration of

Derry–Londonderry (a population of over 107,000) and has specific responsibility to manage and redevelop two former military bases. Mixed-use development is envisaged for both sites and the revitalisation of the riverfront is a key objective. The urban regeneration plan is a response to the ministerial call 'to create and promote a deliverable vision for regeneration of the Derry City Council area, to secure the commitment of all stakeholders to that vision and to pursue single-mindedly its implementation' (www.ilex-urc.org).

While the spatial focus of the plan is on the city centre, it is not intended to be a comprehensive or detailed integration of all spatial planning for the area. Apart from developing the two military sites of Ebrington and Fort George, Ilex works in collaboration with a broad range of public and private sector partners in a sustainable regeneration strategy to exploit the strengths and opportunities of the Derry City Council area.

Ilex is the co-ordinating body for the Integrated Development Fund, which is channelling around £33 million into urban regeneration projects, covering themes of tourism, education, enterprise, entrepreneurship, arts and culture, built heritage, and infrastructure. In recognition of the importance of arts and culture within urban regeneration, a total of £4 million has been earmarked by the North West Challenge Fund. £3.2 million is being spent on capital projects in seven venues in Derry–Londonderry and the remaining £800,000 has been set aside to commission a public artwork within the city, the largest public art commission in Ireland.

Ilex is characteristic of the approaches being taken by URCs throughout the UK, focusing on declining central and industrial areas, and developing and co-ordinating flagship retail, leisure and business projects. Source: www.ilex-urc.org.

URBAN DEVELOPMENT CORPORATIONS (NEW GENERATION)

The UDCs set up in the 1980s and early 1990s (see Chapter 3) had all wound up by 2000 with any of their remaining assets being transferred to English Partnerships, who also took on the responsibility for seeing through any uncompleted projects, such as the development and completion of the Temple Quay site in Bristol. A key commitment in the government's Sustainable Communities Plan (ODPM 2003a) was the creation of four new 'growth areas' in the south-east of England at Ashford, London–Stanstead–Cambridge, Milton Keynes/South Midlands and Thames Gateway. This plan recognised the clear inter-relationship between policy and the means of delivering real changes on the ground. Additional government support was promised for infrastructure and stronger delivery arrangements. These arrangements included the use of specialist delivery vehicles, particularly Urban Development Corporations (UDCs) and URCs. Despite a somewhat mixed previous experience, it has been argued that the UDC model provides a vehicle well suited to the government's desire to link policy and delivery arrangements.

Three post-2003 or fifth-generation UDCs have been established. These are in West Northamptonshire (2003–2010), London Thames Gateway (Lower Lea Valley/Barking/Havering Riverside) (2004–2016) and Thurrock (2004–2016) (ODPM 2003a; 2003b; 2003c; 2004c; 2004d; 2005e). Each UDC is guaranteed direct funding from government to the tune of, for example, £60 million for three years in the case of Thurrock. The legislation remains in place (Local Government Planning and Land Act, 1980) that requires an order to be approved by parliament to set up a new UDC. The powers and role of a UDC are essentially the same, which are securing the regeneration of its area through acquiring, reclaiming and disposing of land; improving buildings and the environment; ensuring the provision of housing and social facilities; ensuring the provision of essential services such as water, gas and electricity; and funding infrastructure projects.

The new generation of UDCs has a more explicit requirement to tackle deprivation and promote social inclusion than their predecessors, in line with wider government philosophies (Schopen 2003; ODPM 2004c; 2004d; Cullingworth and Nadin 2006). Working with local communities and agencies on a wider range of skills and social issues is seen as essential. In policy terms at least, it appears that some of the lessons have been learned from the previous round of 1980s and 1990s UDCs, acknowledging some of their failings (see Chapter 3; Raco 2005a).

In terms of planning powers, the legislation makes provision for the UDC to become the planning authority for the whole or part of its area. It is also possible for these powers to be limited to certain kinds of development. For example, in West Northamptonshire, planning powers are limited to major applications for a series of sites where strategic new growth is planned (ODPM 2004c). In Milton Keynes, where English Partnerships inherited land holdings from the former New Town Development Corporation, a partnership committee has been established by English Partnerships. The committee has planning powers for all major applications within the area of the Milton Keynes partnership over the period 2004–2014, but does not operate as a UDC under the powers of the 1980 Local Government Planning and Land Act (LGA 2005; Cullingworth and Nadin 2006, p. 367).

The London Thames Gateway growth area presents a number of challenges in terms of delivering the government's plans for sustainable urban regeneration and growth (see Raco and Henderson 2006). Covering 18 local authority areas and including three government regions, issues of organisation and co-ordination are paramount (Raco 2005a). The government's response to these delivery issues has been to set up a variety of mechanisms at appropriate geographical and political levels. Overall strategic co-ordination is achieved through a Cabinet Committee chaired by the Prime Minister and a new partnership body, the Thames Gateway Strategic Partnership. Regional planning and co-ordination are achieved through sub-regional partnerships, regional planning bodies and the Government Offices for the three regions. Local delivery is achieved through the two new UDCs (London Thames Gateway UDC and the Thurrock UDC) and seven new regeneration partnerships, which are designated as Urban Regeneration Companies (ODPM 2005e).

UDCs are seen as necessary where the scale and intensity of the task of assembling land and site preparation require an agency with additional powers and the ability to generate additional private sector investment (ODPM 2004c; 2004d). UDCs have always had the central mission of attracting development to their areas, which can be termed 'hard' regeneration. The early UDCs were judged on how many derelict sites had been brought back into development, how many homes had been completed and how many jobs had been created. Thus, their areas of operation often constituted sites that had been abandoned by the market, and which suffered from dereliction, contamination and general unattractiveness.

The government has appeared to recognise some of the failings and lessons of the previous generation of UDCs (see Deas and Ward 2000), particularly surrounding the issue of accountability, by taking a new approach that will include more local representation and substantial local accountability (ODPM 2004b). Partnership working and co-ordination are also seen as essential functions, which place more of an onus on the new UDCs to work effectively with their local authority and other partners than was the case before. The involvement of RDAs and the closer involvement of central government and regional offices in their operation are likely to result in more accountability and ultimately a greater acceptance of their role. There have nevertheless been objections to the establishment of UDCs in some of the growth areas, with some environmental and local amenity groups seeing them as a means of pushing through the development of higher housing numbers demanded by the government (see Raco 2005a; 2005b; Raco and Henderson 2006). However, to date the new generation of UDCs has not been as politically or publicly contentious as the 1980s versions (Raco 2005a).

HOUSING MARKET RENEWAL PATHFINDERS

In contrast to the problems of catering for housing demand in growth areas, declining regions and places in the UK face problems of regenerating whole markets. The Sustainable Communities Plan (ODPM 2003a) provides the government framework for a major programme of action that will tackle the pressing problems of communities across England over a 15–20-year period. One of the key areas forming the basis for the action programme is tackling the problem of low housing demand and housing abandonment. Between 2003 and 2006, £500 million was made available for some of the worst affected areas, known as Pathfinder Market Renewal Areas, with the intention of reversing low demand by 2010. In 2008, there were nine Pathfinder Market Renewal Areas (www.englishpartnerships.co.uk). These were 'Bridging Newcastle Gateshead' (Newcastle and Gateshead); 'Gateway Hull and East Riding' (Hull and East Riding of Yorkshire); 'Transform South Yorkshire' (Sheffield, Barnsley, Rotherham and Doncaster); 'Urban Living' (Birmingham and Sandwell); 'Renew North Staffordshire' (Stoke, Newcastle under Lyme and Staffordshire Moorlands); 'Manchester Salford' (Manchester and Salford); 'Newheartlands' (Liverpool, Sefton and Wirral); 'Partners in Action' (Oldham and Rochdale); and Elevate East Lancashire (Blackburn with Darwen, Hyndburn, Burnley, Pendle and Rossendale).

Partnerships between local authorities and other key stakeholders are developing strategic plans for whole housing markets. The bulk of the Market Renewal Fund was allocated, following negotiation with each pathfinder on the basis of outputs and outcomes included in their strategic schemes. The schemes involve radical and sustained action to replace obsolete housing with modern sustainable accommodation, through demolition, refurbishment and new build. They also seek to ensure other essential requirements of sustainable communities are addressed, such as good quality customer-focused services; good design; and clean, safe, healthy and attractive environments in which people can take pride. These requirements are very much in line with the government's wider urban renaissance and sustainable communities agenda (see Chapter 9).

New initiatives are being developed in East Manchester and Hull, based around URC areas, to deal with empty housing, for example, the Empty Property Initiative (New East Manchester) and the Hull Housing Initiative. Both initiatives have looked to form a partnership between their respective local authority, URC and English Partnerships, which would acquire property within the target areas identified. Acquisitions are carried out by the local authority, under the direction of the URC, within the parameters of an agreed business plan, and seek voluntarily to acquire selected empty properties. English Partnerships was the sole funding partner, and had an initial three-year commitment of £4.2 million plus fees for Manchester, and £4.8 million for Hull.

The Langworthy Project in Salford forms part of the Manchester and Salford Pathfinder area and comprises 3300 houses and 4500 residents. Following discussions between developer Urban Splash and the Seedley and Langworthy Partnership Board, proposals were drawn up for the phased development of 477 back-to-back terraced units in the worst affected area. The majority of the units would be retained and converted to provide a range of duplex apartments, and three- or four-bedroom accommodation with private and/or communal (gated) gardens and controlled vehicular access. Total public sector investment was estimated at £18.1 million, with Salford City Council providing half this sum with English Partnerships supporting the remaining half (www.englishpartnerships.co.uk).

However, there have been significant criticisms of the Housing Market Renewal Pathfinders. The principal contention has been that middle-class residents are being encouraged to move to pathfinder areas and that existing lower-class populations are cast out as part of the problem through a gentrification process (Atkinson 2004b; see Lees et al. 2008; Chapter 11). There is emerging evidence that the pathfinders are spurring private sector investor interest in the areas and inflating house prices based on the assumption that they will receive generous public payments when compulsory purchase and demolition of existing housing occur (Atkinson 2004b; Cameron 2006; Townshend 2006).

BUSINESS IMPROVEMENT DISTRICTS (BIDs)

The government introduced legislation in 2004 to enable cities in the UK to set up Business Improvement Districts (BIDs) under Part 4, Local Government Act 2003 and the Business Improvement District (England) Regulations 2004 (see www.ukbids.org). BIDs were originally introduced in Canada in the 1970s, but have been most commonly used in the US since the 1980s. Despite the general idea, there are wide variations in terms of the scale, budget, role, power and mission of BIDS, and therefore different BIDs have been tailored to specifically local conditions in particular urban areas.

Through the establishment of BIDs, the private sector in essence provides public goods in the city centre. Street cleaning, street furniture and security are provided through a supplemental tax paid by private sector business in the BID which they impose on, administer and spend themselves; in essence a voluntary and hypothecated tax. The private sector takes over some of the functions formerly provided by the state. The growth of BIDs has been fuelled by cuts in public sector spending, competition from out-of-town shopping and entertainment malls, de-industrialisation and the associated decline of the city centre, and the perceived efficiency of the private sector in contrast to public sector bureaucracy. The overarching aim is to boost the local economy of the BID area. BIDs are predominantly found in retail spaces where the businesses have an interest in improving the appearance and safety of an area, and some are located in peripheral industrial estates and business parks.

The National BIDs Advisory Service, which is run by the Association of Town Centre Managers (www.atcm.org) summarises the key elements of setting up a BID. This stated that BIDs represent an investment in the local trading environment through the provision of added value services. BIDs are funded by local businesses through a levy on their rates bill, and therefore all beneficiaries pay. Businesses identify the area and the issues and put together a proposal that should include delivery guarantees, performance indicators and management structure. Businesses must vote in favour of a BID in order for it to be established. Interests of large and small businesses are protected through a system that requires a successful vote to have a simple majority in both votes cast and rateable value of votes cast. The plan voted for has a life span of five years and further proposals have to be reaffirmed through a vote.

The pilot wave of UK BIDs from 2004 sought to test the strategy in a range of locations; establish a required level and nature of managerial competence and skills; produce reliable guidance on operational, training and recruitment issues; identify resources, timescales, mechanisms and measurement needed; test and validate the role of the wide range of partner agencies involved; produce a good practice document and an electronic information bank; define a process that is simple, open and inclusive; build capacity; validate guidance and inform secondary legislation; and identify pitfalls (www.ukbids.org).

To give an idea of the spiralling number of BIDs, 27 were initiated during 2007 compared with 16 in 2006, demonstrating their growing popularity among businesses in town and city centres, on industrial estates or across entire districts. The total stood at 72 in late 2008 and BIDs are now beginning to enter their second five-year terms (Table 5.6; www.ukbids.org). There were a further 13 prospective BIDs in late 2008 across England and Scotland.

CASE STUDY: SWANSEA CITY CENTRE BID

Swansea city centre is the location of the only BID currently in operation in Wales, and is in a city that has been the subject of much research on city centre change, decline and

TABLE 5.6 BUSINESS IMPROVEMENT DISTRICTS

Positive BID Ballot Results (November 2004–August 2008)	*Established*
Kingston First (Kingston Upon Thames)	2004
Heart of London Business Alliance	2004
Better Bankside (London)	2005
Holborn Business Partnership	2005
Coventry City Centre BID	2005
Paddington BID	2005
Plymouth BID	2005
New West End Company	2005
Bedford	2005
Lincoln	2005
Birmingham Broad Street	2005
Bristol Broadmead	2005
Blackpool Town Centre	2005
Keswick	2005
Rugby	2005
Liverpool City Central BID (2nd Ballot)	2005
London Bridge	2005
Reading BID	2005
Winsford Industrial Estate	2005
Bolton Industrial Estates BID	2005
Waterloo Quarter Business Alliance	2006
Camden Town Unlimited	2006
Hainault Business Park Business Improvement District	2006
Ealing	2006
Great Yarmouth BID	2006
Hammersmith	2006
West Bromwich Albion BID	2006
Swansea	2006
Brighton	2006
Ipswich	2006
Teesside Cowpen Estate BID	2006
Hull BID	2006
Altham BID (2nd Ballot)	2006
Retail Birmingham BID	2006
Southern Cross BID	2006
Oldham BID	2006
InSwindon	2007
Cater Business Park	2007
Cannock Chase BID	2007
Coventry City Wide BID	2007
Angel Town Centre BID	2007
Heart of London Business Alliance (2nd Term Ballot)	2007
London Riverside BID	2007
Croydon BID	2007
Erdington	2007
Segensworth Estates BID – Winchester	2007
Segensworth Estates BID – Fareham	2007
Argall BID	2007

TABLE 5.6 CONTINUED.

Positive BID Ballot Results (November 2004–August 2008)	*Established*
E11 BID	2007
Sleaford BID	2007
Truro	2007
Worthing Town Centre BID	2007
Winchester BID	2007
Taunton BID	2007
Blackburn EDZ Industrial Estate BID	2007
Kings Heath BID	2007
Nottingham Leisure BID	2007
Dublin City Centre BID	2007
Longhill and Sandgate BID (Hartlepool)	2007
Derby Cathedral Quarter BID	2007
Halebank Industrial Estate	2007
Astmoor Industrial Estate	2007
New West End Company (2nd Term Ballot)	2007
Dorchester BID Company	2008
Coventry City Centre BID (2nd Term Ballot)	2008
Daventry First	2008
Bathgate BID	2008
Inverness BID	2008
BID Leamington	2008
Clackmannanshire BID	2008
Lancing BID	2008
Falkirk BID	2008
Edinburgh BID	2008
Negative Ballot Results (February 2005–August 2008)	
Maidstone BID (1st Ballot)	2005
Altham BID	2005
Runnymead BID (1st Ballot)	2005
Liverpool City Central BID (1st Ballot)	2005
Southport BID (1st Ballot)	2005
Runnymead BID (2nd Ballot)	2005
Maidstone BID (2nd Ballot)	2006
Malton	2006
Leicester	2006
Chester	2007
Southport BID (2nd Ballot)	2007
Bayton Industrial Estate BID	2007
Colchester BID	2007
Shrewsbury BID	2008
Oxford BID	2008

Source: **www.ukbids.org.**

regeneration. Swansea city centre has experienced continued decline since the 1970s (Tallon et al. 2005) but has recently witnessed significant injections of public and private capital reflected in the National Waterfront Museum, Salubrious Place mixed-use development and the SA1 former docks regeneration, which is ongoing. To complement these physical developments, the City Centre Partnership with support from the City Council and the

Welsh Development Agency, successfully secured the first (and to date only) Business Improvement District in Wales, which is focused on the city centre.

Swansea is the fortieth largest shopping centre in the UK but is ranked fifty-second in the retail rankings, thus is significantly under-performing, mainly as a consequence of competition from out-of-town facilities in the city-region. The city experiences one of the highest levels of out-of-town development per head of population in the UK. Negative perceptions of safety, cost and availability of parking, and the poor array of retailers affect the performance of the city centre (Thomas and Bromley 1996; Bromley and Thomas 1997; Tallon et al. 2005).

The five-year city centre BID was established in 2006 following a 74 per cent positive vote on a turnout of 45 per cent and achieved a first year income levy of £440,000 (www.ukbids.org). The BID aims to highlight the importance of the city centre's shopping role and to enhance the current offer, in addition to improving public transport facilities and public space. It is argued that the Swansea City Centre BID represents a genuine partnership between the public and private sectors, and has established the City Centre Clean Team, City Centre Rangers, the Night Time Exclusion Scheme and re-imaged public car parks (www.swanseacitycentre.com). However, the focus of the BID spending is on what might be regarded as superficial measures designed to enhance the streetscape and economic performance of businesses in the city centre (Plate 5.2). It does nothing to address the causes of social and economic decline and the wider problems of city centre communities and neighbourhoods (see Chapter 10; Symes and Steel 2003).

Similar to the Swansea experience, the BID that was established in central Bristol in 2005 provides an example of how this model aims to bring about the regeneration of a UK retail area (www.bristolbroadmead.co.uk/bid; see also Tallon 2008; Chapter 10). An upgrading

Plate 5.2 Swansea city centre Business Improvement District: upgraded public space

project for the Broadmead shopping area is being implemented as a result of the BID, the fourteenth to be established in the UK. A majority (60 per cent) of business votes were in favour of the BID, in which businesses contribute 1.5 per cent of their rateable value over a three-year period, enabling the scheme to go ahead. A total of over £1.2 million is being reinvested in largely cosmetic improvements to the public space of the streetscape, with the aim of attracting more visitors to the Broadmead shopping area (Broadmead BID Bristol 2005). Overall, these improvements are needed to compete effectively with new out-of-town attractions (see Chapter 10), with rival cities in the region, and with other new centrally located developments (see Tallon 2008).

BIDs aim to contribute to the wider urban renaissance, particularly in city centres, which have long been on the urban regeneration agenda (see ODPM 2002; Chapter 10). Symes and Steel (2003) and Steel and Symes (2005) demonstrate the experience of BIDs in the US in terms of their scale, scope, strengths and weaknesses, which can inform the inception and development of BIDs in the UK context. The experiences of BIDs are uncritically regarded as positive by BID authorities in UK cities, but closer inspection shows that their features resonate with wider academic criticisms of the BID model that are beginning to emerge. These include the potential for the BID to fail as had occurred in the US (Symes and Steel 2003); the increasing privatisation and purification of public space; concerns from community groups regarding the expansion of private sector interests into what had been the responsibility of local government (Steel and Symes 2005); and the focus of relatively limited revenue on cosmetic improvements and marketing rather than on flagship retail projects or a more concerted attack on social and economic problems of the BID localities (see Ashworth 2003; Brown 2008; Cook 2008).

EVALUATING NEW LABOUR'S URBAN POLICIES: CONTINUITIES AND CHANGES

Central government faces a number of urban policy challenges that must be addressed if urban policy is to have a realistic chance of achieving its aims. Although it is arguably premature to draw any firm conclusions about the lasting success or otherwise of New Labour's urban policy under Tony Blair, and his successor Gordon Brown from 2007 (see Chapter 14), six potential obstacles to the lasting success of urban regeneration policy in the UK can be put forward to conclude this final chapter tracing policy from the 1940s to the 2000s (see Atkinson 2003; Johnstone and Whitehead 2004b). Each of these present seemingly intractable challenges and these strongly resonate with the problems of urban policy highlighted at the end of Chapters 3 and 4.

1. The sheer scale and intensity of urban problems such as social exclusion and inequality continue to present a massive and multi-faceted challenge (see Gripaios 2002; Atkinson 2003, pp. 172–173; Dorling et al. 2007; Centre for Cities 2008). This is despite a raft of area-based and social group-based initiatives unveiled and initiated since 1997, which have themselves been criticised (for example, Chatterton and Bradley 2000; Oatley 2000).
2. Evidence continues to show growing regional inequality, especially between the north and south of the UK, in addition to inequalities between and within cities. Two main problems continue to exist in the UK: the first is problem localities within prosperous cities; the second is depressed towns and cities *per se* (Gripaios 2002). It was reported in 2007 that the government was 'in denial' about the north–south divide (IPPR 2007).
3. Despite the rhetoric, confronting urban problems in a joined-up fashion within and between levels of governance poses a considerable and continuing challenge (Atkinson 2003, pp. 167–170). Central government departments including CLG; the Home Office; the Department of Health; the Department for Children, Schools and Families; the

Department for Innovation, Universities and Skills; and the Department for Culture, Media and Sport currently have responsibility for different initiatives focused on urban regeneration and neighbourhood and community renewal. There is in addition the work of organisations such as the Local Government Association's 22 partnerships for regeneration (Cullingworth and Nadin 2006).

4. Encouraging community participation and integration continues to be a difficulty (see Chapter 8) as central and local government is limited in its power to coerce or force communities to engage in neighbourhood urban regeneration initiatives. There is also the danger of 'regeneration fatigue' in particular deprived localities and disillusionment with past, often failed, initiatives (see Atkinson 2003, pp. 170–171; Taylor 2000; 2003).
5. Despite successive attempts by government to address the monumental complexity of urban regeneration policy, it remains as complicated, if not more so, than ever. By 2001, the ODPM was representing 87 policy initiatives in addition to the Urban White Paper commitments illustrating a complex web of initiatives. Johnstone and Whitehead (2004b) argued that the metaphor has evolved from 'patchwork quilt' (Audit Commission 1989) to 'bowl of spaghetti' (a description coming from former regeneration minister Lord Rooker), as demonstrated by Table 5.7. Imrie and Raco (2003b, pp. 14–16) presented an expanded set of tables illustrating this explosion of initiatives with impacts on urban areas, and these comprised a vast number of over 150 initiatives. The Environment and Transport and Regional Affairs committee voiced concerns in 1999 that intervention in urban areas was confused and badly co-ordinated and that there should be fewer, better co-ordinated initiatives locally. The reasoning for complexity is that differing conditions of each area require variety and flexibility in the policy response, tailoring the initiatives to local conditions (Cullingworth and Nadin 2006, p. 379).
6. The managerial and performance indicator culture of New Labour with strict controls, centralised targets, unresponsiveness to local geographical variations (in a similar vein to 1980s Conservative policy), league tables, evaluation and so on, all act against some of the aims of recent policies (Johnstone and Whitehead 2004b; Knutt 2005).

SUMMARY: NEW LABOUR'S URBAN REGENERATION

New Labour's approach to neighbourhood renewal, social exclusion and urban renaissance has undoubtedly addressed some of the inherent weaknesses of policies of the recent past (see Chapters 3 and 4), particularly reflected in the move away from property and economic-dominated approaches, and the involvement of communities in the regeneration process.

However barriers to success still cast some doubt on the government's ability to deliver lasting urban regeneration, especially those relating to social exclusion, community involvement and the complexity of the policy landscape.

Some obstacles are deep-seated 'wicked problems' that have taunted governments for over a century (Harrison 2000), some have been created or least exacerbated by urban and economic policies of the past, and some have been brought about by the actions of New Labour's policies.

The trend that urban policy is to take in the near future is at present unresolved, although indications for the directions of Gordon Brown's post-2007 urban policy are assessed in Chapter 14, as is the Conservative Party's latest thinking.

TABLE 5.7 FROM 'PATCHWORK' TO 'SPAGHETTI': URBAN REGENERATION INITIATIVES COMPARED, 1989 AND 2003

'Patchwork quilt' (1989)	*'Bowl of spaghetti' (2003)*
City Grant	Action Team for Jobs
Derelict Land Grant	Active Community Programme
English Estates	Capital Modernisation Fund (small retailers)
Enterprise Allowance	Children's Fund
Enterprise Zones	City Growth Strategies
Estate Action	Coalfields
Regional Selective Assistance	Community Champions
Section 11	Community Chest
Task Forces and CATs	Community Empowerment Fund
Technical and Vocational Education Initiative	Community Legal Partnerships
Urban Development Corporations	Creative Partnerships
Urban Programme	Crime Reduction Programme
Welsh Development Agency	Drug Action Teams
Work-Related NAFE	Early Excellence Centres
	Early Years Development and Childcare Partnerships
	Education Action Zones
	Employment Zones
	European Regional Development Fund
	Excellence Challenge
	Excellence in Cities
	Fair Share
	Frameworks for Regional Employment and Skills Action
	Health Action Zones
	Healthy Living Centres
	Healthy Schools Programme
	Neighbourhood Management
	Neighbourhood Nursery Centres
	Neighbourhood Renewal Fund
	Neighbourhood Support Fund
	Neighbourhood Wardens
	New Deal for Communities
	New Entrepreneur Scholarships
	Playing Fields and Community Green Spaces
	Positive Futures
	Regional Centres for Manufacturing Excellence
	Safer Communities Initiative
	Single Regeneration Budget
	Spaces for Sport and Arts
	Sport Action Zones
	StepUp
	Street Wardens
	Sure Start
	Sure Start Plus
	Urban Regeneration Companies
	Youth Inclusion Programme
	Youth Music Action Zones

Source: **Weaver 2003.**

STUDY QUESTIONS

1. To what extent do New Labour's urban policies represent a continuity of Conservative policies of the 1980s and 1990s?
2. Do you believe Area Based Initiatives (ABIs) are effective in dealing with urban problems? If not what measures do you believe would be more effective?
3. Evaluate New Labour's approach to combat social exclusion.
4. 'A step change or a step back?' (Raco 2005a). Discuss this statement by comparing the 1980s and 1990s generation of Urban Development Corporations with the post-2003 Urban Development Corporations in the UK.

KEY READINGS ON NEW LABOUR'S URBAN POLICY

Atkinson (2003) 'Urban policy', in Ellison and Pierson (eds.) *Developments in British Social Policy 2* [brief overview of the early years of New Labour's urban policies and philosophies]

Cullingworth and Nadin (2006) *Town and Country Planning the UK* [Chapter 10 focuses on urban policies and regeneration and links with town and country planning policy, much of it covering the post-1997 period]

DETR (Department of the Environment, Transport and the Regions) (2000a) *Our Towns and Cities: The Future – Delivering an Urban Renaissance* [the 2000 Urban White Paper remains the government's overall policy approach to urban regeneration]

Imrie and Raco (eds.) (2003a) *Urban Renaissance? New Labour, Community and Urban Policy* [key text on the early years of New Labour's philosophies and approaches to neighbourhood and community renewal]

Johnstone and Whitehead (eds.) (2004a) *New Horizons in British Urban Policy: Perspectives on New Labour's Urban Renaissance* [critical analysis of the approaches to urban policy in the late 1990s and early 2000s]

ODPM (Office of the Deputy Prime Minister) (2003a) *Sustainable Communities: Building for the Future* [recent urban regeneration policies have been developed in the context of the 2003 Sustainable Communities Plan]

Smith, Lepine, and Taylor (eds.) (2007a) *Disadvantaged By Where You Live? Neighbourhood Governance in Contemporary Urban Policy* [a collection exploring the neighbourhood both as a sphere of governance and as a point of public service delivery under New Labour]

KEY WEBSITES ON NEW LABOUR'S URBAN POLICY

Business Improvement Districts: www.ukbids.org [the National BIDs Advisory Service contains latest news on BIDs, case study examples and advice on setting up BIDs]

CLG (Communities and Local Government): www.communities.gov.uk [the government department with responsibility for urban regeneration in the UK; contains archived reports and evaluations, and an overview of current policies]

Core Cities Group: www.corecities.com [a network of the eight major regional 'Core Cities' in England outside of London]

English Partnerships: www.englishpartnerships.co.uk [the national regeneration agency for England supporting sustainable growth; contains news, information and urban regeneration case study examples]

European Structural and Cohesion Funds: www.berr.gov.uk/regional/european-structural-funds/index.html [information at the Department for Business, Enterprise and Regulatory Reform on funds available for 2007–2013]

Homes and Communities Agency: www.homesandcommunities.co.uk [national housing and regeneration agency, formed in 2008]

Local Government Association: www.lga.gov.uk [a voluntary organisation which promotes the interests of English and Welsh local authorities, a total of almost 500, representing over 50 million people and spending around £74 billion a year on services]

Planning Resource: www.planningresource.co.uk [up-to-date planning and policy information of relevance to contemporary urban regeneration]

Regen.net: www.regen.net [latest regeneration news, updated daily]

Regional Development Agencies: www.englandsrdas.com [links to the nine RDAs and their current activities]

Social Exclusion Task Force: www.cabinetoffice.gov.uk/social_exclusion_task_force [originally the Social Exclusion Unit, now subsumed within the government's Cabinet Office]

Urban Regeneration Companies: www.urcs-online.co.uk [source of information, best practice guidance, resources and case study examples of URCs]

Urban Task Force: www.urbantaskforce.org [the independent update and progress report on the 1999 Urban Task Force report]

SECTION III

Cities in Transition: Themes and Approaches

6 Urban competitiveness

INTRODUCTION

As illustrated by Section II, policy initiatives by public and private bodies of different kinds have played a key role in shaping the changes taking place in cities. The main focus of Section II was on examining the policies that have been adopted by central government to influence the social, economic and physical development of cities. The principal focus of Section III is on examining the policies and strategies that have been pursued by cities themselves, focusing particularly on improving economic competitiveness and tackling social disadvantage and exclusion.

As was demonstrated in Chapter 1, the decline of cities was closely tied to economic decline and therefore strategies for economic growth underpin urban regeneration in the UK as elsewhere globally. This chapter identifies some key approaches to urban economic regeneration pursued by cities themselves, in the context of urban policy initiatives covered in Section II. First, the links between globalisation and urban competitiveness are identified and examined. Second, the connections between competitiveness and place competition are examined, as are the links with urban place marketing and urban entrepreneurialism. Third, some key strategies pursued in UK cities to secure competitive advantage are introduced. To investigate competitive advantage and the city, it is also necessary to investigate the drivers behind the growing emphasis on the 'competitiveness' of places, the meaning of 'competitiveness' when applied to cities, indicators for measuring a city's 'competitive advantage', and the requirements or conditions essential for competitive success. Evidence is drawn from a major UK project on cities, competitiveness and cohesion undertaken in the late 1990s and early 2000s, and from other recent macro-studies such as the State of English Cities and Competitive European Cities. Case studies from UK cities illustrate these approaches, which have also been followed in cities internationally such as Chicago, Paris, Barcelona, Bilbao, Rotterdam, Berlin, Frankfurt and Melbourne (see for example, Bianchini and Parkinson 1993; Kearns and Philo 1993; Couch et al. 2003).

COMPETITIVENESS, PLACE COMPETITION AND URBAN REGENERATION

Porter (1998, p. xii) commented that 'there was no accepted definition of competitiveness'. It clearly concerns the performance of economies and can at one level be equated with the trajectory of the economy in terms of variables such as value added, employment or incomes of the population. If inputs into the productive process are defective, deficient or costly, the competitiveness of a locality will suffer (Begg 2002b). These factors determine a significant part of the costs of doing business in a particular location. Begg (2002b) argued that the most competitive place is not necessarily the one with the lowest costs, but perhaps one that raises the living standard of its residents and workforce, representing a social and economic regeneration of a place. A definition of competitiveness for the purposes of urban regeneration can be taken as 'the degree to which [a city] can, under free and fair market conditions, produce goods and services which meet the test of international markets, while simultaneously maintaining and expanding the real incomes of its people over the long term' (Begg 2002b, p. 3).

Parkinson and Boddy (2004, p. 1) contended that since the 1990s cities have been viewed as 'dynamos of the UK national economy, vital to the competitiveness of "UK plc", rather than economic basket cases as they were sometimes portrayed'. Cities are now regarded as economic assets rather than urban liabilities reflected in the renewed investment in cities and growing attraction of urban life, both of which are vital to economic competitiveness (ODPM 2004b; 2006a).

GLOBALISATION AND THE COMPETITIVE CITY

Globalisation is one of the principal reasons behind the need for cities to compete, and the impact of globalisation on competitiveness is important in understanding urban change and regeneration strategies. In the age of footloose global capital flows and globalisation, cities at the global level have to become ever more competitive to secure future growth and the regeneration of urban economies. To this end, cities seek to ensure competitive advantage over their regional, national, European and global counterparts. The networks of inter-urban competition in which individual cities find themselves caught have increased spatially and numerically, and individual cities are subject to fewer protective measures and structures than has previously been the case (Short and Kim 1999; Hall 2006).

This has implications for urban regeneration in terms of attracting business, residents and tourists to particular cities. Cities have to fight to secure competitive advantage through attracting internationally footloose capital (see Hall 2006, pp. 82–84). Place and local context matter in terms of competitive strength, and some cities offer a better mix of attributes for business and business success than others (Begg 1999; 2002a; Boddy and Parkinson 2004; Buck et al. 2005).

To investigate urban competitiveness empirically in the context of economic globalisation, a major £4 million UK research project undertaken on cities, competitiveness and cohesion between 1997 and 2002 sought to improve an understanding of how cities develop and mobilise distinctive economic assets in order to secure competitive advantage, and how these processes impact upon and are influenced by social cohesion and environmental quality (Begg 2002a; Boddy and Parkinson 2004; Buck et al. 2005). The study investigated competitiveness in four integrated case studies of London (see Gordon et al. 2004), Manchester and Liverpool (see Harding et al. 2004), Glasgow and Edinburgh (see Turok et al. 2004), and Bristol (see Boddy et al. 2004). In addition, 19 thematic studies were undertaken covering themes such as the knowledge economy, housing, ethnicity, enterprise, leadership and social capital (Boddy and Parkinson 2004).

The Glasgow and Edinburgh study (Turok et al. 2004) looked at the contrasting performance and very different trajectories of these two neighbouring Scottish cities and the relationship between them. Over the last two decades, Glasgow has experienced major economic decline, the effects of which continue to be apparent, but more recently has witnessed a degree of urban renaissance in selective areas. By contrast, Edinburgh has emerged as a highly economically successful city. Economic success was found to support some level of cohesion in terms of inclusion, equality, social order and stability, provided the success is broad-based enough to create a sufficient range and quality of jobs suitable for the local population. However, Turok et al. (2004, pp. 27–29) found that there was no evidence that social cohesion (see Chapter 8) was fundamental to short-term economic growth. External forces of national and international economic relationships and business connections were found to impact on competitiveness and cohesion. Broadening and expanding the economic base of cities was seen as key to providing sustainable jobs and incomes for a higher proportion of residents and to generate the resources to fund public services. Industrial diversity and balance, rather than narrow specialisation, are required in order to reflect the broad assets of cities. There is the need for a range of job opportunities to meet the diverse social structure and skills of the population. Finally, Turok et al. (2004) argued that physical assets including supply of suitable land and premises, the quality of the environment, and the state of the public and private transport system could hamper economic development in cities more than elsewhere.

Harding et al. (2004) looked at economic competitiveness and social cohesion in the neighbouring cities of Liverpool and Manchester in north-west England. Both cities suffered major economic decline and job losses in an underperforming region relative to the UK in general. Unlike Liverpool, Manchester was found to have capitalised more effectively on opportunities to boost its competitive strength.

By contrast, the studies on competitiveness in Bristol (Boddy et al. 2004) and London (Gordon et al. 2004) are both focused on cities in prosperous southern England, and both demonstrate considerable competitive strength. Bristol, although not immune from the negative impacts of wider economic forces, has exhibited considerable adaptive capacity and competitive strength over recent decades (see also Lambert and Smith 2003; Tallon 2007a). Bristol performs as a relatively successful and growing city-region experiencing continued growth in population, jobs and investment (Boddy et al. 2004).

A number of explanations have been put forward as to why Bristol is currently a relatively competitive city (Boddy 2003a; Boddy et al. 2004). The city benefits from being located in the flourishing south of the country in a relatively prosperous region. The effect of the global city of London, a less than two-hour car or train journey away, makes Bristol an advantageous place for business. The transport infrastructure facilitates motorway links to London and its airports and the south-east, and rail facilities connect to national and international networks (Lambert and Smith 2003). The logic of location for business has therefore contributed to Bristol's relative success. Traditional locational factors remain important with available sites and property, accessibility, and a wide labour market catchment for technical and professional staff extending along the M4 corridor (Boddy et al. 2004). Related to this, the city and surrounding areas have a good quality of life reputation and image as a place to live (Boddy 2003d; Boddy et al. 2004).

In terms of the local economy, the city's historical inheritance, especially aerospace manufacturing, the absence of a major heavy industrial legacy, and a diverse economy, meant that Bristol did not suffer to the same degree as many other cities in the UK (such as Liverpool and Manchester) during the de-industrialisation process. Boddy (2003d) argued that adaptive capacity and diversity of the economy rather than economic clusters or key sectors of the economy have contributed to the long-term economic buoyancy of the city. Although there are a number of specialised clusters within high technology, financial services, and the cultural and media industries (Bassett et al. 2003), they are small-scale features of sectors dominated by a small number of relatively large employers. Diversity rather than specialisation appears to drive relative economic success. Overall, the city-region has a collection of assets that are attractive to a wide range of employers and potential inward investors (Boddy 2003b; 2003d; Lambert and Smith 2003; Boddy et al. 2004).

The case study of competitiveness in London represents a unique example because of its scale, complexity and intense internationalisation (Gordon et al. 2004). London's inherited strengths are particularly advantageous for the climate of the new competitive, international and post-industrial economy. At a national and European level, London has emerged as a strikingly successful city over the last two decades (Gordon et al. 2004), although one in which there are sharp inequalities (see Hamnett 2003a). Internationalisation and flexibility have had significant positive impacts on the London economy, and in common with the other case study examples, social cohesion has had a limited influence on the city's competitiveness (Gordon et al. 2004).

COMPETITIVENESS AND THE STATE OF THE ENGLISH CITIES

In the 2000 Urban White Paper, *Towards an Urban Renaissance* (DETR 2000a), central government made a commitment to commission a report on the progress and performance of English cities after a five-year period, as an update to an earlier State of the Cities report (DETR 2000b). The independent State of the Cities report provided a comprehensive audit of urban performance in England and a review of the impact of government policies upon cities (ODPM 2006a). The main themes covered were social cohesion, demographics, economic competitiveness and performance, liveability, and governance and the impact of policy. The report represented an authoritative statement of the state of English cities, the opportunities and challenges being faced in an international context, and the policy steps required to further develop the economic competitiveness of cities.

The main findings were summarised in ODPM (2006b) and the following key messages emerged for urban competitiveness:

- Many English cities have improved their economic competitiveness through restructuring and finding new market niches, most strikingly exhibited by the urban renaissance of many city centres.
- The urban renaissance must be sustained and widened both within cities and across the country.
- The gap between cities in the north and west, and cities in the south and east of England remains a challenge to be addressed by regional economic and urban policy.
- Lessons for economic competitiveness can be learnt from European cities, such as decentralising decision-making to regional and local levels of government.
- Entrepreneurial, local leadership appears crucial in helping to find new economic futures for cities, their businesses and residents.
- Cities are critical to the successful delivery of wider government policy such as greater investment of public resources in the mainstream programmes which impact upon cities; greater recognition and focus upon the economic potential of cities and the policy levers to encourage it; greater willingness to address regional imbalances; recognition of the importance of sustainable communities; greater focus upon city-regions and collaboration across regions; and growing willingness to simplify and reduce demands and constraints upon local and regional government.
- Government departments and mainstream policies should continue to focus on cities as crucial determinants of regional and national economic performance. The balance of powers between national, regional and local government should allow English cities to

benefit from the freedoms, resources and responsibilities found in the more successful European and American cities.

The key overall message is that England's cities are now better placed than at any time since the end of the nineteenth century to become engines of national advance.

COMPETITIVE EUROPEAN CITIES

A further related recent in-depth study on urban competitiveness sought to assess the performance of the English Core Cities outside of London (Birmingham, Bristol, Leeds, Liverpool, Manchester, Newcastle, Nottingham and Sheffield) relative to their European counterparts. The English Core Cities have shown signs of an economic renaissance since the late 1990s as demonstrated by the State of the English Cities report (ODPM 2006a). However, there remain concerns that they are not competing sufficiently economically in the national context; are falling behind London; lack the right powers and resources to improve their performance; and do not make as great a contribution to the national economic welfare as comparable cities in continental Europe (ODPM 2004b; 2004k). The messages identified based upon the experience of the most successful European cities can help increase the economic competitiveness of UK cities, regions and the national economy.

Key findings were synthesised from the Competitive European Cities research, which have implications for UK cities and competitiveness (ODPM 2004k):

- The English Core Cities have witnessed a renaissance since the late 1990s, but economic challenges remain.
- Urban competitiveness can be measured by economic diversity, skilled workforce, connectivity, strategic capacity to implement long-term development strategies, innovation in firms and organisation, and quality of life.
- Not all continental or English cities do equally well or equally badly in every aspect of competitiveness. Core Cities have improved their performance since the late 1990s, but many cities lag in the eyes of international investors, and lag behind their competitors in terms of GDP, innovation levels, educational levels, connectivity, social cohesion, quality of life, political capacity, and connections with their wider territories.
- Successful European cities considerably outperform their national GDPs, but with the exception of Bristol, Core Cities lag significantly behind the national average. If the Core Cities could match European competitiveness, the national economic impact would be enormous.
- Structural characteristics of competitiveness mean that cities that performed well in the mid-1990s still do so, although cities can improve their competitiveness as demonstrated by cities such as Helsinki, Barcelona and Madrid.
- The framework set by national government is highly significant as exemplified by the impact on successful provincial cities of even limited decentralisation in France over 20 years.
- Continental cities have responsibilities for a wider range of functions that affect their economic competitiveness than their English counterparts. The mix varies, but their combination of powers and resources seems to make continental cities more proactive, entrepreneurial and competitive.
- European governments recognise that cities are in a relationship with each other in their national system and develop policies that make this explicit. This has shaped their investment policy in transport, higher education, and location of research and development facilities. In the UK, there has been little sense of the relative roles and contributions of different cities and how they impact upon each other.

- France and the Netherlands are the two most centralised countries and have paid most attention to cities. They are attempting to build better working relationships between national government and urban areas. There need to be greater levels of trust between national and city governments to encourage city economic competitiveness.
- Large urban areas have substantial assets in hard and soft infrastructure, which gives them the potential to be successful, but not all large cities are successful. The successful cities were often found to be the larger cities in Europe and certainly the largest in their national system. The Core Cities in England therefore have potential as targets for government strategies.
- No successful regions were found which did not have successful cities at their core. Regions that performed well were those where the Core City performed well and vice versa. Many national and regional governments have recognised the contribution that cities make to regional economic performance, and there is an imperative to develop strategies, policies and instruments that bring Core Cities and their economic hinterlands together.
- There is a need for a national policy for regions that takes a strategic view of the appropriate relationships between different parts of England, and appreciates the impact of government machinery, policies and resources upon them. This would form part of the wider debate about the best way of improving the economic competitiveness of the English urban, regional and national systems.
- Economic competitiveness does not necessarily drive out social cohesion. The successful cities of Europe have the most skilled and better-educated workforces and the highest performing economies have the lowest rates of unemployment. The social agenda is critical in European cities, and this is not incompatible with an economic growth strategy.
- Cities operate within a set of powerful structural, economic, social, physical and institutional constraints but are not powerless to shape their economic trajectories.

Elements and conditions seen as crucial for urban economic competitiveness and success across all UK cities have emerged from research including the *State of the English Cities* and *Competitive European Cities* reports. These will be highlighted and exemplified in the remaining chapters of Section III on themes in contemporary urban regeneration in the UK.

TRANSPORT STRATEGIES AND COMPETITIVE CITIES

Competitive advantage to a large extent relies on location and accessibility as revealed by the case studies within the cities, competitiveness and cohesion project (Boddy and Parkinson 2004). In the post-war period, public and private transport links have seen considerable investment to facilitate the movement of capital and people to the more peripheral parts of the UK and within urban areas. These have been characterised in recent decades by motorway links (for example, the M25 orbital road), bridges (such as the Second Severn Crossing), rail links (including the Jubilee Line extension in London) and airport expansion (exemplified by the recently completed £4 billion Heathrow Terminal 5) (see Cullingworth and Nadin 2006, Chapter 11). Conversely, transport infrastructure problems in urban areas can negatively impact on a city's competitive edge and obstruct urban regeneration (see Pacione 2005, Chapter 13). Transport infrastructure improvements were essential to the realisation of London's Docklands regeneration (Hall 1992) for example, as examined in Chapter 3.

A contemporary example of a major proposed transport infrastructure project predicated on enhancing competitiveness and contributing to the regeneration of deprived communities is London's Crossrail project. This £16 billion scheme, approved in 2007 and due to be completed by 2017, aims to link Maidenhead and Heathrow in the west with the City, Docklands, Essex and Kent in the east. Some view this as the most important new transport infrastructure project in the UK for 30 years and as vital to the country's future economic prosperity by benefiting the economy to the tune of £30 billion (www.crossrail.co.uk). It is estimated that the scheme will create 30,000 jobs, and that it will contribute to London's financial reputation and competitiveness in the face of increasing competition from European cities such as Paris and Frankfurt. However, counter-arguments reflect those of wider criticisms of 1980s-style regeneration, especially regarding who would benefit from the project, and whether public spending in areas such as the East End of London would be reduced and reallocated.

The role of transport in urban regeneration in the UK is further explored and exemplified in Chapters 9 to 12, in terms of its role in sustainable urban regeneration, and its place within retail, housing, and leisure and cultural regeneration projects.

EXPLAINING AND CRITIQUING URBAN COMPETITIVENESS

In the wider context of globalisation and competitiveness, and informed by the four UK case studies and research evidence on the economic state of cities, five key drivers behind place competition can be identified. First, the globalisation of investment flows means that cities are ever more tied up with and interdependent on each other (Short and Kim 1999). Footloose inward investment in cities provides an unstable basis for economic growth and urban regeneration, because it is able to switch location to a much greater degree than manufacturing or heavy industrial investment. Hall (2006, p. 83) pointed out that inter-urban networks of competition have become increasingly international. The choices of location available to decision-makers are now potentially international because of the global nature of markets, and therefore cities now find themselves competing with each other at a global level (see Short and Kim 1999). Cities are also coming into competition with an increasing number of other cities because of the decline of specialisation within city economies and because urban regeneration strategies across cities of the UK, Europe and North America have been remarkably similar, furthering inter-urban competition.

Second, the decreasing importance of the nation state in regulating the economy since the 1970s means that city authorities in partnership with business have an increased interest and role to play in attracting and regulating economic development (Griffiths 1998a). Protective structures and agreements have been eroded including government regional policy, trade agreements between countries, military force or occupation and systems of empire (Hall 2006). Central government regional economic policy has been replaced by grants from European and national levels of government, which themselves are often allocated on a competitive basis (Oatley 1998a; see Chapter 4). These types of awards to cities have consequently been another driver of inter-urban competition.

Third, the rise of new markets such as urban tourism and culture (see Chapter 12) has become another component of inter-city competition. Cities throughout the UK, Europe and North America have replicated a small number of what are regarded as essential components of 'successful' and competitive cities such as waterfront redevelopments, convention centres, sports stadia and flagship urban regeneration projects. This has heightened inter-urban competition as a larger number of cities are competing with each other to lure tourists to increasingly similar tourist experiences (see Griffiths 1998a).

Fourth, the shift from managerial urban government to competitive and market-based forms of public service delivery has been another driver behind the process of competition between urban places, synonymous with the rise of the post-modern city (see Hambleton 1998; Chapter 1). The rise of urban entrepreneurialism has driven competition between localities, as authorities have sought to promote alternative forms of economic growth to replace cutbacks in central government funding for local authorities. City authorities have become increasingly concerned with risk-taking and growth-oriented activities to facilitate urban regeneration (see Hall and Hubbard 1996; 1998; Griffiths 1998a).

Fifth, there has been the emergence of flexible production methods and local interdependencies since the 1970s, associated with the post-Fordist era (see Mole 1996; Oatley 1998b). Under the mass production regime, plants and machinery were set up to produce enormous quantities of identical products, which was fairly inflexible. With the development of new computer-based technologies, it has become possible to organise production in new, more flexible ways. The relative importance of different factors of production has shifted. Raw materials now make up a smaller and smaller share of the final value of products and it is information, as the outcome of human ingenuity and creativity, that makes up a bigger and bigger share. This footloose aspect of the new economy is subject to inter-urban competition to add to a city's economic success.

The combination of these five processes has driven place competition since the 1970s. However, although this analysis appears to make sense at the level of an individual firm, it is more problematic when applied to cities. This is because cities do not represent single corporate entities, and control is dispersed across the city (see Short and Kim 1999). Cities do not produce commodities as such in the way that an individual firm does. In addition, it is hard to define who exactly an individual city's competitors are. How to judge city competitiveness and success is problematic (Begg 2002b), as this can be defined using multiple criteria as shown by the Competitive European Cities research (ODPM 2004b). Such issues are reflected in wider academic debates and contrasting perspectives on the applicability of city competitiveness ideas. Indeed, Deas and Giordano (2002, p. 191) argued that 'the notion of "urban competitiveness" remains a somewhat nebulous one, its conceptual turbidity rendering its measurement notably problematic'.

Krugman (1996a; 1996b) argued that enterprises compete, and not localities. This can be seen as a 'geosceptic' argument in that cities as an entity do not compete with each other. However, Porter (1998) claimed that localities do compete by establishing suitable conditions for enterprises to locate. These conditions include a skilled labour force, efficient communications infrastructures, flexible land and property markets, high standards of liveability, and responsive local government (Porter 1998).

Deas and Giordano (2002) put forward the ways in which competitive advantage might be measured. A distinction can be made between competitive assets linked to contextual factors, and competitive outcomes related to economic performance. Four broad types of asset can be used to measure competitive advantage in cities, namely economic, policy, the environment and social assets. Two broad types of outcome can be identified as firm-based and area-based. However, these indicators do not account for image factors and their impacts on place as discussed later. The choice of valid measurement indicators is selective, and in practice the distinction between many of the indicators becomes blurred.

Harvey (1989) further contended that the conditions that are important depend on the market in which a city is attempting to compete. These different kinds of markets comprised production investments (for example, in the social and physical infrastructure), consumption expenditures (for example, creating high-end retail, cultural and entertainment quarters), command and control functions (for example, investing in communications infrastructure), and state spending (for example, attracting funding for urban regeneration projects) (see Griffiths 1998a, pp. 42–43).

Governance factors are seen as a key condition for enhancing city competitiveness. The 'organising capacity' or 'institutional capacity' of city regions is vital (van den Berg et al. 1997; van den Berg and Braun 1999). Competitive cities have the power to be able to anticipate, respond to and cope with changing intrametropolitan and extrametropolitan relations (van den Berg et al. 1997). This reflects a shift in emphasis from 'hardware', which was important in the modern city through to the 'soft infrastructure' of the post-modern city (van den Berg and Braun 1999).

Further questions and debates indicate the problematic notion of urban competitiveness as a vital component of urban regeneration (see Begg 2002a; Boddy and Parkinson 2004; Buck et al. 2005). One question relates to how secure competitive advantages are and how quickly advantage can be established or eroded. The question of accountability in the context of new informal governance models of the entrepreneurial city is another issue. The question of who the real beneficiaries of inter-urban competition are has been a matter of some debate. Business and the private sector more generally appear to gain from a competitive city through an advantageous business environment, whereas city authorities benefit from urban and economic regeneration stimulated by business growth. Evaluating the benefits of inter-city competition on local communities is more problematic, particularly if success bypasses excluded communities and neighbourhoods within entrepreneurial and 'competitive' cities. Finally, questions surround whether 'competitive' is the same as 'successful', and whether there are alternative routes for cities wishing to regenerate their economies. These could include developing city networks and encouraging values of collaboration and co-operation between cities to share good practice and promote policy development (see, for example, Dawson 1992; Griffiths 1995b) (see Chapters 7 and 8).

COMPETITIVE ADVANTAGE AND THE CREATIVE CLASS

Competitive advantage has been closely linked with the emergence of the 'creative class' in cities. This influential thesis is associated particularly with the American economist Richard Florida (2002; 2005). He argued that the basis of economic success is the capacity to attract and retain creative class talent, as 'access to talented and creative people is to modern business what access to coal and iron once was' (Florida 2002, p. 6). The creative class is synonymous with the quaternary sector of the economy, which includes knowledge-generating professionals as distinct from wider service sector employees in areas such as retail and leisure. Florida (2002) identified a 'super creative core' of the creative class consisting of those who are employed in the fields of computer software, mathematics, engineering, university lecturing, and art, media and design.

Florida argued that creative people drive growth and competitiveness in a city. Key factors in cities attracting this group are cultural and associated with diversity, openness and tolerance, and the environment. The most successful creative cities are those that demonstrate high levels of talent, technology and tolerance (Florida 2002). Florida measured these factors by various indices. The talent index indicates the proportion of people with a university degree. The techpole index measures the proportion of innovation and technology in a city by comparing the rate of growth of its high-tech sector with the average. The gay index measures the percentage of same sex households living in a city, which Florida (2002) argued, reflects tolerance. The bohemian index measures the proportion of writers, artists, musicians, actors and painters living in a city. A combination of these four indices feed into a 'creative index' for a city. In the US, creative cities are revealed to include Washington, Boston, Austin, Seattle and San Francisco.

Local governments often do the wrong things in attempting to attract the creative class, for example, through encouraging investment in commercial malls, tourist developments and sports stadia rather than fostering authentic edgy, mixed districts, and facilities such as wine bars, libraries and nightclubs which feed the interests of the creative class. 'Creatives' move from standardised lifestyles in conventional city landscapes comprising shopping malls, office parks, call centre jobs, suburban life and a 9 to 5, 50-year career. They are attracted to interesting jobs and flexible working patterns in vibrant cities where eccentrics and non-conformists fit into a diverse and tolerant community in centres with a sense of place and authenticity.

A number of critics of Florida's creative class thesis have argued that it is an insufficient basis for urban competitiveness and that attention is deflected away from lower social groups (Malanga 2004; Jacobs 2005; Kotkin 2005; Peck 2005). Malanga (2004) argued that Florida's theories are based on thin evidence, and that creative cities are not shown to sustain success over the longer term. In addition, Malanga criticised Florida for encouraging city authorities to produce fabricated and idiosyncratic art scenes at a cost to the taxpayer. Further, Malanga (2004), Jacobs (2005) and Slater (2006) were critical of creative cities as gentrification is promoted (see Chapter 11). Here, the incoming and growing creative classes displace working-class inhabitants. Higher rates of income disparity are also evident in creative cities. The focus on the creative class also neglects the role of service sector workers who sustain the creative class (Jacobs 2005). Finally, the numbers of people who make up the creative class in a city are relatively small in number compared with the wider economy, and might therefore represent a small contribution to a city's growth and competitiveness. Pursuit of the 'Floridisation' of cities should be treated with caution (see also Chapter 12).

PLACE MARKETING AND URBAN ENTREPRENEURIALISM

Linked with the intensification of the globalisation process and the emergence of place competition have been strategies of place marketing as an element of urban entrepreneurialism. Griffiths (1998a) illustrated that the marketing of urban places, and the emphasis on the projection of deliberately crafted images to external audiences and local populations, have been one of the key features of the entrepreneurial mode of urban governance that came to prominence since the 1970s (Harvey 1989). This shift to entrepreneurialism came as a response by individual cities to the collapse of the post-war social democratic consensus, which was characterised as a managerialist form of governance (see Chapters 1 and 7). Since the emergence of urban entrepreneurialism, publicity and advertising campaigns have been important ingredients in the process of place marketing, which seeks to improve urban competitiveness.

Urban place marketing (also referred to as 'selling the city', 'city marketing', 'place promotion' and 'urban boosterism') is closely linked to urban regeneration and change, and to the negative images that have become associated with many cities following the de-industrialisation of their economies (see Ward 1998). Principal players in the urban economy have had to become more entrepreneurial in order to compete in the post-industrial economy (Hall and Hubbard 1996; 1998). Central to the strategies that contemporary cities employ is the promotion of a city as a great place to do business, to live, to play and to visit (see Hall 2006, Chapter 6). The images of cities that are created for place marketing often draw on the physical changes that urban regeneration projects have achieved. However, as Hall (2006) indicated, images that are produced for place promotion strategies are fabricated and draw selectively on gratifying aspects of a city's past and present while ignoring other less savoury elements. In the era of place promotion and competition, it is perception rather than reality that is crucial. The manipulation of perception has been termed 'imagineering' reflected in the conception of the 'imagined city'.

Place images developed to enhance urban competitiveness are typically simplified, generalised, stereotypical impressions that people have of any place or area. Place images exaggerate certain physical, social, cultural, economic or political features, while at the same time excluding others (Hall 2006, p. 78). Urban images can be cleverly manipulated and transformed by city authorities to encourage investment or to attract visitors, concomitant with urban regeneration aspirations. Hall (2006) noted that these urban images could be formed in five main ways. City authorities seek to create images of places through slick marketing and promotional campaigns. Media coverage of events in places, which become the prevailing impressions of those places, also contributes to images of particular cities. Further, satire and visual comedies; personal experience; and hearsay and reputation add to images of cities.

Many former industrial cities such as Los Angeles, Liverpool, Birmingham and Manchester have suffered the stigma of a bad image both presently and historically. These negative images can persist and become exaggerated and ingrained. In response it is now clear that cities are increasingly engaging with and spending on place marketing as an integral part of urban regeneration strategies. This commodification of the city, that is treating the city as a commodity or a sellable product, has become a requisite strategy in economic development and competition to lure external investment into the city (Griffiths 1998a). Current approaches to place marketing characteristically endeavour to reimage or reinvent the city by promoting local cultures and traditions to stress the city's uniqueness and attract tourists (see Chapter 12).

Place marketing is a key to transforming city image and thus enhancing competitive advantage, and can be defined as 'the conscious use of publicity and marketing to communicate selective images of specific geographical localities or areas to a target audience' (Ward and Gold 1994, p. 2). Philo and Kearns (1993, p. 3) observed that the practice of selling places:

> entails the various ways in which public and private agencies, local authorities and local entrepreneurs . . . strive to 'sell' the image of a particular geographically-defined 'place', usually a town or city, so as to make it attractive to economic enterprises, to tourists and even to inhabitants of that place.

In essence, place promotion is concerned with attracting investment or visitors to places. It has been argued that place marketing

> is the principal driving force in urban economic development in the 1980s and will continue to be so in the next decade . . . The logic that more jobs make a better city is giving way to the realisation that making a better city attracts more jobs (Bailey 1989, p. 3)

– a sentiment which remains valid.

To this end, cities in the UK have engaged in place marketing to overcome negative imagery, to combat the consequences of deindustrialisation, and in response to the changing roles of cities and an associated 'crisis of urban representation' (Hall 2006). The purposes of place marketing are fourfold: economic (wealth and job creation), social (to create a quality of life image), cultural (to enhance local pride, identity and community to act as a boost for the local population), and political (to enhance identity and a sense of belonging).

Since the 1970s, cities have actively promoted what they have to offer through an array of guides, brochures, media, videos and DVDs, posters, literature, websites, tourist offices, libraries, commercial information services, specialist magazines, city slogans and city logos. Promotional material from any city reveals a range of similar themes being conveyed. Hall (2006, pp. 88–93) distilled these as consisting of geographical centrality and accessibility; images of industrial heritage and high-technology industry; images of business and flagship office developments; images of lifestyle, culture and quality of life; and images of post-modern environments mixing spectacular and historic architecture, attractive suburbs, and proximity to the countryside.

Short and Kim (1999, pp. 99–106) developed four categories for images promoted by place marketing campaigns in western cities. The first, 'wannabe world cities', are that group of cities lagging just behind the three global cities, and to enhance their competitiveness they bid to stage major global sporting events and develop iconic or signature buildings. Second, a group of cities promote an image exclaiming 'look, no more factories!' where the emphasis is on the new rather than the old, depicting fashionably post-modern places, and consumption rather than production. The third 'city for business' category positively markets location, workforce, quality of life, transport, business climate, education and research. A final cluster of cities aims at 'capitalising culture' and seeks to exploit cultural capital, the historic feel and the festival package.

A famous case study example of urban place marketing and competition in practice is London's Docklands (Crilley 1992a; 1993; see Chapter 3). Physical regeneration was undertaken to promote a positive image of the area (Brownill 1999). The area was promoted as a place to invest, a place for business, a place for the affluent to live and play, and a tourist destination. This involved a major media advertising campaign during the 1980s, emphasising the spectacular architecture and the historical legacy of the area (see Crilley 1992a; Hall 2001, p. 138).

In Birmingham, Hall (2006, p. 86) showed that since the 1980s urban landscapes have been shaped to appeal to potential consumers, and have become more entangled with place marketing, through a process of 'investment marketing'. Recently, urban regeneration has been in the vein of Birmingham's brand of a diverse, multi-cultural and global city, as reflected in the spectacular Bull Ring mixed-use redevelopment (see Chapter 10). In Birmingham, as elsewhere, urban image is becoming increasingly central to physical, economic, social and cultural development (Hall 2006).

However, it is Glasgow that is widely regarded as the UK pioneer in urban reimaging (Pacione 2001). To counter a negative image based on industrial decay, unemployment, violence, trade union militancy and poor environmental quality, Glasgow embarked on a concerted marketing campaign from the early 1980s. This commenced with a cartoon character (Mr Happy) and a slogan (Glasgow's Miles Better), and the strategy advanced to focus on the arts and culture as a means of increasing tourism and conveying the image of Glasgow as a post-industrial city. In 1988 the city hosted one of five Garden Festivals in the UK, in 1999 it was designated City of Architecture, and most significantly Glasgow was designated European City of Culture in 1990 (see Boyle and Hughes 1994). The promotion of urban image involved physical reconstruction, re-emphasis of neighbourhood identity, and the creation of a 24-hour city. Culture was employed as a tool of city marketing in 1990, and the success of this year was built upon the combination of external marketing alongside the promotion of local activities. As with the 'Miles Better Campaign' of the 1980s, the achievements of 1990 can be put down to a skilful balance between satisfying the demands of the local community and extending the possibilities of improving the city's image in the wider world (see Pacione 2001, p. 323; Mooney 2004). Glasgow's version of European City of Culture was primarily concerned with the use of culture for urban marketing and tourist promotion (see Chapter 12).

It is difficult to generalise about the success or the influence of place promotion campaigns in attracting new business or tourism to cities (Griffiths 1998a; Hall 2006). Griffiths (1998a) argued that place marketing is highly ideological in that it is predicated on the manipulation of images, meanings and perceptions. Imagineers construct images of cities that are often marketed to a small, wealthy elite with the needs of less well-off people being ignored. Positive images of the city created in promotional campaigns and in urban regeneration have acted as 'masks' hiding the reality of wider urban problems (Harvey 1989; Griffiths 1998a; Hall 2006).

Distributional consequences in terms of shifts from social and economic inequalities to developing flagship buildings and attempting to compete with other cities have been demonstrated for a range of UK examples. These include Birmingham's International Convention Centre built at a cost of £180 million just 200 yards from the deprived Ladywood area (Hall 2006, p. 143). At the same time, local government was cutting spending on services (Griffiths 1998a, p. 55; see Loftman and Nevin 1994; 1995).

Place marketing is intrinsically a highly speculative venture. Investments in major entrepreneurial projects such as flagship developments, large-scale spectacles or signature events are gambles, which might reap no reward despite significant expenditure at the local level. Only a small number of cities can host the Olympic Games, or be designated European Capital of Culture, and there is a limit to the number of convention centres required for a given locality (Harvey 1989; Griffiths 1998a).

Image may be less important than assumed; other factors can be more important, such as quality of office space and quality of workforce for business. Image may be more important to attracting tourists and business tourism than to attracting businesses (Young and Lever 1997; Hall 2006). Anything innovative is quickly replicated by other cities meaning that to a significant extent cities end up displaying an identical array of attractions and facilities 'making sameness' (Griffiths 1998a).

Despite doubts over its effectiveness and some of its seemingly inequitable outcomes, the sheer amount of place promotion is likely to ensure its continuation (Hall 2006). If cities fail to advertise their claims, they are likely to be completely ignored, especially in the context of intensifying competition between cities for public and private investment in the increasingly globalised economy.

Overall, urban regeneration approaches taken by cities to improve competitiveness are many and varied. Some of the principal thematic means of undertaking regeneration inextricably linked to competitiveness are explored in later chapters, and include mixed-use developments of housing, retailing, office developments, and leisure and cultural attractions. These thematic approaches have emerged in a wider context of creating competitive localities through the new urban entrepreneurialism (Griffiths 1998a), the introduction of competition and contracting in UK local government (Hambleton 1998), and the growing importance of partnership and leadership at the local level (Stewart 1998).

SUMMARY

A collection of often conventional assets contributes to competitive advantage, and the drivers behind place competition include globalisation, the rise of new markets and the shift to competitive forms of service delivery.

Competitiveness is increasingly important to cities across the world in the face of intensifying globalisation and competition for business, residents and tourists.

Cities attempt to contribute to competitiveness through a variety of strategies including mixed-use developments, which are explored in later chapters in Section III.

Competitiveness is a problematic concept when applied to cities and does not necessarily contribute to social cohesion.

Place marketing is an entrepreneurial strategy that cities have deployed to entice footloose global capital, and transforming the image of a city is an essential ingredient of urban regeneration strategies.

Images of the city created by marketing campaigns are highly selective and the benefits and problems associated with these approaches are not evenly distributed.

STUDY QUESTIONS

1. Discuss whether cities are 'competing' with each other in the post-industrial global economy.
2. What types of strategies have been adopted by cities to enhance their 'competitive advantage'?
3. 'The emerging geography of the Creative Class is dramatically affecting the competitive

advantage of regions' (Florida 2002, p. 243). Explain and critically evaluate this claim with reference to Richard Florida's principal theories.

4. Discuss the contention that place marketing involves the creation of fabricated images that divert attention and city spending away from urban problems and disadvantaged communities.

KEY READINGS ON URBAN COMPETITIVENESS

Begg (ed.) (2002a) *Urban Competitiveness: Policies for Dynamic Cities* [an overview of recent thinking on competitiveness and cities in the UK]

Boddy (ed.) (2003a) *Urban Transformation and Urban Governance: Shaping the Competitive City of the Future* [a focus on competitiveness using the case study of Bristol, a relatively successful city in the early twenty-first century, placed in a wider theoretical context]

Boddy and Parkinson (eds.) (2004) *City Matters: Competitiveness, Cohesion and Urban Governance* [contains summaries of the projects carried out under the ESRC research programme on Cities: Competitiveness and Cohesion]

Buck, Gordon, Harding, and Turok (eds.) (2005) *Changing Cities: Rethinking Urban Competitiveness, Cohesion and Governance* [a collection of pieces based on the research carried out for the ESRC Cities: Competitiveness and Cohesion programme]

Gold and Ward (eds.) (1994) *Place Promotion: The Use of Publicity and Marketing to Sell Towns and Regions* [international perspectives on geographical marketing through the advertising and promotion of cities, districts, regions and entire countries for tourism, investment, industrial relocation, development and migration]

Oatley (ed.) (1998a) *Cities, Economic Competition and Urban Policy* [explores the shift towards competitive approaches in urban policy in the 1990s, including a chapter on place marketing]

Urban Studies vol. 36, no. 5/6 (1999) [special issue on urban competitiveness, marketing and the need for organising capacity containing 14 articles]

KEY WEBSITES ON URBAN COMPETITIVENESS

Competitive European Cities: www.communites.gov.uk [report of the state of England's core cities and their competitiveness compared with European cities; full report and urban research summary available]

Core Cities Group: www.corecities.com [this group is a network of England's major regional cities: Birmingham, Bristol, Leeds, Liverpool, Manchester, Newcastle, Nottingham and Sheffield; contains information and research reports]

State of the English Cities (2006) report: www.communities.gov.uk [a comprehensive audit of urban performance in England and a review of the impact of the government's urban renaissance policies on cities; full report and urban research summary available]

7 New forms of urban governance

INTRODUCTION

This chapter aims to develop an understanding of the political processes through which urban regeneration is delivered, developing some ideas covered in Chapter 6 surrounding competitiveness. The chapter first relates governance with general approaches to urban regeneration. It then situates urban governance within the wider shift towards a post-Fordist economy. It then illustrates the decline of the state and its implications for urban policy. The related moves from local government to local governance, and from urban managerialism to urban entrepreneurialism, are then analysed. The role of inter-city collaborative networks in dealing with urban economic and social problems is then explored. Further, the relative importance of different levels of government and their relationship with contemporary urban regeneration in the UK are looked at, particularly the regional dimension and the increasing profile of the city-region. Knowledge of new forms of urban governance is requisite to an understanding of how contemporary central government urban policy and city-level initiatives are formulated and delivered. These ideas underpin the content of many of the other chapters in the textbook in Section II and in the later chapters of Section III.

CONTEXT: URBAN REGENERATION AND URBAN GOVERNANCE

Since the 1970s, several basic attempts at urban regeneration have been closely linked with theories in economics and urban governance. Relevant terms and ideas here are growth poles, the multiplier effect, agglomeration economies and the trickle-down effect. Different governmental actors and arrangements have been to varying degrees important in the field of urban regeneration since the 1970s (see Section II). Initiatives have been state-led, market-led, or partnerships between the public and private sectors. Strategies have been top-down or state-led, bottom-up or community-led, or based on property-led regeneration.

Growth poles relate to the way in which firms locate near to other firms with which they do business. Growth is not uniform over a region or country, and key locations attract more investment. Cities compete to become the growth pole and once a city becomes the growth pole, its competitors can suffer long-term decline (see Chapter 6). Related to growth poles, good agglomeration economies contribute to the development of a city as a growth pole; in theory, cities that attract firms attract more firms. Successful cities have features such as a labour supply, markets, services, a transport hub, face-to-face contact, firms in the same line, firms in linked businesses, and proximity to government. The multiplier effect theorises that growth triggers more growth. Through the construction sector, growth creates more jobs and investment, supplier linkages develop whereby firms do business with each other, and more spending power is generated for services thus creating further jobs through the multiplier effect. The fabled trickle-down effect asserts that money earned is spent in retail and entertainment environments, which creates jobs for the unskilled in the service sector (see Chapter 3).

State-led regeneration can take the form of top-down or bottom-up interventions. Top-down urban regeneration has historically dominated approaches to solving the urban problem. As demonstrated by much of Section II, central government devises urban regeneration strategies led by the public sector, and these are typified by large-scale schemes and long-term programmes, often on a macro-economic scale. Government plans what it views as best for a city, which can be based on the skills base of a city as was historically the case for cities such as Pittsburgh in the US and Sheffield in the UK. The government acts as a co-ordinator for any private sector investment that must fit the 'master plan' for the city. Bottom-up growth, by contrast, is characterised by local organisations in a city operating on a small, micro-economic scale. Bottom-up regeneration can be focused around local government or community groups. Individual streets and local businesses may be targeted, and 'supplier linkages' are fostered within a city.

Market-led regeneration works where the government provides the general framework for intervention through tax breaks, grants and infrastructure improvement. This encourages the private sector to invest, and then lets the free market operate (see Chapter 3). No 'master plan' is instigated; the market dictates with the private sector at the lead and there is a minimal role for government. Third-way regeneration is characterised by public–private partnerships and a high degree of co-operation between the public and private sectors. Business investment is co-ordinated by government with the groups working closely together (Bailey et al. 1995).

Property-led regeneration approaches (see Chapter 3) viewed the built environment as the key to regenerating failed cities. Government and more often the private sector inject money into infrastructure, the remediation of brownfield sites, refurbishment of existing stock, new housing, new office and retail space, and more urban green space. The aim has been to transform the image of a city and to create a growth pole, as exemplified by the Canary Wharf development in central London (Hall 2006, Chapter 7).

FROM FORDISM TO POST-FORDISM AND THE CITY

Fordism and its successor developed as part of a body of ideas known as regulation theory and is associated with the key writings of Aglietta (1979) and Lipietz (1987). In the urban studies literature, Amin and Thrift (1995); Goodwin and Painter (1996); Jessop (1995) and Painter (1995) among others have linked regulation theory to changing urban areas and governance. The approach has its roots in Marxist analysis, which argues that the way in which a society works is shaped very strongly, or even determined, by the mode of production that is dominant within it. A mode of production is not merely about technology, but is also associated with social organisation and social relations. There is a necessary historical sequence in the relationship between different modes of production, which historically have been organised along class lines. Feudalism gave birth to capitalism, which is currently the dominant mode of production. Different modes of production have different class relations stemming from different forms of exploitation. Under capitalism, the Marxist analysis argues that exploitation tends to take a subtle, hidden form. The evolution of capitalism has had profound implications for social relations at all spatial levels from the global to the local. Consistent with other class-based modes of production, capitalism is self-destructive due to internal, structural contradictions. These internal contradictions generate periodic crises, in which intense and painful processes of restructuring occur, such as de-industrialisation. The fundamental role of the state is to stabilise the system, so the state is not neutral in class terms. The state intervenes in the market in all economic systems.

The core argument behind regulation theory is that during the capitalist era, certain historical periods can be identified in which particular sets of arrangements emerge to manage or reconcile the breakdown tendencies of that stage of capitalist development. These sets of arrangements have been termed modes of social regulation, and the mode of social regulation at work from the 1920s until the 1970s in countries such as the US and the UK was known as Fordism (see Mackinnon and Cumbers 2007). As mentioned in Chapter 6, Fordism took on a set of characteristics that underpinned economic production in the UK and the wider western world from the 1920s until the 1970s (see Stoker and Mossberger 1995; Griffiths 1998a; Table 7.1). Fordism was named after Henry Ford and the automobile industry in the US in the early twentieth century. Fordism characterised much of industry in the period since 1945 but had its origins in the car manufacturing industry of Detroit. It can be viewed as a product of the US dominance and mass consumerism of the era, and came to be the dominant mode of production of the UK economy.

Fordism was dominated by intensive accumulation or mass production. Another characteristic was monopolistic regulation by the state realised through protective regimes and a high degree of intervention by the state. In terms of its politics, Fordism was essentially protectionist; had a key role for industrialists; and was founded on negotiation around tensions between industrialists and workers reflected by labour disputes and union strength.

From the 1970s, a 'crisis in Fordism' became apparent (see Harvey 1982). Regulation theory suggests that there is an inherent paradox between the tendency of Fordism towards instability, crisis and change, and its ability to coalesce and stabilise around a set of institutions, rules and norms, which serve to secure a relatively long period of economic stability. In the 1970s, there was a lull in the Fordist economy related to the production of oil, which fuelled the Fordist economic system. Regulation theorists suggest that this was a general crisis of institutional forms that had come to guide the post-war economy, rather than being specifically oil-related. The 1970s slow-down was deemed as a symptom of a crisis in Fordism, whatever the specific explanation.

Since the 1970s a number of changes to the mode of social regulation occurred associated with structural economic change and the de-industrialisation process (see Chapter 1). Boyer (1990) suggested four reasons for the structural crisis in Fordism. First, productivity gains decreased as a result of the social and technical limits of Fordism exemplified by worker resistance to the Fordist organisation of work and increasing difficulties in balancing ever longer and more rigid production lines. Second, the expansion of mass production led to an

TABLE 7.1 CHARACTERISTICS OF FORDISM

Economic life dominated by large firms
Firms nationally oriented
Mass production methods characterised the production process
Mass consumption and standardised commodities dominated
Commodities geared to the model of the standard nuclear family
Fordism was supported by the national state through Keynesianism, a universalist welfare policy, and equity

increasing globalisation of economic flows, which made national economic management ever more difficult. Third, Fordism led to growing social expenditure, because the relative cost of collective consumption increased due to the inapplicability of mass production methods in this area, leading to inflationary pressures and distributional conflicts. Fourth, consumption patterns gradually transformed towards a greater variety of use values; new demands were at odds with standardisation, the basis of economies of scale, and they were unable to be easily satisfied through mass production methods.

Globalised capital has been a driving force for structural change, which has resulted in a number of different characteristics of post-Fordism (see Amin 1994) (Table 7.2).

This transition to post-Fordism is reflected in the shift from the modern to post-modern city, as described in Chapter 1 (see Table 1.2) and the remainder of this chapter focuses on the changing governance arrangement in cities and the implications for urban regeneration strategies. The move to post-Fordism since the 1970s represents a transition from one distinct phase of capitalist development to another (see Amin 1994). One aspect of this new phase has been the emergence of the information age (see Graham and Marvin 1996). There is a debate, however, on whether there are fixed points of transition, or whether this is more fluid.

Characteristics of the post-Fordist city strongly embody some of the more general defining features of the post-Fordist period. The post-Fordist city has witnessed an increasing engagement of local government in local economic development. There has also been a restructuring and subordination of social consumption. In terms of urban governance, the post-Fordist city has seen an expansion of local political action, the rise of public–private partnerships, and the emergence of urban governance rather than government (see Graham 1992). Associated with rising levels of crime and fear in urban areas has been the emergence of the 'fortress city', particularly in the US and to a lesser degree in the UK (Christopherson 1994; see Chapter 11). The fortress city can be seen in a wider context of the loss of direct productivity activities from western countries, rising unemployment and urban dereliction, the reordering of urban environments as consumer products, and consumption-driven development and politics. The post-Fordist city is a managed environment as illustrated by place marketing (see Chapter 6), post-modern urban spaces and developments, and security-led urban planning.

THE CHANGING ROLE OF THE STATE IN URBAN REGENERATION

Changes in urban governance are connected to wider processes of the restructuring of state power. The nation state consists of a set of institutions (legislative, executive and judicial), which serve the whole of society with sovereign authority over a defined territory (see Dearlove and Saunders 2000). The role of the state in the economy and society in general has evolved through a process, which can be divided into three distinct phases (see

TABLE 7.2 CHARACTERISTICS OF POST-FORDISM

Economic life dominated by multi-national corporations with global flows of resources
Customised products and niche markets
A polarised labour force
Surveillance and the segregated city
A neo-liberal state characterised by deregulation, privatisation and competitive principles
An emphasis on entrepreneurialism at the local level

Abercrombie and Warde 2000, Chapter 13; Dearlove and Saunders 2000; Giddens 2006, Chapter 20). Each of these phases is linked to the role and nature of government involvement in public policy and more specifically, for example, in urban regeneration programmes.

The first phase can be characterised as an era of 'the limited state'. During the early nineteenth century, the state in the UK undertook a very limited range of functions compared with the standards of the early twenty-first century. There was no health service, no system of unemployment benefits, and state schools were virtually non-existent. Families and communities had to fend for themselves. There was only a very embryonic police force and there were few controls on land development and limited intervention by way of activities seeking to regenerate urban areas. There were no nationalised industries and the level of democracy was also very limited as only male property owners had the right to vote.

The second period, which can be traced from the late nineteenth century until the middle part of the twentieth century, was one in which state powers expanded considerably for several reasons. The problems that were thrown up by urbanisation and city living were major factors in this expansion. In the rapidly growing cities, living conditions for most people were wretched, diseases could travel very easily, and social disorder could not easily be contained by traditional institutions, such as the church. It therefore became clear to the ruling class that society and the economy could no longer be left to market forces and mutual aid. Some form of organised public intervention started to appear necessary. It also became apparent to industrialists and military leaders that an educated and healthy working class would help increase the efficiency of production, and guarantee better quality recruits for the armed forces, who were needed in greater numbers as the British Empire spread globally. The concentration of people in factories and city neighbourhoods also fuelled popular demand for greater industrial and political democracy, which in turn led to stronger demands for improved public service provision.

Another important factor in the expansion of state activity was the experience of severe economic depression and mass unemployment in the 1920s and 1930s. An influential economist of the time, John Maynard Keynes, put forward the theory that economic depressions could be avoided if governments were prepared to take steps to increase aggregate demand. They could do this, Keynes argued, by deliberately raising public spending above the level of tax revenues. According to Keynes, it was not necessarily a positive endeavour for the government to try always to 'balance the books'. In some circumstances, 'deficit financing' by spending more than was being raised was justified if the economy needed a boost to take it out of recession. Although it took a number of years and the experience of the Second World War for Keynes's ideas to become generally accepted, they eventually became the 'conventional wisdom' for the management of the economy, and provided intellectual support for the continued expansion of state activity. Illustrating the overall scale of state activity, government expenditure as a proportion of Gross Domestic Product (GDP) peaked during the two World Wars, reaching 46 per cent in 1918 and 61 per cent between 1942 and 1944. Although government expenditure fell back to between 34 per cent and 38 per cent during the 1950s, this still represented a much higher proportion than the pre-war level of around 25 per cent. It was during the period of the mid-1940s to the mid-1960s that government spending on urban areas focused on the post-war reconstruction of housing and city centres, which is generally regarded as the beginning of the current understanding of 'regeneration' (see Chapter 2).

By the 1960s, the UK had a more or less fully developed welfare state. Among its main features were the following.

- The state provided an extensive set of public services, to which all citizens had access on the basis of need, rather than on the basis of ability to pay. In the main, welfare services were intended to be 'universal' (open to everyone), rather than 'selective' (aimed primarily at the poor). Welfare provision consisted of the five main areas of health care; education; income support and pensions; housing; and personal social services including social workers and youth workers.

- These public services were mainly paid for out of general taxation and were mostly free 'at the point of delivery'. There were, for example, no charges for visits to a doctor or dentist.
- Taxation followed a broadly progressive pattern, rather than being in the form of a flat tax, with significantly higher tax burdens being carried by wealthy individuals and high earners than by poorer and middle-income households.
- The state was expected to play a major role in the management of the economy. In particular, it was seen as having responsibility for keeping unemployment low. The state's economic role also included direct public ownership of many key industries, such as gas, electricity, coalmines, post and telecommunications, railways, and water supply. It also included a network of controls over the actions of private companies, such as the system of town and country planning controls established in the late 1940s (see Chapter 2).

From the 1950s until the 1970s, this arrangement had the broad support of all the main political parties. Consequently, this period is sometimes referred to as the period of the 'social democratic consensus'. There were disagreements, for example, over exactly which industries should be in public ownership, and over charges for certain welfare services, but these were mostly over details rather than fundamental principles. By the mid-1970s, as the welfare state continued to expand, government expenditure had climbed to nearly 50 per cent of GDP.

The third phase of state activity in the society and economy of the UK witnessed the post-war consensus fracturing by the end of the 1970s, when Thatcher's Conservative government was elected (see Chapter 3). The Thatcher government was fundamentally opposed to the ideas on which the Keynesian social democratic consensus had been based. There followed a period in which public services were scaled down and made more selective; controls over business activity were weakened; industries were denationalised; taxation was shifted away from the rich; greater emphasis was placed on social control through the police and the courts; and the state sought to limit its economic management role to that of keeping down inflation (see Chapter 3 for the implications of these changes for urban policy).

Since 1997, New Labour has retained many of the characteristics of the 1980s and 1990s Conservative governments. There has been a concern to keep taxation low; key industries have not been renationalised; there has been an enthusiastic promotion of partnerships between the public and private sectors; and there have been cutbacks on certain types of benefit, such as student grants. However, New Labour's governmental priorities have differed from those of the Conservatives in a number of ways. For example, the introduction of tax credits for working families has helped to tackle child poverty. There have also been attempts at more integrated 'joined up' approaches to tackling entrenched problems of social disadvantage and social exclusion (see Chapters 1 and 5). There are contrasting perspectives about whether New Labour represents a continuation of the era that began with the Thatcher government, or the beginning of a new phase in terms of the state's involvement in the economy and society, and by implication in urban regeneration policy (see Giddens 1998; 2000).

Throughout the 1980s and early 1990s, as the government took measures to reduce the share of GDP devoted to public expenditure, the proportion spent fell from 46 per cent in 1981 and 1982 to 38 per cent in 1988. After a slight rise in the early 1990s due to an economic downturn, the proportion had fallen back again to 37 per cent by 1999, but has more recently increased under New Labour's spending on public services to 43 per cent in 2006.

General government expenditure in 2006/2007 amounted to £552 billion (Self and Zealey

2007). A closer inspection of figures on government expenditure by activity has implications for activities related to urban regeneration. The key area of spending related to 'transfer payments' such as family benefits, disability allowances, unemployment benefits and pensions, all aimed to reduce poverty and social exclusion. The biggest five areas of public service spending were social protection (38 per cent), health (16 per cent), education (12 per cent), defence (7 per cent), and public order and safety (5 per cent), which together accounted for 78 per cent of the total (Summerfield and Gill 2005). Direct urban policy spending is dwarfed by these figures for mainstream activities accounting for around 0.1 per cent of total public spending.

There are strong justifications for state activity in society, but there are sharply differing views about what the state should do and how it should be done. For example, some people believe that the state should take a major responsibility for making sure that people's needs are met; others believe that society works best if individuals are left to take responsibility for their own destiny. Some people believe that taxation should be borne mainly by the well off; others believe that distribution from the rich to the poor is not ethically justified, and may even be harmful. Some people believe that public services should cater for all sections of society; others believe that public services are only needed for those who can not cater for themselves, and so on. One way of making sense of these differing views is by distinguishing between a number of underlying 'political philosophies' that have strongly influenced successive urban policy periods as demonstrated in Section II. The key political philosophies for the purposes of seeking to understand the state's involvement in urban regeneration are conservatism, liberalism, socialism and social democracy (see Giddens 2006; Heywood 2007).

Conservatism argues that social order, or the stability of society as a whole, is the most important priority for a government. This is the reason why conservatives attach such value to tradition, and to institutions such as the family, religion, the monarchy and the police. For conservatives, state action should be concerned with supporting, not supplanting, these institutions. Conservatism also argues that societies are naturally hierarchical and that democracy has its place in a modern society, but is not a supreme virtue.

Liberalism, in its original 'classical' form, argues that the freedom of the individual is the most important principle. The main role of the state, therefore, should be to protect individual liberty. Liberals argue that state action should not be too intrusive on people's freedom. For example, liberals have traditionally been advocates of low taxation, and of the minimum of regulation or 'red tape', as exemplified by the approaches in 1980s urban policy (see Chapter 3). Liberals tend to be critical of welfare provision, as they believe it creates 'dependency'.

Socialists believe that capitalist societies, such as the UK, cater for the interests of the rich and powerful, and do not serve properly the mass of the population. In their view, capitalism is based on exploitation. It is the labour of working people that produces a society's wealth, but working people do not get the full fruits of the wealth they create. Socialists are also critical of capitalism because it is not democratic. Key decisions, for example about investment and disinvestment, are made by a small, powerful, circle, yet those decisions potentially affect large numbers of people, in different parts of the world. Socialists have, therefore, traditionally argued for 'common ownership' of the means of production.

Finally, social democracy can also be thought of as 'modern', as distinct from 'classical', liberalism. It is the philosophy of the mixed economy and perhaps is the political philosophy most reflected in post-1997 approaches to urban policy in the UK (see Chapter 5). In keeping with the liberal tradition, social democrats recognise the benefits of markets in fostering innovation, efficiency and choice and therefore want private business to thrive. However, they believe that markets do not regulate themselves effectively and that intervention by the state is necessary. Social democrats also believe that social disadvantage is damaging for the whole of society, and is usually not the fault of the poor themselves. Welfare provision by the state, therefore, is seen as a necessary element of advanced societies. Social democrats believe that state power should be used to 'humanise' capitalism to make it work better for people. This is sometimes referred to as the idea of the 'social market economy'.

It is useful to bear in mind that political parties are not always guided by the political philosophies that their names would suggest. The political ideas of the Conservative Party under Margaret Thatcher were based more on liberalism than conservatism. The political ideas of the Liberal Democrat party today are based more on social democracy than on liberalism in its classical form. The impacts of these philosophies on the implementation of urban regeneration initiatives are evident from the approaches and philosophical underpinnings illustrated by Section II.

There is currently a debate surrounding the role of the state in society and the economy, which can be applied to the field of urban regeneration in the UK. There are many different dimensions that are being questioned, which include such issues as:

- The role of the public sector versus the role of the private sector
- The role of the public sector versus the role of the voluntary and community sector
- The role of the national level of the state versus the role of local, regional and city-regional levels of the state
- The role of the national state versus the role of supra-national institutions (such as the European Union).

THE SHIFT FROM LOCAL GOVERNMENT TO LOCAL GOVERNANCE

The nature of the contrasts between the second and third phases of state activity points to another issue of importance to the practice of urban regeneration in the UK. This has been termed as a shift from government to governance (for more discussion, see, for example, Cochrane 1993; Andrew and Goldsmith 1998; Di Gaetano and Strom 2003; Geddes 2006). 'Government' can be defined as:

> the activity of the formal governmental system, conducted under clear procedural rules, involving statutory relationships between politicians, professionals and the public, taking place within specific territorial and administrative boundaries. It involves the exercise of powers and duties by formally elected or appointed bodies, and using public resources in a financially accountable way (Stewart 2003, p. 76).

'Governance' is a term used to indicate:

> [the] shift away from direct government control of the economy and society via hierarchical bureaucracies (government) towards more indirect control via diverse non-governmental organisations. It represents a broader approach to urban management and is often associated with the declining power of local government (Pacione 2005, p. 670).

Stewart (2003, p. 76) stated that it is:

> [a] much looser process often transcending geographical or administrative boundaries, conducted across public, private and voluntary/community sectors through networks and partnerships often ambiguous in their memberships, activities, relationships and accountabilities. It is a process of multi-stakeholder involvement, of multiple interest resolution, of compromise rather than confrontation, of negotiation rather than administrative fiat. Transaction costs are minimised, trust maximised, collaborative advantage extracted.

The process of the shift from government to governance has been represented as the 'hollowing out of the state' (see Jessop 2004), and corresponds with the changes experienced in the 1980s when unelected agencies and the private sector assumed a greater role in urban regeneration in particular, and society and the economy more generally. State functions have been lost outwards to the market, upwards to supra-national institutions, and downwards as localities become more significant political agents. The governance of urban areas now embraces the public, private and voluntary and community sectors (Community Development Foundation 1996; see Chapter 8). This thesis was initially developed by Harvey (1989), who argued that the 'managerial' approach to urban governance typical of the 1960s gradually gave way to 'entrepreneurial' forms of action in the 1970s and 1980s, a trend that has developed and been entrenched in the 1990s and 2000s. Harvey (1989, p. 11) argued that essentially 'the task of urban governance is, in short, to lure highly mobile and flexible production, financial, and consumption flows into its space'.

FROM MANAGERIALISM TO ENTREPRENEURIALISM

The third phase of the evolution of the state is closely linked with the shift from managerialism to entrepreneurialism (Harvey 1989; Wood 1998), or from the Keynesian state to the competitive state (Short and Kim 1999, Chapter 9) or 'neo-liberal' state (Ward 2003b). As discussed in Chapters 3 and 6, entrepreneurialism as a mode of urban governance resulted from the response by individual cities to the collapse of the Fordist social democratic arrangements that had underpinned the economic expansion of the three decades following the Second World War (Griffiths 1998a). The demand-side buoyancy of the Fordist era made possible the steady expansion of state expenditure on social investment and collective consumption. This made possible the spread of 'managerial' forms of urban governance (Harvey 1989; Hall and Hubbard 1996; 1998).

Managerialism is 'an analytical approach that focuses upon the influence of managers on access to scarce resources and local services' (Pacione 2005, p. 671), in this case predominantly those within government at national and local levels. This mode of urban governance was defined by the three main characteristics of an emphasis on the allocation of state surpluses, rather than on the attraction of private investment flows; the dominance of bureaucratic organisational forms in the delivery of public services, rather than the more flexible, less formalised, organisational approaches that were being adopted by the private sector; and the dominance of a social welfarist ideology, as distinct from the private sector values of wealth generation and competitive success (Griffiths 1998a, p. 42).

During the 1970s recessions and the concomitant erosion of the economic basis of managerial governance, national governments such as in the UK were facing increasing pressure to cut urban social spending. At the same time, the contraction of manufacturing industry in most of the UK's urban centres and the fall in male-dominated full-time secondary sector employment, were generating a collapse of the 'structured coherence' in their local social relations of production, consumption and reproduction (Harvey 1985; Goodwin 1993; Griffiths 1998a). Expectations that had formerly been taken for granted such as a job for life, the relative durability of occupational skills and the support mechanisms provided by formal and informal social networks, gradually disappeared from the social landscape (Griffiths 1998a). Local political cultures often based on a 'labourist' coupling of paternalism and universalism that had bolstered and legitimised managerialist governance became destabilised. As Griffiths (1998a, p. 42) argued, 'deprived of its economic and cultural foundations, managerialism gave way (unevenly, and often conflictually) to new approaches'.

Entrepreneurialism is the new mode of urban governance that has emerged from the crisis of managerialism (see Short and Kim 1999). It can be defined as 'a perspective on urban development that views the city as a product that needs to be marketed. This marketing approach and emphasis on restructuring the city to appeal to global business assign pre-eminence to economic interests in the decision-making process of urban planning' (Pacione 2005, p. 669). Drawing on the example of Baltimore's Harborplace, Harvey (1989) argued that entrepreneurialism had three defining characteristics. First, it is centred on public–private partnerships, whereby local government attempts to attract external sources of funding, new direct investment or new employment sources. Second, the activities of these partnerships are speculative in nature, with the risk resting on the public sector. Third, the effects of entrepreneurial projects are not only associated with the localities in which they are located.

Entrepreneurialism is predicated on a competitive quest for new sources of economic development in response to a collapsing manufacturing base and a growing internationalisation of investment flows. It has also been distinguished by the emergence of different organisational forms and institutional processes from those that characterised the managerialist era. Entrepreneurial regimes of urban governance have typically involved the formation of alliances and partnerships between public and private sector bodies (see, for example, Hutchinson 1994), together with some degree of displacement of institutional processes based on democratic representation. Entrepreneurialism has usually been associated with an ideological shift away from public service criteria, towards an acceptance of social and spatial inequalities (Hall and Hubbard 1996).

As illustrated in Chapter 6, entrepreneurialism is inextricably linked with the rise of inter-city competition and urban competitiveness measures. Four 'basic options' for urban entrepreneurialism identified by Harvey (1989) encompassed production, consumption, command and control functions, and attracting state surpluses. Within these categories, a wide array of entrepreneurial strategies can be followed by cities, although Griffiths (1998a) pointed out that cities do not have a free choice in selecting an entrepreneurial strategy. The strategic choices made by city leaders occur in conditions beyond their own choosing, and the options open to any individual city will depend on the place it occupies within the global urban system (Hall and Hubbard 1998; Short and Kim 1999). There is a powerful trend towards greater unevenness in urban fortunes stemming in part from the interactions that can and do occur between different strategies. For example, the pursuit for command and control functions will be assisted if the city is already an internationally recognised centre for arts and consumption (Griffiths 1998a). Entrepreneurialism is founded on speculation, risk-taking and competition, which inevitably results in winners and losers.

The displacement of managerial by entrepreneurial forms of urban governance (Table 7.3) in the context of the growing power of global capital and the de-industrialisation of former manufacturing centres, has not just reinforced uneven development between cities. The period since the 1970s has also witnessed a sharpening of social inequalities and divisions within cities, as middle income jobs lost through de-industrialisation are not replaced, and state action at all levels has been redirected away from social support and towards economic priorities consistent with the shift from government to governance (see, for example, Hamnett 1996; 2003a).

TABLE 7.3 CHARACTERISTICS OF ENTREPRENEURIAL GOVERNANCE

Government works in partnership with business
Tangible and visible symbols of urban regeneration
Pro-development agenda
Anti-democratic decision-making
Outcome-based decision-making model in urban regeneration and planning
Facilitation of consumption

Along with the shift to entrepreneurialism as an economic growth strategy has been the rise of the entrepreneurial city (see Hall and Hubbard 1996; 1998; Short and Kim 1999, Chapter 9). Painter (1998) provided a range of meanings involved in the notion of the rise of the entrepreneurial city including the city as a setting for entrepreneurial activity, increased entrepreneurialism among urban residents, a shift from public sector to private sector, and a shift in urban politics and governance from the management of public services towards the promotion of economic competitiveness, as discussed in Chapter 6.

Short and Kim (1999, p. 119) identified around 20 well-documented entrepreneurial cities in the urban studies literature that had deployed entrepreneurial strategies to facilitate urban regeneration. In the US, these included Los Angeles (Fulton 1997), New York (Fainstein 1991) and Syracuse (Short et al. 1993). Demonstrating a range of urban regeneration approaches, entrepreneurial cities in the UK included Birmingham (Hubbard 1996a; 1996b; Loftman and Nevin 1998), Bristol (Bassett 1996), Cardiff (Imrie et al. 1995), Glasgow (Boyle and Hughes 1991; 1995; Booth and Boyle 1993; Loftman and Nevin 1996; Paddison 1993), Liverpool (Parkinson and Bianchini 1993), London (Brownill 1994; Fainstein 1991), Manchester (Cochrane et al. 1996; Lawless 1994; Loftman and Nevin 1996; Peck and Tickell 1995), and Sheffield (Lawless 1994; Loftman and Nevin 1996; Raco 1997).

More recent additions to the research base on entrepreneurial cities in the UK include prestige projects in London (Imrie et al. 2009); Peck and Ward (2002), Williams (2003) and Ward (2003a; 2003b) on Manchester's urban redevelopment; Unsworth and Stillwell (2004) and Ward (2003b) on Leeds; Couch (2003) on Liverpool's transition; Ward (2003b) on Birmingham's continued entrepreneurial rejuvenation; and Punter (2007) and Hooper and Punter (2007) on Cardiff's mega-project regeneration. Each of these analyses shows how urban regeneration initiatives in UK cities are strongly tied in with the wider context of urban entrepreneurialism and governance structures. Thematic approaches to urban regeneration undertaken in these city examples are highlighted in Chapters 10–12. These entrepreneurial urban regeneration schemes have typically included mega projects, high-end housing, shopping malls, new leisure and cultural developments, convention centres, sports stadia, public art, and iconic urban design strategies.

INTER-CITY COLLABORATIVE NETWORKS

Associated with the shift towards urban entrepreneurialism, and the pursuit of policy responses and innovations to attempt to deal with the economic and social issues faced by cities, has been the development of inter-city collaborative networks (see Cappellin 1991; Parkinson 1992; Robson 1992). These have developed in the context of contradictory dynamics that have been at work in terms of governance at the European level. First, the need to maintain sovereign national states and sovereignty-sharing institutions has been set against the rise of sub-state-level actors (see John 2000). Second, the emphasis on territoriality or geographical places has been eroded by the growing prominence of the fluid logic of networks and flows. These conflicting dynamics have had implications for cities which have taken a much more 'international' outlook and acted as entrepreneurial agents in competition with other places (Short and Kim 1999). These developments reflect the dynamics of fragmentation and territoriality. In response, cities have started entering into collaborative relationships of one type or another with other places. This reflects, by contrast, the dynamics of integration and networks.

The forerunners of networks between urban areas were the town and city twinning schemes of the post-war period. For example, the city of Bristol has been twinned with the seven cities of Bordeaux in France (1947), Hannover in Germany (1947), Oporto in Portugal (1984), Tbilisi in Georgia (1988), Puerto Morazan in Nicaragua (1989), Beira in Mozambique (1990) and Guangzhou in China (2001). A wider variety of collaborative network models has emerged since the 1980s that has been much more sophisticated, politically astute and proactive.

First, networks can exist within a national territory such as the Core Cities Group in the UK, which was established in 1995 as an association of the major regional cities in England (see Chapter 6). Initially, the cities comprised Birmingham, Bristol, Leeds, Sheffield, Liverpool, Manchester and Newcastle, to which Nottingham was added in 2001. The main purpose of the group is to highlight the critical economic role of major cities, and to enhance their competitive advantage in an international context (see Chapter 6; Comedia 2002; ODPM 2004b; 2004i; Murray 2007). The group is given added importance with the involvement of key players such as council leaders and chief executives. Its central unit in Manchester comprises a director and a policy officer. There are eight work streams consisting of transport and connectivity; innovation; skills and employment; sustainable communities; culture and creative industries; climate change; finance and industry; and governance and partnerships. Annual summits report on each on these themes, which symbolise the key contemporary thematic issues of urban regeneration in the UK (www.corecities.com).

A second type of network brings together cities of a similar size and is active across a wide spectrum of issues. An example is Eurocities, which was founded in 1986 as a network of 'major European cities' with populations of over 250,000. Founder members were Barcelona, Birmingham, Frankfurt, Milan and Rotterdam, and its headquarters is in Brussels. The network grew out of a desire to promote an 'urban' dimension in the policies and structures of the EU. Its two primary aims are to give cities a voice at the European level, and to increase practical co-operation between cities. A subsidiary aim is to reach out to cities beyond the current EU boundaries to support them in their progress towards democratic government and a market economy (see Griffiths 1995b). The membership of the Eurocities collaborative network has grown from 60 cities in 1994 to 130 in 2007 in over 30 countries (www.eurocities.org).

Co-operation between cities takes place at three main levels comprising forums, working groups and projects. Seven forums encompass culture (based in Nottingham), economic development (Budapest), environment (Gothenburg), knowledge society (Bologna), mobility (Copenhagen), social affairs (Rotterdam), and co-operation (The Hague). Each forum has established a number of working groups of 20 to 40 members. Examples of working groups within the seven forums are creative industries (culture), urban regeneration (economic development), waste management (environment), e-government (knowledge society), road safety (mobility), urban security (social affairs), and governance (co-operation). Each of these working groups works on an array of projects.

Other types of networks between cities include those which can extend across national boundaries (see Church and Reid 1996; Williams 1996); networks focused on a particular policy area such as Car Free Cities, Telecities (www.telecities.org) or Cities of Culture (see García 2004b; Griffiths 2006); and networks based on campaigning issues such as the Climate Alliance of European Cities.

In terms of the wider trends towards new forms of urban governance, coalitions and associations between cities offer major benefits to members. These benefits embrace the sharing of information and ideas to help cities address their problems, providing a platform for lobbying and influence, and the creation of frameworks for better strategic thinking. They also act as forces for integration and cohesion across the extended Europe, and help to counteract the self-defeating impacts of place competition (see Chapter 6). However, cautionary issues include membership exclusivity of certain cities, difficulties in transplanting lessons to other contexts, issues of legitimacy and accountability, and the management of tensions between networks of cities and national governments.

THE GROWING IMPORTANCE OF THE REGION AND CITY-REGION

The regional agenda in the UK has been strengthened since New Labour came to power in 1997, in particular through the establishment of the nine Regional Development Agencies in England in 1999, following devolution of powers to the other constituent parts of the UK. These oversee economic development and regeneration; business efficiency, investment and competitiveness; employment; skills; and sustainable development (www.englandsrdas.com; see Chapter 5). These were complemented by the creation of the Government Offices for the English Regions, which represent the 11 Whitehall departments of the Cabinet Office; Communities and Local Government; Department for Business, Enterprise and Regulatory Reform; Department for Children, Schools and Families; Department for Culture, Media and Sport; Department for Environment, Food and Rural Affairs; Department for Work and Pensions; Department for Transport; Department of Health; Home Office; and Ministry of Justice (www.gos.gov.uk).

Consistent with the emergence of the regional agenda has been a renewed interest in the concept of the city-region. The city-region is a long-established urban growth and development model within urban geography. It is an area focused on the major employment centre in a region and encompassing surrounding areas, for which it acts as the primary high-order service centre (Pacione 2005, p. 91). This form of urban development and unit of governance has recently been used as an appropriate description of monocentred urban areas of up to a million people. It has been applied to the hinterlands surrounding the major core cities of the UK such as in Manchester, Bristol and Edinburgh (see Bramley and Lambert 2002; Boddy 2003b; Boddy and Parkinson 2004), as well as in continental Europe (see Herrschel and Newman 2002). City-regions reflect the polycentric nature of contemporary urban development, which has developed through edge-city types of development, or through linked networks or urban nodes in a more extended sub-regional context in the UK (Cooke et al. 2002).

In recent years, the notions of the city-region and neighbourhood (see Chapter 5) have gained increasing prominence in both academic and policy fields, and both are frequently seen as 'natural units' for analysis and policy focus in terms of addressing a range of problems facing urban areas (Atkinson 2004a; 2007; Goodwin 2004). An argument made since the mid-1990s has been that the city-region, in addition to the neighbourhood, is an appropriate unit of governance at the sub-regional level. Across Europe, under the growing pressures of globalisation and the apparent decline of the nation state, the city-region has increasingly been defined as the focus for economic development policies linked with urban competitiveness (see Chapter 6; Atkinson 2007, p. 65). This is because it is important to urban regeneration and competitiveness that the policy context and the planning framework in particular, along with infrastructure provision, administrative and political structures, relate to exurban developments (Boddy 2003d, pp. 93–94).

Urban regeneration and city-region-wide social and infrastructural developments have been hindered by administrative fragmentation and competition between local authorities within a single city-region. This is perhaps most clearly exemplified by the example of the Bristol city-region, which consists of four local authorities struggling to reconcile views relating to issues such as management of growth, new housing development and transport infrastructure schemes (see Boddy 2003d; Lambert and Smith 2003; Stewart 2003; Tallon 2007a).

Despite its recognised importance at the European level (see Le Gales 2002, pp. 156–159) and in the UK (Atkinson 2007), 'the city-region level of governance remains relatively underdeveloped in terms of its institutional capacity to effect change and in terms of how it relates, politically and democratically, to the constituent administrative units (for example the local authority and the neighbourhood) that make up a city-region' (Atkinson 2007, p. 77). Another unresolved contemporary problem in terms of urban policy identified by

Atkinson (2007) is how the regional level of governance characterised by Regional Development Agencies and Government Offices (see Chapter 5) link to the city-region, local authorities and neighbourhoods in a strategic sense. The focus continues to be on the local authority level in terms of urban governance and policies.

SUMMARY

From the 1920s until the 1970s, the Fordist economy was underpinned by government activity and the welfare state. Since the 1970s, the post-Fordist economy has emphasised entrepreneurialism at the local level.

The changing role of the state in society and the economy has paralleled the transition to post-Fordism. Since the late 1970s, the state has been 'hollowed out' and a greater role has been afforded to the private sector and non-governmental organisations.

There has been a change in the nature of urban governance from managerialism to entrepreneurialism, driven by the situation in which cities found themselves during the post-Fordist era.

Changes in the purpose of governance have witnessed a focus on competitiveness, marketing and image management, and forms and structures of urban governance exemplified by horizontal networks and partnerships.

At one end of the spectrum has been the focus on the neighbourhood as a level of governance, and at the other end has been the recognition of the importance of the regional and city-region levels of governance.

The entrepreneurial city is vividly symbolised by urban regeneration mega-projects as each competes with one another for investment, jobs, skilled workers and visitors.

There is a high degree of inter-linkage between the various transitions, which have impacted upon cities and urban regeneration strategies since the 1970s.

Thematic approaches and examples of urban regeneration demonstrated in Chapters 10 to 12 symbolise the transitions towards post-Fordism, governance and entrepreneurialism, all elements of the so-called post-modern city.

STUDY QUESTIONS

1. Explain and evaluate the concept of post-Fordism as a contribution to understanding the recent phase of urban change.
2. Explain the distinction between government and governance, and discuss the causes and consequences of the shift in emphasis from government to governance at the city level.
3. The 'entrepreneurial city': myth or reality? Discuss this claim with reference to urban regeneration developments in UK cities.
4. Some decisions about public services are made nationally (by central government); some are made locally (for example, by local authorities, police authorities, regional development agencies and health trusts). What arguments are there for and against locally based decision-making?
5. Assess the forming of collaborative networks between cities to share good practice and promote policy development.
6. Evaluate the role of the city-region as a spatial unit of urban governance.

KEY READINGS ON NEW FORMS OF URBAN GOVERNANCE

Boddy and Parkinson (eds.) (2004) *City Matters: Competitiveness, Cohesion and Urban Governance* [most of this collection relates the changing nature of urban governance to issues of economic competitiveness and social cohesion]

Cars Healey, Madanipour and de Magalhaes (eds.) (2002) *Urban Governance, Institutional Capacity and Social Milieux* [a collection of papers exploring recent urban governance]

Cities Journal [look at some of the recent 'City Profiles' on UK cities for case studies of new forms of urban governance and entrepreneurial regeneration, including Williams 1996; Haywood 1998; Ellis and McKay 2000; Pinch 2002; Tiesdell and Allmendinger 2004; Tallon et al. 2005; Tallon 2007a]

Hall and Hubbard (eds.) (1998) *The Entrepreneurial City: Geographies of Politics, Regime and Representation* [a collection of pieces exploring the make-up of the entrepreneurial city]

Harvey (1989) 'From managerialism to entrepreneurialism: the transformation in urban governance in late capitalism', *Geografiska Annaler* 71B: 3–17 [a seminal early piece on the nature of the shift from managerialism to entrepreneurialism]

Healey, Cameron, Davoudi, Graham and Madanipour (eds.) (1995) *Managing Cities: The New Urban Context* [contains discussions of institutional capacity, urban governance and post-Fordism]

Oatley (ed.) (1998) *Cities, Economic Competition and Urban Policy* [explores the shift towards competitive approaches in urban policy, and analyses the move towards entrepreneurialism and the role of partnerships]

Smith, Lepine and Taylor (eds.) (2007a) *Disadvantaged by Where You Live? Neighbourhood Governance in Contemporary Urban Policy* [contemporary focus on the neighbourhood as a sphere of governance and a point of public service delivery under New Labour]

KEY WEBSITES ON NEW FORMS OF URBAN GOVERNANCE

Office of National Statistics (ONS): www.statistics.gov.uk [useful resource for central government spending figures and the make-up of the public sector more generally; includes the annual Social Trends survey]

Social Trends: www.statistics.gov.uk/socialtrends [published annually, this draws together social and economic data from a range of government departments and other organisations and paints a broad picture of UK social trends and change]

8 Community and regeneration

INTRODUCTION

An increasingly important component of recent urban regeneration has been the involvement of communities in driving forward the regeneration of their area. This chapter looks at the debates surrounding this issue and at the nature of the involvement of communities in regeneration programmes. Community involvement in urban regeneration has emerged in a context of moves towards more entrepreneurial governance in which cities are keen to have the contribution, participation and approval of local communities, and to sell strategies to communities for reasons of legitimacy.

This chapter will first track the history of community involvement in urban regeneration initiatives from the late 1960s through to the latest policy developments on neighbourhoods under the New Labour government. It will then examine why it is that community engagement is considered so important for urban policy today. The third part of the chapter will examine the experience of community engagement in neighbourhood renewal partnerships and the challenges that this poses for the future of urban regeneration in the UK. The fourth section examines the importance of the social economy as an element of community regeneration within the UK.

There has been a strong emphasis in recent government policy on neighbourhoods and community engagement (see, for example, DETR 1997; ODPM 2005a; 2005g; 2005h; CLG 2006b; 2007e; 2007f; see Chapter 5). During the development of the 2006 Local Government White Paper (CLG 2006b), David Miliband, then Minister for Communities and Local Government, spoke of the need to re-engage citizens with the institutions of government and proposed a 'double devolution' from the centre to the local authority and beyond that to the neighbourhood. Although this commitment was watered down by the time the Local Government White Paper was published, giving citizens and communities a greater say remains a key theme. The emphasis in urban regeneration policy on neighbourhoods and community engagement is not new. Community has been a recurrent theme in government policy since at least the late 1960s (see Chapter 2).

COMMUNITY IN PUBLIC POLICY: A HISTORICAL PERSPECTIVE

The beginning of the story of community involvement in public policy corresponds with the emergence of a distinct urban policy in the 1960s when a series of high profile area-based initiatives were introduced to tackle the problems in what were seen as 'deprived' communities. These initiatives, the key of which were examined in Chapter 2, included Education Priority Areas, Housing Action Areas, General Improvement Areas, national Community Development Projects, Inner Area Studies, the Comprehensive Community Programme, and the Urban Programme, which directed smaller amounts of money into a greater number of areas (see Taylor 2003).

The need to address area-based deprivation was partly triggered by structural change. Urban unrest and racial tensions intensified in the 1960s. However, an additional stimulus was the discovery that poverty had not been abolished by the creation of the welfare state as had been expected. The work of Peter Townsend focused attention on poverty, while the work of Michael Young and others demonstrated that the redevelopment programmes of the 1960s, built to replace the slums, were disrupting rather than creating new communities (see Cullingworth and Nadin 2006). A third stimulus for a new policy drive was the Skeffington report in 1969, which recommended more public participation in planning (Taylor 2003).

The nature of the response at this time was partly informed by existing programmes in the US, particularly the US War on Poverty. In addition, the interest in community development as part of the response was also influenced by experience in the British Commonwealth, where 'community development' was being introduced to prepare former colonies for independence.

Taylor (2003) argued that there were two main rationales behind this emphasis on community and participation in the late 1960s. The first assumed that the problem lay in communities themselves. Community needed to be 'restored' in the more disadvantaged areas from which it had been lost, either because housing programmes had displaced community ties or because of moral breakdown, particularly what a leading Conservative thinker (Sir Keith Joseph) defined as cycles of deprivation, which continued down the generations in some families and communities. This restoration of community had been attempted through the New Towns and slum clearance programmes. The second rationale assumed that the problem lay with the state and public services, and that there was a need to address the lack of co-ordination between services in the rapidly expanding local government organisations of the time.

Other thinkers at the time were critical of both of these rationales. The Community Development Projects focused on 12 areas across the country, and combined action with research teams in each locality, which was another idea borrowed from the US War on Poverty. The research from this programme was highly critical of the focus on community development because of its colonial antecedents and of the area-based approach for its assumption that the problems of these areas lay in communities themselves. Instead, a series of research reports highlighted global economic trends in capitalism of which the most disadvantaged areas were invariably the casualties (see Chapters 1 and 2). This research argued against the consensus approaches of the late 1960s and early 1970s, and promoted a more radical, confrontational and quasi-Marxist approach to the structural causes of poverty. It was perhaps not surprising that the government sponsors allowed this particular initiative to fade away, although other parts of the policy package were to last much longer. The oil crisis of the mid-1970s resulted in an economic recession, and increased pressure on working-class communities. In response, the focus shifted to job creation and training programmes, both of which recognised the economic nature of area-based disadvantage. There was to be more limited encouragement of community enterprise.

The arrival of the Thatcher government with its landslide victory in 1979 brought with it a more radical economic approach, which relied on economic incentives to bring industry and development into poorer areas such as docklands (see Chapter 3). The assumption was that wealth generated by economic development would 'trickle down' to the poor through job creation (see, for example, Healey et al. 1992). However, research soon showed that this fabled 'trickle-down effect' did not work in practice. Poorer populations were displaced, bypassed or abandoned by wealth creation (see, for example, Imrie and Thomas 1999a). Industries often moved on once the subsidies ran out, and at a local level there was resistance to the market-driven agenda. New urban left local authorities sought to gain support for their services and their role, although subject to continued attack by central government, by decentralising services, providing new opportunities for participation, and funding marginalised groups. For example, women's units and race equality units spread in many cities and many local authorities employed community workers. These authorities sought to build a so-called 'rainbow coalition' of resistance to the Thatcherite agenda through municipal socialism. However, by the end of the 1980s, many local authorities were reeling in the face of financial constraints and the dismantling of public sector provision in favour of services contracted out to the private sector. Community development continued during the 1980s, but this was in opposition to, rather than in partnership with, central government.

The departure of Thatcher and the arrival of the Major government in 1990 led to some softening of the market-led agenda. The aggressive 'economistic' analysis of the 1980s was gradually replaced by the beginnings of an emphasis on consumerism and community control of services, community involvement, and partnership with emphasis on a citizens' (consumers') charter. The softening of market excesses was exemplified by the flagship initiative of the administration, City Challenge, which demanded partnership working and consultation with communities (see Chapter 4; MacFarlane 1993; McArthur 1995; Davoudi 1995; Lowndes et al. 1996). However, this involvement was expected over a tight six-week bidding timescale, which was totally unrealistic for any genuine community involvement (Davoudi 1995). At the same time, an emphasis on consumers in policy led to an encouragement of tenant involvement in management. In the public housing sector, tenants were given the right to vote on having other landlords, although the response was often that tenants preferred to stay with local authorities. For example, Estate Action finance to improve housing on run-down estates was dependent on the inclusion of measures to encourage tenants to become involved in management, hence giving them a stake in social housing and reducing 'dependency' (Stewart and Taylor 1995; Hastings et al. 1996; Taylor 1998). Partnership was also encouraged in bids for the Single Regeneration Budget challenge fund, which replaced City Challenge (see, for example, Oatley 1998d). Central government encouraged voluntary and community sector-led bids, although these were few and far between in the early rounds. However, in later rounds this was given a higher priority, which was further entrenched under the New Labour administration.

NEW LABOUR POLICY TOWARDS COMMUNITIES AND REGENERATION

Tackling social exclusion was a key theme of the incoming New Labour government (SEU 1998; 2000; 2001; Griffiths 1998c), drawing upon the European experience (Madanipour et al. 1998; Alden and Thomas 1998; Atkinson 2000a). Community involvement became a prominent theme in the plethora of policies introduced to tackle social exclusion which included the continued SRB programme, Sure Start for under-fives, the New Deal for Communities and the National Strategy for Neighbourhood Renewal, a 20-year strategy (see Chapter 5; Community Development Foundation 1997; DETR 1997; Joseph Rowntree Foundation 1999; DTLR 2000; Lepine et al. 2007). In 1998, the then Prime Minister Tony Blair, in the preface to a report by the Social Exclusion Unit, proclaimed 'Too much has been imposed from above, when experience shows that success depends on communities themselves having the power and taking the responsibility to make things better' (SEU 1998, p. 2). The emphasis here was on giving power to communities, but also on the responsibilities of these communities. The approach was reminiscent of the 1970s Community Development Projects. The extent of community involvement in the late 1990s and early 2000s was seen as a watershed and was unusual at the time.

The Social Exclusion Unit's report led to the development of a National Strategy for Neighbourhood Renewal, which was based on an unprecedented two years' consultation with communities as well as other players. Policy development over this period included 18 Policy Action Teams involving civil servants and others in the field. People with experience of living and working in disadvantaged communities were also seconded into government departments and a National Community Forum, made up largely of residents of disadvantaged communities, was set up to advise government on the implementation of the strategy. At the local level, the forerunner of the National Strategy for Neighbourhood Renewal, the New Deal for Communities, required communities to be at the heart of renewal. Communities were also expected to be central players in Neighbourhood Management Pathfinders, which were established to join up services at neighbourhood level and 'bend mainstream budgets' to serve these areas better (Smith et al. 2007a). Faith was placed in bottom-up regeneration initiatives.

As Tony Blair said in his foreword to the consultation document on the National Strategy for Neighbourhood Renewal (SEU 2000, p. 5), 'Unless the community is fully engaged in shaping and delivering regeneration, even the best plans on paper will fail to deliver in practice'. The strategy required community involvement not only in the special area-based initiatives but also in Local Strategic Partnerships (LSPs). These brought together the major players at local level to encourage 'joined-up' working especially in response to those cross-cutting issues that had proved resistant to previous approaches. In the 88 priority areas for neighbourhood renewal (those local authority areas containing the 10 per cent most disadvantaged wards in England), the LSP was charged with developing a neighbourhood renewal strategy that could then draw down additional neighbourhood renewal funding (Keith 2004). Funding was made available to support the involvement of the voluntary and community sector in LSPs in these areas through three related programmes comprising the Community Chest, the Community Learning Chest and the Community Empowerment Fund.

Through the combination of the various programmes shown in Table 8.1, central government provided limited funding to support and encourage small-scale community activity on the ground through the Community Chests. The government also put in place a requirement that communities should be major players in special initiatives to turn around deprived areas, as witnessed in policies such as New Deal for Communities (which pre-dated

TABLE 8.1 THE ACTION PLAN

New Deal for Communities
Neighbourhood Management Pathfinders
Neighbourhood Wardens
Neighbourhood Renewal Fund
Local Strategic Partnerships
Community Participation Programmes
Community Chests, Community Learning Trust, Community Empowerment Fund

the National Strategy for Neighbourhood Renewal), the Neighbourhood Renewal Fund (targeting the 88 most deprived local authority areas), and the Neighbourhood Management Pathfinders (36 set up in two waves from 2002 for seven years). In addition, the government provided funding to support community involvement in strategic decision-making at local authority-wide level. The National Strategy for Neighbourhood Renewal also supported a three-year Neighbourhood Wardens pilot project.

The four goals of the National Strategy for Neighbourhood Renewal in terms of community involvement (ODPM 2005a) were to enhance:

- Social capital: developing capacity, confidence and skills;
- Social inclusion and cohesion: networking, cohesion and co-ordination;
- Governance: developing an effective voice, and involving communities;
- Service delivery: influencing public service delivery giving communities an effective voice.

These developments needed to be viewed in the context of other wider emerging policy trends in relation to local government, community development and planning. The four parallel developments encompassed modernising local government, democratic renewal, civil renewal and a comprehensive third sector policy. The Local Government Modernisation Agenda sought to improve local leadership both politically and administratively, in order to counter criticisms that initiatives represented a 'bowl of spaghetti' (Johnstone and Whitehead 2004b). One outcome of this was the Local Area Agreement, which became one of the centrepieces of local government reform. A Local Area Agreement is defined as 'a three-year agreement, based on local Sustainable Community Strategies, that sets out the priorities for a local area. The agreement is made between Central Government, represented by the Government Office (GO), and a local area, represented by the lead local authority and other key partners through Local Strategic Partnerships (LSPs)' (www.communities.gov.uk, 2008). All local authorities had developed Local Area Agreements by the mid-2000s.

A second development has been the promotion of democratic renewal to give the government greater legitimacy among the population in an era of falling voting turnout. This has been attempted through initiatives like citizens' juries, whereby local people are invited to contribute to discussions on problems such as anti-social behaviour. The government has more recently pursued democratic renewal beyond disadvantaged neighbourhoods.

A third parallel policy agenda in the 2000s has centred on civil renewal to encourage people and government to work better together, which to an extent has taken on a moralistic tone. This was the idea of David Blunkett when he was Home Secretary, but has now been transferred to the CLG, who described civil renewal as 'about people and government, working together to make life better' (see Home Office/ODPM 2005). 'It involves more people being able to influence decisions about their communities, and more people taking responsibility for tackling local problems, rather than expecting others to' (www.communities.gov.uk, 2008).

Community cohesion has also been a significant related agenda in the wake of racial tension, the rise of the extreme right British National Party, and the London bombings of 2005 to tackle 'racism, extremism and hate and promoting inter-faith activity and a shared sense of belonging' (www.communities.gov.uk, 2008). Social cohesion, allied with economic development, has been prominent in approaches to urban regeneration since the late 1990s (see Chapter 6; Kearns and Forrest 2000; Forrest and Kearns 2001; Boddy and Parkinson 2004, Chapters 13, 16, 20 and 21).

The government has developed a comprehensive third sector policy that incorporates the voluntary sector, the community sector and the faith sector. An Office for the Third Sector has been created in the Cabinet Office illustrating its position at the heart of government.

In the field of planning, housing market renewal has its own implications for urban regeneration (see Chapter 5). A key feature has been the promotion of mixed communities in order to introduce greater tenure, and hence social diversity to social housing estates that are characterised by multiple deprivation (see Bailey et al. 2006; Roberts 2007). This has been most vividly illustrated through the Housing Market Renewal Pathfinders in northern English cities (Atkinson 2004b).

Neighbourhood renewal policy has more recently taken another turn, with the 2006 Local Government White Paper (CLG 2006b). Devolution is now the ambition and central government funding for much of the neighbourhood renewal agenda has now been devolved to local authority level within the localism agenda (see ESRC 2007). This is part of a more comprehensive approach to local government reform and is now an ingredient of the Local Area Agreement.

In the late 2000s, provisions for giving citizens and communities a greater say included:

- Devolution of neighbourhood renewal funding;
- The Local Government White Paper incorporating local charters, community calls for action, participatory budgeting, asset transfer, and citizens' juries;
- Duty to involve guidelines published;
- A Community Engagement Action Plan including a National Empowerment Partnership, a Take Part campaign, a national network of empowering authorities, and community kitties;
- A strong emphasis on voice in third sector policy with investment in community sector infrastructure and community anchors.

The policies and language in the area of community and local government constantly change, as do ministers responsible, making this a dynamic and confusing policy landscape. Illustrating the constant evolution of policy seeking to involve communities in the fields of regeneration, planning and housing, the government's draft legislative programme for 2008/2009 included a Community Empowerment, Housing and Economic Regeneration Bill with the purpose of creating greater opportunities for community and individual empowerment; reforming local and regional governance arrangements to promote economic regeneration; and continuing the government's programme of housing reform (see www.commonsleader.gov.uk). Such legislation would clearly have an impact on how communities engage with urban regeneration.

THE IDEOLOGY OF COMMUNITY AND ITS IMPORTANCE TO URBAN REGENERATION

On its election in 1997, New Labour committed itself to look for a 'third way' between the market and the state (Tiesdell and Allmendinger 2001; Imrie and Raco 2003a; Johnstone and Whitehead 2004a). Community was a central theme inspired by sociologist Amitai Etzioni's communitarianism (see, for example, Etzioni 1997), and more recent ideas of social capital as promoted by Robert Putnam (1993; 1995; 2000; 2002). Etzioni argued that the state was distant from the individual, and looked to local communities for the solution. However, this work has been criticised for being overly moralistic and idealistic.

In Putnam's analysis, the norms and networks resulting from systematic face-to-face association enable participants to act together more effectively to pursue shared objectives. Putnam (1995) contended that social capital represented features of social life including networks, norms and trust that enable participants to act in this way. Social capital in essence refers to social connections and related norms and trust. Putnam saw voluntary and community associations such as sports clubs, co-operatives, mutual aid societies, cultural associations and voluntary unions as the key source of social capital. Putnam's studies in

Italy associated membership of such 'horizontally ordered' and mutual organisations with effective local government. They generated the trust, reciprocity and capacity for civic engagement, which is seen as essential to the effective functioning of modern democracy. Resources or social capital can be gained from networks. 'Vertically ordered' organisations, on the other hand, such as the Mafia, the Roman Catholic Church and some political parties, created dependency, and Putnam (1993) argued that they were negatively associated with good government. Critics of Putnam have argued that networks are unevenly distributed and exclusive.

Social capital as promoted by Putnam has three forms. Bonding social capital forms between people who are alike and in close relationships. Bridging social capital builds ties across communities with different interests such as ethnic communities. Linking social capital establishes ties between communities and power holders. Some define social capital as something that individuals have, for example, networks that give them access to opportunities and resources. Others define social capital as something that is a collective resource (see Kearns 2003). Kearns (2003) demonstrated the confusion regarding its definition but synthesised its components as being social networks used by people; social norms adhered to in people's behaviour, and in particular whether or not these norms are widely shared; and levels of trust people have either in their neighbours, in people in general, or in the institutions of government. Either way, a range of studies suggest that social capital is associated with better government, lower levels of crime, improved health, economic growth and thus has the potential to regenerate deprived urban communities (PIU 2002; Halpern 2004).

Kearns (2003) argued that the New Labour government's dialogue around neighbourhood and community regeneration shifted away from social exclusion and economic development towards social capital, reflecting a 'soft' goals agenda. Tiesdell and Allmendinger (2001, p. 921) further contended that New Labour's 'Third Way' approach to neighbourhood regeneration included 'a greater recognition of the concept of "social capital", in the form of local partnerships and bottom-up approaches and more generally in strategies, policies, and initiatives that integrate the "people and communities" with a "bricks and mortar" dimension'.

Social capital has featured prominently in New Labour's approach to urban regeneration (see Taylor 2000; Forrest and Kearns 2001). New Labour has maintained that social capital can contribute to community cohesion and promises multiple social outcomes in terms of reduced crime, improved health, better labour market outcomes, higher educational attainment and more effective democracy (Kearns 2003, pp. 55–59).

The emphasis on community and the development of social capital is not confined to the UK. It is a global phenomenon viewed as a new magic wand in response to:

- the breakdown of moral cohesion in an increasingly fragmented post-modern society, and the emergence of an 'underclass';
- part of the more general move from government to governance, in recognition of the limits to both state resources and legitimacy in an increasingly complex global environment (Stoker 1998, see also Somerville 2005);
- increasing demands on welfare provision and an emphasis on active citizenship, and responsibilities as well as rights of individuals – community is seen as a supplement or alternative to the state and its limited resources;
- the breakdown of belief in democracy and a need to gain legitimacy for government at the local level through increasing the number of people voting in elections.

More pragmatically, it enhances community involvement, in the following ways:

- brings resident knowledge to bear in finding solutions to issues that have proved difficult to resolve from the outside;

- encourages communities to buy into and own solutions thus making them more integrated and sustainable (see Anastacio et al. 2000); residents are more likely to protect their local area if involved in the locality;
- contributes to people's own welfare through developing skills, social capital and so on, thus reducing demand on the welfare state.

Community involvement is seen as intrinsically worthwhile, but whether community can deliver on these expectations remains to be seen. However, it is important to remember that community and social capital have their 'dark side' and cannot be seen as a panacea for urban regeneration. First, communities are defined by 'them' as well as 'us'; in this sense they exclude as well as include (see Taylor 2002; Kearns 2003). Strong communities can operate on the basis of distinguishing between 'insiders' and 'outsiders' (Giddens 2000). Strong ties within disadvantaged neighbourhoods often exist, but ties are absent with those outside the neighbourhood. Second, wider structural reasons for exclusion are ignored. Prescribing community to populations that have been excluded by mainstream society appears hypocritical. It could be argued that the expansion of choice in housing and schools has concentrated those who cannot choose increasingly in disadvantaged pockets, as communities cannot turn the decline of their neighbourhoods around on their own. Third, the strong ties with romantic ideals of community are not consistent with the way in which society now functions, and community can be both oppressive and excluding.

Kearns (2003, pp. 45–46) added some notes of caution regarding the downside of social capital as a method of neighbourhood and community regeneration. First, economic performance can be undermined by sectional interests relating to relationships with associations, such as trade unions, which are inefficient. Second, the 'old boy' networks of the middle class maintain social class barriers and inhibit social inclusion across groups; the advantages of the more extensive and diverse social networks of advantaged groups can exacerbate social inequality, representing social capital as a 'club good' rather than a 'public good'. Third, the strong bonding of social capital of poorer groups may exacerbate community conflicts and limit advancement where bonding social capital within groups far exceeds bridging social capital between groups. Fourth, some groups that have strong bonding social capital and are also spatially concentrated might become insular and disconnected by desire and by default. Fifth, strong communities can be oppressive and conformist, restricting routes out of poverty and exclusion. Sixth, and finally, Kearns (2003) argued that social capital through networks and norms can be used to promote damaging behaviour such as crime, truancy and drug addiction, thus damaging community welfare, health and education.

Kearns (2003, p. 39) contended that despite social capital arising as a means to the end of social inclusion and as a way of tackling social exclusion, it has its upsides and downsides. In terms of social capital as promoted by New Labour's urban regeneration policy, Kearns (2003) concluded that it can be appreciated far more for its positive elements than its negative elements.

THE EFFECTIVENESS OF COMMUNITY INVOLVEMENT IN URBAN REGENERATION

There has been more scope for communities to get involved in urban regeneration since the late 1990s than ever before. There has been a corresponding explosion in the language associated with community and partnership. However, reconciling the different cultures of communities and government takes time to develop, and there are pitfalls as well as opportunities associated with community involvement (see Taylor 2007). Partnerships between the public sector and communities are often characterised by stereotypes about other people at the table and a lack of trust, which is difficult to resolve. In the face of the complexity of the issues, the effectiveness of community involvement and partnership has been problematic (see Atkinson 1999a; Goodlad et al. 2005). Despite repeated government commitments to community engagement, most research suggests that community involvement in partnerships over the past forty years or so has yet to deliver (see, for example, Colenutt and Cutten 1994; Mayo 1997; Cebulla et al. 2000). Many communities still feel marginalised and unable to influence the decisions that affect their areas or the services they use in the realisation that past involvement has not delivered. There are four key factors responsible for this (see Taylor 2002; 2003; 2007), comprising barriers between community and local power holders, partner capacity, community capacity, and the role of central government.

The first issue relates to barriers in the form of local power holders. The 'rules of the game' dictate that communities are usually expected to adapt to government cultures. It is government bodies and professionals that usually define the rules of the game, who plays and on what terms (Taylor 2003). As a result many communities feel they are in a game they do not understand. They often find themselves in the wrong game too, since decisions are usually taken elsewhere within pre-existing power networks or the corridors of power. Communities do not have these informal links into the policy-making process, only a formal legitimating role. In addition, only those who can hit the ground running are able to engage with the process, as there is rarely any 'induction' or attempt to make conventional ways of doing things more accessible to local communities (Taylor 2002; 2007). Expectations are often unrealistic and timescales are too short.

Research by Taylor (2003) illustrated some of the problems faced by voluntary, community and user organisations experienced in partnerships. These related to adapting to the rules of the game, demonstrated by the following quotes: Who adapts?:

> In all the work I've been involved in, it's us who have to put effort into reaching the council's level . . . they never come down to ours.

Understanding unfamiliar systems:

> With so many men in suits, it was difficult to find the courage to speak up. Sometimes you went along determined to say something this time, but somehow the meeting would be over and you wouldn't have opened your mouth.

A discouraging response:

> some never regained their confidence after making their initial comment, as they were 'made to feel' they had 'said something stupid' or 'at an inappropriate time during the meeting' or 'under the incorrect agenda item'.

On representation:

> The officers pay more attention to questioning whether we are representative or not than they do to what it is we actually say.

A second topic of concern relates to partner capacity. Many studies suggest that there is resistance to the community involvement agenda in many parts of the public sector, particularly from professionals and from local councillors. Against a background where the powers of both have been eroded by increasing central government control since the 1980s (see Chapters 3 and 7), public sector players often feel threatened by any emphasis on community. Despite the rhetoric, power tends to stick at the local authority level rather than devolving down to the neighbourhood level to any significant degree (see Taylor 2003). Even where there is political will and leadership, many of the targets that come down from the centre take precedence at middle management level. In terms of Arnstein's (1969) famous ladder of participation in policy making, too much community involvement is at the manipulation or therapy level rather than at the partnership, delegated power or control rungs through tenant management, for example (see Figure 8.1). At the same time, few public sector players are encouraged or trained to work in new ways. The rhetoric is not backed up by practical incentives and as central government devolves powers, it has fewer and fewer levers with which to ensure that its community empowerment agenda rolls forward at local authority level.

A third issue regarding the effectiveness of community involvement in urban regeneration relates to community capacity. New policies since 1997 have given communities the instruments and the right to challenge public authority, through offering new triggers and resources at neighbourhood level. However, studies suggest that effective community empowerment depends on more than structures and triggers or even rights. There has been a strong emphasis on capacity building in government policy, and research suggests that effective involvement in policy requires a strong foundation in community activities and organisations on the ground, as well as a variety of ways into engagement.

There are many challenges, however, in organising communities to engage effectively and contribute to urban regeneration (see Jones 2003). It requires time and resources if engagement is to move beyond the few, and probably realistic expectations of how many and to what extent people will engage. Some of the challenges relate to the diversity within communities, which raises questions as to how to mediate between the many different interests and communities in any one neighbourhood; there is no 'one' community in a given neighbourhood. The competitive nature of funding can often drive wedges between communities and cuts across calls for more collaboration and cohesion. There is a tension between leadership (see Purdue et al. 2000) on the one hand and widespread participation on the other.

Another problem found by Taylor (2003; 2007) in work on participation is that there is a great deal of mistrust of government in communities and a fear of being co-opted into government agendas. While some residents are overcoming this to work productively with power holders, others get stuck in opposition or find it difficult to look beyond their immediate concerns and think strategically. These problems are not insoluble but take time and resources to resolve.

A fourth aspect limiting the success of community involvement in urban regeneration has been central government. In a policy environment where the emphasis since the late 1990s

Degrees of citizen power	Community control
	Delegated power
	Partnership
Degrees of tokenism	Placation
	Consultation
	Informing
Non-participation	Therapy
	Manipulation

Figure 8.1 The ladder of community participation
Source: **Arnstein (1969; 1971).**

has been on evidence-based policy (DTLR 2000; Harrison 2000; Grimshaw and Smith 2007), it has been difficult to provide the evidence that community involvement does make a difference. This makes the policy commitment to community involvement vulnerable in an era of evidence-based policy. However, there is some evidence emerging that programmes like the New Deal for Communities and Neighbourhood Management Pathfinders (see Chapter 5) have led to improvements in terms of community involvement.

The sheer pace of policy change means that it is a moving target that is being measured, and in any case all the evidence suggests that community involvement takes time to work effectively (see Taylor 2003). In the past, communities have been put off by excessive central government controls and heavy regulation relating to the way that new money is spent and the way progress is monitored through audits and targets. The devolution agenda suggests that this micromanagement might be reduced in the future. The opportunities for communities to close the power gap have clearly opened up in recent years. However, questions remain about how far, once central government relinquishes the controls, local public bodies will commit to the community empowerment agenda, and how far central government is prepared to push the agenda if they do not. In short, it can be argued that the rhetoric continues to far outstrip the practice.

MEASURES TO PROMOTE THE SOCIAL ECONOMY

In parallel with and linked to the emergence and promotion of new forms of community involvement in urban regeneration partnerships, recent efforts have sought to build and promote the social economy as a way of tackling problems of urban social disadvantage and exclusion. The social economy has become an important element of the social and economic policy agenda of the UK government and the European Union (see Oatley 1999; Amin et al. 1999; 2002; Spear 2001; Moulaert and Ailnei 2005; Moulaert and Nussbaumer 2005). Community-based credit unions and barter-based cashless exchange systems have emerged, and signalled the development of 'community economies' designed to help communities marginalised by structural economic change (Hall 2006, p. 160).

Principal among measures within the social economy have been Local Exchange Trading Systems (LETS), credit unions and community social enterprise (Pacione 2005, pp. 347–350). The social economy is not part of the private sector or the public (state) sector and is sometimes referred to as the 'third sector' of the economy, which is 'not for personal profit'. The activities within the social economy are undertaken by organisations and groupings that are part of civil society, and therefore not part of the state, with the primary aim of providing a social or environmental benefit of some kind. The term is used intentionally to link with economic regeneration rather than referring to any particular values or expertise that an organisation might bring such as community development, care, the environment and so on. A social enterprise is seen as entrepreneurial rather than just as a service, campaign, trust or organisation (BCC 2001).

The spheres of activity of the social economy or third sector include building housing, exchanging skills and services, providing loans, running cafés and restaurants, running leisure activities, recycling furniture and providing transport. The sector has made a valuable contribution including the provision of jobs and new routes into employment, capacity building, local democracy, and the supply of goods and services to local communities (see Taylor 2003; 2007).

The social economy exhibits a number of characteristics in terms of organisation, forms and modes of operation. Goods and services are sold or exchanged without money. They can be highly localised organisations or serve a wider area, and they can have formal constitutions or operate more informally. Examples of the social economy include community-owned businesses; local self-help or interest organisations such as development trusts; most 'public good' purpose trusts and national initiatives; and user, worker and marketing co-operatives and other mutual societies, except the building society and insurance sectors (BCC 2001).

Development trusts, as an example of community-owned and -led organisations, have proliferated since the 1990s. They use self-help, trading for social purpose, and ownership of buildings and land to facilitate social, economic and environmental benefits in their community. They operate in both urban and rural areas, often in neighbourhoods that have experienced the most severe effects of economic decline. They are independent bodies, but work in partnership with the public sector, private sector and with other community groups. They are community 'anchor' organisations, which deliver services and facilities, find solutions to local problems, and help other organisations and initiatives succeed (see www.dta.org.uk).

The three principal categories of social economy activity in the UK that aim to promote social cohesion and address social disadvantage in urban areas are LETS (Local Exchange Trading Systems), credit unions and social enterprises.

LETS are systems for exchanging skills, services and goods between members, without the use of money. They are 'a community attempt to develop a local alternative to the capitalist economy by creating a complementary form of social and economic organisation that uses a non-commodified local currency to generate trading activity among members' (Pacione 2005, p. 671). Members register the skill or service they can provide, and indicate its value in the currency of the scheme. Examples of services could include dog walking, fixing computers, gardening, tutoring, decorating or providing garden produce. Units of currency are transferred by the administrator of the scheme. The benefits of LETS for local disadvantaged communities are that they reduce dependence on cash, retain benefits in the locality thus reducing 'leakage', and create social networks that contribute to social capital and social cohesion (see Williams 1996a; 1996b; 1996c; 2000a; O'Doherty et al. 1999).

Credit unions were introduced in Ireland in the 1950s and are financial co-operatives in which savings and loan schemes are operated by the users. Members make a regular saving, and become eligible to take out loans at low rates of interest. There are approximately 400 credit unions in the UK, which are generally situated in low-income districts. The benefits of credit unions are that they encourage regular saving habits, give access to affordable loans, break the grip of exploitation by loan sharks, and strengthen civil society.

Social enterprises, also known as 'community enterprises' and 'social businesses', are trading organisations that operate as businesses. However, their ownership and control rest with a local community or trust, and the surpluses are ploughed back into the scheme to enhance their activities. Social enterprises are committed to high ethical standards.

Social enterprises, community-owned businesses, co-operatives, local self-help groups and development trusts are seen in the city of Bristol as having a vital role in economic development and social cohesion (BCC 1997; 2001). In 2001, the social economy in Bristol was estimated to be worth around £223 million covering over 1100 organisations. The largest social economy 'business' had a turnover of £6 million with the average being around £10,000. Around 9000 people were employed in the sector and a further 20,000 volunteers were engaged in social economy activity. Activities of organisations within the social economy in Bristol encompassed advice, care, housing, property management, health, transport, manufacturing, environment, culture, recreation, sport, community and economic development (BCC 2001).

An example of a successful social enterprise in Bristol is the Sofa (Shifting Old Furniture Around) Project, which was set up in 1982, and operates across the whole city-region. The enterprise collects and repairs unwanted furniture and appliances, and sells them on to people on low incomes. The scheme was re-launched in 2001 with funding from the EU and English Partnerships. The scheme operates from a purpose-built factory in the inner city and consists of showrooms, repair workshops, offices, and space for other organisations to use. The Sofa Project has a fleet of four vans and 15 full-time employees. It receives around 40,000 donated items of furniture a year, which are sold to over 7000 low-income households (www.sofaproject.org.uk).

The social economy as a whole can contribute to urban regeneration in a range of ways. LETS, credit unions and social enterprises are capable of meeting the needs of those who are not well catered for by the private or public sectors. They give people the chance to build their skills, self-confidence and independence, and they create jobs (Centre for Local Economy Studies 1996; Gosling 2001). The social economy can help build social capital by knitting together fragmented communities and fostering leadership capacity (Williams 2000b; Williams and Windebank 2000a; 2000b). This is very much in tune with New Labour's 'third way' philosophy. The social economy can be supported at the city level by local authorities by setting up social economy development units, providing funding support and premises, carrying out social economy audits, and through publicity and awareness-raising activities.

Despite its contribution to urban regeneration, the social economy faces a number of difficulties and challenges. There is limited knowledge of the social economy, which suffers a poor image and profile and has limited appeal to people with professional and managerial skills. Staff are often underpaid and not well trained. In practice, it is often hard to secure the involvement of members of excluded communities in the social economy. Profitable niches within the sector become attractive to the private sector, which is counter to the philosophy of the social economy. There are complex legal frameworks within which social economy activity has to operate, and access to funding from public sector sources is difficult.

SUMMARY

Community involvement in public policy has a long history dating back to the 1960s.

New forms of community involvement in urban regeneration partnerships began to emerge in the 1990s, and these have evolved and become more entrenched since the late 1990s.

Community involvement in urban regeneration is seen as intrinsically worthwhile as a response to moral breakdown, increasing demands on welfare provision, the breakdown of democracy, and because local knowledge can be utilised to enhance the sustainability of projects.

There are pitfalls as well as opportunities associated with the involvement of communities in urban regeneration.

The trend towards community involvement is linked with changing modes of urban governance and political action as discussed in Chapter 7.

Measures to promote the social economy have included Local Exchange Trading Systems (LETS), credit unions and social enterprises.

STUDY QUESTIONS

1. Outline the benefits and difficulties associated with community participation in urban regeneration initiatives.
2. With specific reference to the issue of community participation, consider the failings of urban entrepreneurialism.
3. Partnership has become a pervasive feature of contemporary urban governance. Discuss the reasons for this and explore how it is possible to achieve successful community involvement in local regeneration partnerships.
4. Explain what is meant by 'social capital', and discuss the approaches that might be taken to build social capital in disadvantaged urban neighbourhoods.

5. 'Social capital, then, has its upsides and its downsides' (Kearns 2003, p. 46). Discuss this assertion with reference to post-1997 urban regeneration policy.
6. How important is 'social capital' to the economic prosperity of cities?
7. Evaluate the role of the social economy in neighbourhood regeneration.

KEY READINGS ON COMMUNITY AND REGENERATION

Kearns (2003) 'Social capital, regeneration and urban policy', in Imrie and Raco (eds.) *Urban Renaissance? New Labour, Community and Urban Policy* [critiques the idea of social capital and evaluates its use in contemporary urban policy]

Taylor (2003) *Public Policy in the Community* [traces and evaluates the involvement of communities in public policy since the 1960s]

KEY WEBSITES ON COMMUNITY AND REGENERATION

Association of British Credit Unions: www.abcul.org [information and news on credit unions in England, Scotland and Wales, and the site of the Association of British Credit Unions Ltd, the main trade association for credit unions]

Capacity Builder: www.capacitybuilder.co.uk [community development and social policy information resource for academics, students and those responsible for community development in the UK]

Communities and Local Government: www.communities.gov.uk [government department with current responsibility for communities; includes evaluations of SRB, NDC and Neighbourhood Management Pathfinders]

Community Development Foundation: www.cdf.org.uk [non-departmental public body resource for intelligence, guidance and delivery on community development in the UK]

Co-operative Assistance Network Limited: www.can.coop [training, consultancy and development work for co-operatives, social enterprises and credit unions]

Involve: www.involve.org.uk [an independent organisation which aims to put people at the heart of decision-making; contains resources on neighbourhoods and community involvement]

Joseph Rowntree Foundation: www.jrf.org.uk [independent charitable organisation which carries out research on issues including neighbourhood renewal, urban regeneration and community participation]

New Economics Foundation: www.neweconomics.org.uk [independent think-tank that aims to improve quality of life by promoting solutions that challenge conventional thinking on economic, environmental and social issues]

New Start Magazine: www.newstartmag.co.uk [weekly magazine focusing on regeneration, sustainable communities and social policy in the UK]

Social Enterprise London: www.sel.org.uk [organisation that works with individuals, enterprises, organisations and government to facilitate social enterprise in London]

Urban Forum: www.urbanforum.org.uk [umbrella body for community and voluntary groups with interests in urban and regional policy, especially regeneration]

Young Foundation: www.youngfoundation.org.uk [a centre for social innovation combining practical projects, the creation of new enterprises, research and publishing; includes resources on neighbourhoods and community involvement]

9 Urban regeneration and sustainability

INTRODUCTION

Sustainable development has become a central concept in all discussions of contemporary urban regeneration in the UK. This chapter explores the key economic, social and environmental aspects within sustainable urban development. The chapter first looks at the relationship between cities and sustainability. It then briefly looks at the contested definitions of sustainability and sustainable communities. The chapter then analyses the debates surrounding the extent to which urban and economic development and regeneration can be reconciled with sustainable development. It then examines the nature of sustainable cities and communities before outlining the links between urban regeneration and sustainability. Key contemporary approaches to sustainable urban development and regeneration are then presented, focusing on the compact city debate, and sustainable urban planning and design. This is followed by an examination of urban environmental management systems in the UK, exemplified by brownfield site regeneration. Issues linking urban regeneration and sustainability have been key to an understanding of urban policies and the economic, social and environmental problems faced by cities.

CITIES AND SUSTAINABILITY

The urban regeneration agenda linked to sustainable development includes areas such as housing, communities, local governance, climate change, energy consumption, economics, construction, design, health, land use planning, natural resources and environmental limits, waste, transport, education, and young people. Sustainable development principles are apparent at the neighbourhood, local, regional, national, European and international policy levels, and especially focus on cities (see Giradet 1996; 1999; 2003; 2004).

Concerns to regenerate the central city in particular have evolved in the UK since the 1970s following the adverse effects of de-industrialisation, decentralisation and suburbanisation as discussed in Chapter 1. By the 1990s, this concern had translated into government policy, built around the promotion of a combination of uses including commercial, leisure and residential (DoE 1996a; Ravenscroft et al. 2000). It is challenging to define, and even more difficult to assess or evaluate, the intertwined goals of urban regeneration and sustainability, and there are contradictions inherent in current urban policy. Current planning policy espousing sustainability includes DETR (2001) on transport; CLG (2006a) on housing, ODPM (2005c) on town centres and ODPM (2005b) on sustainable planning.

Since the late 1990s in the UK, there has been an explosion in the debate among academics, politicians, pressure groups and the media surrounding sustainability at local, regional and global scales. Much of this debate has focused on the city, which has been identified as a building block in the path towards a more sustainable world (Hall 2006, p. 153). Cities have continued and will continue to affect fundamentally the development of the environment, and the environment should affect the development of cities (see Hall 2006, Chapter 9; Pacione 2005, Chapter 30).

Questions of sustainability are not purely environmental in their concerns, as it is increasingly apparent that many economic processes and social forms are equally unsustainable. For example, it became apparent in the late 1990s that whole inner city estates in some northern cities were in a state of terminal and irreversible decline and required demolition (see Atkinson 2004b; Cameron 2006; Townshend 2006). Out-migration from many peripheral and former industrial cities also poses a threat to the future viability of some cities in the UK, Western Europe and North America (see Hamnett 2005; Leunig and Swaffield 2008). Sustainability has therefore been applied to both the human and environmental dimensions of the urbanisation process. Not only is the economic and social viability of neighbourhoods, cities and their regions in doubt, but their environmental viability is in addition uncertain and the processes of urbanisation as they are currently proceeding pose a series of threats to the global environment (Hall 2006, Chapter 9; Pacione 2005, Chapter 30).

On the one hand, cities represent a threat to the environment, as they are major contributors to global environmental problems including pollution, resource depletion and land consumption. Cities cover just 2 per cent of global land surface but contain over 50 per cent of the world's population (see Hamnett 2005; Pacione 2005). Cities consume 75 per cent of the world's resources and generate a majority of the world's waste and pollution (Blowers and Pain 1999, p. 249). On the other hand, the environment can be viewed as a threat to cities as the environmental problems generated by cities are felt most severely within cities (Blowers and Pain 1999). Environmental problems such as pollution and its symptoms are long-established and increasingly apparent aspects of urban life. Hall (2006) argued that the environmental consequences and costs of urbanisation impact unevenly on different social groups, in the same way that the de-industrialisation process has been spatially and socially uneven. This is because environmental problems tend to impact most severely on the most vulnerable groups in urban society.

DEFINING SUSTAINABILITY AND SUSTAINABLE CITIES

Unsustainability implies that in the future, development will be compromised or threatened as environmental capacity is reached or environmental limits are breached (Hall 2006, p. 155). Unsustainability can be apparent at a variety of scales from the local to the global, and can be most striking at a local level. Concepts of sustainability have assumed a central place within urban development models since the 1990s (Hall 2006). The most commonly cited definition of sustainable development is that put forward by the Brundtland World Commission on Environment and Development in 1987, which argued that it is 'development that meets the needs of the present without compromising the ability of future generations to meet their own needs'. Although useful, the difficulty of this definition is that 'needs' are not absolute. Further, Blowers and Pain (1999, p. 265) pointed out that 'what may be regarded as needs in the cities of the North would be luxuries in those of the South'.

The definition is a starting point although an idealistic and impractical definition of sustainable development. Frey (1999) identified the characteristics of a sustainable city (Table 9.1).

Rogers (1997, p. 169) further argued that the sustainable city is:

- just (justice, food, shelter, education, health and hope are fairly distributed; all people participate in government);
- beautiful (art, architecture and landscape spark the imagination and move the spirit);
- creative (open-mindedness and experimentation mobilise the full potential of the city's human resources and allow a fast response to change);
- ecological (minimises its ecological impact; landscape and built form are balanced; buildings and infrastructures are safe and resource-efficient);
- fully accessible (information is exchanged both face-to-face and electronically);
- compact and polycentric (protects the countryside, focuses and integrates communities within neighbourhoods and maximises proximity);
- diverse (a broad range of overlapping activities creates animation, inspiration and fosters a vital public life).

TABLE 9.1 CHARACTERISTICS OF A SUSTAINABLE CITY

i. Physical properties of the city
   - Containment
   - Densities to support services
   - Adaptability

ii. Provisions of the city
   - Readily available, good quality public transport
   - Reduced and dispersed traffic volumes
   - Cycle provision
   - Pedestrian walkways
   - Hierarchy of services and facilities
   - Access to green space

iii. Environmental and ecological conditions
   - Low pollution, noise, congestion, accidents, crime
   - Available private outdoor space
   - Range of green space habitats for nature
   - Existence of local, non-fossil fuel energy generation, including micro-generation
   - Existence of urban farms
   - Evidence of local food consumption
   - Small ecological footprint

iv. Socio-economic conditions
   - Social mix that reduces social stratification
   - Degree of local autonomy
   - Degree of local economic self-sufficiency, including local firms
   - 'Liveability' – good quality of life

v. Visual-formal quality
   - Positive 'image' to city and its constituent parts
   - Sense of 'centrality'
   - Sense of 'place' (distinctiveness)

Source: **Based on Frey (1999, pp. 32–33).**

BOX 9.1 SOME DEFINITIONS OF SUSTAINABLE CITIES AND COMMUNITIES:

'[a city that] enables all its citizens to meet their own needs and to enhance their well-being without damaging the natural world or endangering the living conditions of other people, now or in the future' (Giradet 2003, p. 9).

'Sustainable communities are places where people want to live and work, now and in the future. They meet the diverse needs of existing and future residents, are sensitive to their environment, and contribute to a high quality of life. They are safe and inclusive, well planned, built and run, and offer equality of opportunity and good services for all' (www.communities.gov.uk, 2008).

Similarly, 'smart growth' represents 'a set of planning techniques designed to achieve more sustainable development, that includes infill development, revitalisation of existing neighbourhoods, mixed-use developments, environmental preservation, and integrated regional transport and land-use planning' (Pacione 2005, p. 674).

The Communities and Local Government department (www.communities.gov.uk, 2008) outlined eight components of a sustainable community that clearly resonate with the wider approaches to urban regeneration:

1. Active, inclusive and safe (fair, tolerant and cohesive with a strong local culture and other shared community activities) – sustainable communities offer:
   - a sense of community identity and belonging;
   - tolerance, respect and engagement with people from different cultures, background and beliefs;
   - friendly, co-operative and helpful behaviour in neighbourhoods;
   - opportunities for cultural, leisure, community, sport and other activities, including for children and young people;
   - low levels of crime, drugs and antisocial behaviour with visible, effective and community-friendly policing;
   - social inclusion and good life chances for all.
2. Well run (with effective and inclusive participation, representation and leadership) – sustainable communities enjoy:
   - representative, accountable governance systems which both facilitate strategic, visionary leadership and enable inclusive, active and effective participation by individuals and organisations;
   - effective engagement with the community at neighbourhood level, including capacity building to develop the community's skills, knowledge and confidence;
   - strong, informed and effective partnerships that lead by example (for example, government, business, community);
   - strong, inclusive, community and voluntary sector;
   - sense of civic values, responsibility and pride.
3. Environmentally sensitive (providing places for people to live that are considerate of the environment) – sustainable communities:
   - actively seek to minimise climate change, including through energy efficiency and the use of renewables;
   - protect the environment, by minimising pollution on land, in water and in the air;
   - minimise waste and dispose of it in accordance with current good practice;
   - make efficient use of natural resources, encouraging sustainable production and consumption;
   - protect and improve bio-diversity (for example, wildlife habitats);

- enable a lifestyle that minimises negative environmental impact and enhances positive impacts (for example, by creating opportunities for walking and cycling, and reducing noise pollution and dependence on cars);
- create cleaner, safer and greener neighbourhoods (for example, by reducing litter and graffiti, and maintaining pleasant public spaces).

4. Well designed and built (featuring quality built and natural environment) – sustainable communities offer:
   - sense of place – a place with a positive 'feeling' for people and local distinctiveness;
   - user-friendly public and green spaces with facilities for everyone including children and older people;
   - sufficient range, diversity, affordability and accessibility of housing within a balanced housing market;
   - appropriate size, scale, density, design and layout, including mixed-use development, that complement the distinctive local character of the community;
   - high quality, mixed-use, durable, flexible and adaptable buildings, using materials which minimise negative environmental impacts;
   - buildings and public spaces which promote health and are designed to reduce crime and make people feel safe;
   - accessibility of jobs, key services and facilities by public transport, walking and cycling.
5. Well connected (with good transport services and communication linking people to jobs, schools, health and other services) – sustainable communities offer:
   - transport facilities, including public transport, that help people travel within and between communities and reduce dependence on cars;
   - facilities to encourage safe local walking and cycling;
   - an appropriate level of local parking facilities in line with local plans to manage road traffic demand;
   - widely available and effective telecommunications and Internet access;
   - good access to regional, national and international communications networks.
6. Thriving (with a flourishing and diverse local economy) – sustainable communities feature:
   - a wide range of jobs and training opportunities;
   - sufficient suitable land and buildings to support economic prosperity and change;
   - dynamic job and business creation, with benefits for the local community;
   - a strong business community with links into the wider economy;
   - economically viable and attractive town centres.
7. Well served (with public, private, community and voluntary services that are appropriate to people's needs and accessible to all) – sustainable communities have:
   - well-performing local schools, further and higher education institutions, and other opportunities for lifelong learning;
   - high quality local health care and social services, integrated where possible with other services;
   - high quality services for families and children (including early years child care);
   - good range of affordable public, community, voluntary and private services (for example, retail, fresh food, commercial, utilities, information and advice) which are accessible to the whole community;
   - service providers who think and act long-term and beyond their own immediate geographical and interest boundaries, and who involve users and local residents in shaping their policy and practice.
8. Fair for everyone (including those in other communities, now and in the future) – sustainable communities:
   - recognise individuals' rights and responsibilities;

- respect the rights and aspirations of others (both neighbouring communities, and across the wider world) also to be sustainable;
- have due regard for the needs of future generations in current decisions and actions.

Reflecting these central government aspirations, three principal New Labour 'sustainable communities' policies relating to urban regeneration can be identified in the late 2000s – the Sustainable Communities Plan, growth areas and housing market renewal (see also Chapter 5; Raco 2007a). The Sustainable Communities Plan (ODPM 2003a) set out a long-term programme of action for delivering sustainable communities in both urban and rural areas. It aims to tackle housing supply issues in the south-east of England, low demand in other parts of the country, and the quality of public spaces. The plan includes not just a significant increase in resources and major reforms of housing and planning, but a new approach to what is built and how it is built. Growth areas have been identified in the south-east of England, the UK's largest region with a population of around 8.1 million, which is 13.5 per cent of the total population (www.communities.gov.uk). It is among the fastest growing regions, and key issues faced relate to housing supply, affordability of housing, and transport. The government has identified four growth areas in Ashford, London–Stansted–Cambridge–Peterborough, Milton Keynes and the South Midlands, and the Thames Gateway (see Chapter 5). Housing market renewal policies have sought to address low demand for housing leading to market failure, which has emerged as a problem since the late 1990s. It affects around 1 million homes and threatens to undermine the urban renaissance now being experienced in some cities. Nine Housing Market Renewal Pathfinder projects have been established to tackle the most acute areas of low demand and abandonment in parts of the north and Midlands in England (see Chapter 5; Cameron 2006; Townshend 2006).

The role of urban regeneration is clear among these components, criteria and related policies, and is exhibited in approaches explored elsewhere in Sections II and III. Delivering sustainable development should involve planning for the long term, fully integrating economic, social and environmental factors into decision-making and considering impacts beyond the local area. The 2005 UK sustainable development strategy (HM Government 2005) adopted five principles of sustainable development. These comprised living within environmental limits; ensuring a strong, healthy and just society; achieving a sustainable economy; promoting good governance; and using sound science responsibly. These principles aim to underpin an integrated and balanced approach to sustainable development.

RECONCILING ECONOMIC REGENERATION AND ENVIRONMENTAL PROTECTION

There are two key aspects of the relationship between economic development, as a key feature of urban regeneration activities, and sustainable development (Hall 2006, Chapter 9). First, there appears to be a conflict between short-term economic development and longer-term environmental needs (Blowers 1997; Mohan 1999). Governments in the developed world aim primarily to secure continued economic growth (Jacobs 1997; Myers 2006). In the UK, the government has five macroeconomic objectives for the whole economy, which are closely replicated across the developed world. Myers (2006) summarised these as economic growth, which is measured by Gross Domestic Product (GDP) (the value of goods and services produced in an economy) in billions of pounds, and how much it has increased or decreased as a percentage from the previous year or quarter; full employment as measured as a percentage of the workforce or as a number which closely follows economic growth up or down; price stability as reflected in price inflation which is measured by an index and expressed as a percentage; a balance of payments between imports and ex-

measured by an amount in billions of pounds; and environmental protection, a newer objective added in the 1990s, which has difficulties associated with its measurement, as there are a plethora of measures including the carbon footprint, air quality and congestion.

The benefits for the population of the government striving for economic growth are rising living standards, as measured by increasing levels of personal consumption. However, there appears to be an incompatibility between individual lifestyles characterised by increasing levels of personal consumption and long-term collective interests in environmental sustainability (Blowers 1997). The problem becomes even worse when acknowledging that the aspirations of many developing world countries are to emulate the standards of living in the developed world. The Stern Review (HM Treasury 2006) on the economics of climate change highlighted the effect of climate change and global warming on the global economy, and found this to be a significant threat to economic growth.

One outcome of the current regime of capitalist development and flexible accumulation is a sharpening of levels of social polarisation both within, and between, cities on a global scale (see Chapter 1). Such inequalities are inherently unsustainable socially, but also environmentally, and provide a significant barrier to achieving any level of sustainable development. Poverty and social exclusion in cities are inextricably associated with the generation of environmental degradation (Hall 2006). Two of the government's macroeconomic objectives therefore appear mutually incompatible.

Rather than seeking to restructure the fundamental nature of the economy, the response of governments has been to argue that rising levels of economic development and personal consumption are compatible with sustainable development and urbanisation, a position referred to as 'ecological modernisation' (Hall 2006). Economic growth embraces anything that contributes to increasing GDP, and specifically anything that contributes to increasing economic activity in the construction or property sector, such as the value of output figures, the number of transactions, employment and productivity – some key aims of urban regeneration. Environmental protection includes anything that contributes to promoting a more efficient use of resources, or to not harming the natural environment, including increasing energy efficiency levels in new housing, landfill tax, Section 106 agreements, planning gain and eco-towns. Such measures directly target environmental problems stemming from economic development and urban growth.

Governments have therefore aimed to 'green' capitalism and urban development by introducing various measures in planning and business that have attempted to secure sustainability. Measures adopted or proposed by the government have included internalising the costs of environmental externalities by incorporating them into the costs of production. However, Hall (2006) noted that businesses might continue to generate environmental externalities by paying for the 'right' to pollute (see Haughton and Hunter 1994), and that these measures might not severely impact on profits as costs can be passed on to the consumer. Corporate adoption of green strategies is voluntary and thus uneven. Corporations have enacted measures such as the production of environmental statements or policies and the use of recyclable materials in packaging (Hall 2006). However, in an increasingly globalising economy, without international agreement and regulation, such measures are likely to be ineffective as business might simply relocate to countries where the costs of environmental legislation on production are low or absent.

Alternative paths towards sustainable economic development, although having a limited impact globally, have had major impacts within specific localities and include moves towards locally responsive and responsible enterprise (see Chapter 7). These have included the development of locally owned small businesses such as community food co-operatives (Haughton and Hunter 1994). Such businesses provide alternatives to the large supermarket chains and out-of-town locations that are inaccessible to deprived neighbourhoods, which are often reliant on poor public transport services (see Chapter 10). Community-based credit unions and local exchange trading systems have also emerged to assist marginalised urban communities (see Chapter 7).

Hall (2006, p. 160) argued that the reconciliation of economic development and sustainable development does not seem possible without some reconstruction of what constitutes economic development. Such reconstruction would need to incorporate some aspect of long-term collective needs, both economic and environmental, rather than short-term individual demands. It seems highly unlikely at anything beyond the most local of scales. Such community initiatives suggest that the most appropriate scale at which alternative concepts of economic development and regeneration might be pursued is the neighbourhood scale (Rudlin and Falk 1999; Barton 2000; Hall 2006). This certainly appears more practical and realistic than seeking to engineer massive structural change to urban forms. It is also a recognition that centralised systems of government have failed to deliver economic and social regeneration to neighbourhoods (see Taylor 2003; Chapter 8), and seem unlikely to deliver environmental sustainability in the future. Neighbourhood-level planning has the potential to deliver solutions to a neighbourhood's own problems as well as providing the seedbed for sustainability at greater scales (Hall 2006).

ECOLOGICAL FOOTPRINTS AND CITIES

The sheer scale of the impacts of cities as the focus of economic development upon the environment is conveyed well by the concepts of the ecological footprints of cities and the global hinterlands of cities. Maintaining contemporary cities requires that they draw upon and impact upon vast areas of land and water from beyond their own immediate geographical hinterlands. Cities draw resources, building materials, food, energy and so on from all over the world (Hall 2006). They also disperse pollution and waste globally. Wackernagel and Rees (1995) developed the idea of the 'ecological footprint' to represent the environmental impact of cities in terms of the amount of land required to sustain them. An ecological footprint is defined as 'the land area required to supply a city with food or timber products and to absorb its waste' (Giradet 1999, p. 27). Along similar lines, Rees (1997, p. 305) defined the ecological footprint of cities as 'the total area of productive land and water required on a continuous basis to produce the resources consumed and to assimilate the wastes produced by that population, wherever on Earth the land is located'.

Currently, the concept of the ecological footprint suggests that city living is unsustainable. It is estimated that if the entire Earth's population consumed the resources at the rate of a typical resident of Los Angeles, it would require at least three planet Earths to provide all the material and energy they consume (see Fokkema and Nijkamp 1996; Lees 2002). Similarly, London's ecological footprint equates to 125 times larger than the city's surface area; the London model needs two Earths to sustain the city. Bristol's ecological footprint is 191 times the size of the city, and if everyone on Earth had the same lifestyle and used the same quantity of resources as the city's residents, the city would need three Earths to sustain itself (BCC 2003). Finally, the ecological footprint of Vancouver in Canada is estimated as 180 times larger than the city's surface area. The idea of the ecological footprint conveys the disproportionate impact of cities upon the environment. Clearly, if sustainable urban development is to be achieved, reducing the ecological footprints of cities is an imperative. Cities must become more sustainable in order to protect the environment and people's quality of life, within which urban regeneration has a key part to play.

SUSTAINABLE CITIES AND COMMUNITIES

Much of the recent thinking on urban regeneration is also central to the idea of sustainable urban development and planning (Couch and Dennemann 2000; Naess 2001). Blair and Evans (2004, p. 1) defined sustainable development by stating that it aims 'to connect environmental, economic and social welfare by thinking and managing for the longer term'. Although the term sustainable development dates back to the 1970s, it was not until the 1990s that it was commonly applied to cities (Hardoy et al. 2001, p. 339), and linked with government policy for urban regeneration (DoE 1994c).

Approaches to sustainable development in cities at first tended to concentrate on spatial and ecological aspects, adopting the narrower 'ecological modernisation' perspective (Blowers and Pain 1999). Even the Urban White Paper (DETR 2000a) focused principally on environmental sustainability. This narrower approach usually translates into a specific and practical focus for sustainability objectives, which concern minimising air and water pollution, the depletion of energy resources, and any adverse effects on the living environment. A sustainable city will have an efficient solid waste disposal system involving recycling, and there will be limited reliance on the private car and instead a greater use of walking, cycling and public transport.

The concept of the compact city, which is discussed later in the chapter (see, for example, Breheny 1992; 1993; Haughton and Hunter 1994; Low et al. 2000; Raemaekers 2000; Pacione 2005, pp. 618–619), is a part of the sustainable city vision that limits its ecological footprint, and advocates the merits of urban containment as a way of reducing energy consumption and pollution (Breheny 1995; Blowers and Pain 1999; Hall 2006). The higher urban densities of the compact city encourage an improved public transport system, enhanced access to facilities, while reducing social segregation (Burton 2000). Urban housing, which is developed with sustainability in mind, will consume less land, generate fewer private car journeys, use existing urban resources and conserve energy. Better use of urban land will reduce the pressure for new housing on greenfield sites outside urban areas, and will simultaneously revitalise city centres by bringing people back to the centre to make use of the services and jobs that are already available (Edgar and Taylor 2000; Pacione 2005). Moreover, lower travel-to-work rates will reduce pollution and congestion. These reductions in travel to work will only be possible if housing development does not occur to the extent of threatening rather than supporting the economic functions of the central city (Bromley et al. 2005).

Such ideas about urban sustainability were brought together and extended in the 2003 government plan *Sustainable Communities: Building for the Future* (ODPM 2003a). The plan revolved around the three core sustainability aims of a healthy environment, a prosperous economy and social well-being (Power 2004, p. 10). In the list of key requirements of sustainable communities, a flourishing local economy, which provides jobs and wealth, is the very first requirement (ODPM 2003a). Among other requirements, a conception of sustainable communities in which local people, organisations and businesses play a key role in both planning and 'long-term stewardship' (ODPM 2003a, p. 2) is now emphasised alongside the more traditional concerns about minimising the use of resources. Social well-being as well as social and democratic inclusion and safety (see Raco 2007b) are seen as essential elements of sustainable communities.

In terms of housing development, the plan incorporated the New Labour urban renaissance ideas of 'higher density, greater use of brownfield land and existing buildings, higher quality, more environmentally sensitive design and more mixed communities' (Power 2004, p. 5). Most of these recommendations seem to have been accepted almost uncritically. However, Bromley et al. (2005), Bailey et al. (2006) and Roberts (2007), for example, have demonstrated that the concept of mixed communities can be controversial and contested (see Chapter 11). However, the plan for sustainable communities specifically includes the requirement that there is 'a well-integrated mix of decent homes of different types and tenures to support a range of household sizes, ages and incomes' (Power 2004, p.6).

LINKING URBAN REGENERATION AND SUSTAINABILITY

The emphasis within most urban regeneration policies has tended to be on economic rather than environmental or social regeneration (Couch and Dennemann 2000). However, for example, the recent promotion of city centre living to meet environmental and social 'sustainability' aims, as well as supporting economic regeneration, is a key strand of urban regeneration policies, and indicates how the two dimensions of regeneration and sustainability are closely interrelated (Chanan et al. 1999; see Chapter 11).

An assessment of the effects of urban regeneration policy also reveals the linkages with sustainability. Roberts and Sykes (2000, pp. 298–299), drawing on the British Urban Regeneration Association's criteria for the Best Practice Awards, suggested an approach to assessment that specifically includes the environmental sustainability of the scheme alongside aspects such as financial viability, and the contribution to economic regeneration, to community spirit and cohesion. More recently, the Sustainable Development Commission have pulled together ideas of sustainable regeneration in a report entitled *Mainstreaming Sustainable Regeneration: A Call to Action* (2003). The report stated that, as a first action point, sustainable development principles should be at the heart of regeneration policy and practice (Sustainable Development Commission 2003, p. 4). A further action point stressed that 'housing and construction should be regarded as a major opportunity to embed sustainable development in regeneration' (Sustainable Development Commission 2003, p. 5). Although energy efficiency measures are at the forefront of thinking, social, economic and environmental impacts are all emphasised (see www.sd-commission.org.uk).

Energy efficiency measures are usually at the heart of attempts to cost the introduction of sustainable urban regeneration schemes. Most research on sustainability avoids quantification of the costs and benefits other than for the energy savings to be derived from changes to transport use and type, building design, and recycling (Bromley et al. 2005). In terms of central city housing regeneration, a short-term costing that focuses on the greater residential development costs of a brownfield site where land values are high, might neglect the long-term environmental benefits to accrue from reducing urban sprawl and greenfield development (Bromley et al. 2005). In the same way, costing residential development cannot easily factor in the advantages of incorporating affordable housing alongside more profitable, high-end private developments, because the benefits of social mix and equity do not have a monetary value. To satisfy sustainability goals, the exclusionary effects of high-class housing development should be reduced and tenure diversification promoted, whether or not the short-term assessment of costs and benefits suggests this is inappropriate. Such difficulties of financing for a long-term future are emphasised in broader terms by Blair and Evans (2004), and are recognised in the government's guiding principles for sustainable development (ODPM 2003a). The chapter now moves on to discuss ways in which sustainable urban regeneration has been pursued through some of the key contemporary urban planning and design policies.

THE COMPACT VERSUS DISPERSED CITY DEBATE

Lees (2002) argued that the city is 'choking to death' due to (sub)urban sprawl, congestion and air pollution. As such, low-density spacious suburban development is no longer considered the remedy for urban and inner city problems. Planners now argue that the way to make cities more sustainable is to make them more compact and more urbanised (see

Blowers and Pain 1999). In contrast to suburban living, more compact settlement patterns conserve open space and enable residents to use energy-efficient mass transit public transport systems. In this way the densification of urban settlements is said to provide a solution to local-scale environmental problems, such as traffic and air pollution from car use, but also to help reduce greenhouse gas emissions and the impact of cities on the global climate. Since the mid-1990s, city centre regeneration, densification, infilling and brownfield development have been encouraged by central and local government policy. This has been implemented through focusing retail (see, for example, Lowe 2005), office (see, for example, Barlow and Gann 1993; Heath 2001) and housing development (see, for example, Seo 2002; Tallon and Bromley 2004; Bromley et al. 2005; 2007; Barber 2007) in central mixed-use and sustainable environments (Urban Villages Group 1992; Coupland 1997; Evans 1997).

There has been much debate over the question of whether there is an optimal city size in terms of sustainability. Debate generally associates increasing city size with unsustainability (Hall 2006). Increasing city size tends to be associated with increasing energy use per capita and larger cities are more dependent upon external environmental resources. Larger cities are also associated with greater environmental problems such as air and water pollution. However, as Hall (2006) summarised, some have argued against this and say that large settlements might be more sustainable as decentralisation can potentially decrease commuting distances (particularly where employment also decentralises to suburban locations), large cities are characterised by relatively low car use, and large cities have the most developed public transport systems. Large cities also have more advanced political and managerial infrastructures giving them greater political resources that can be mobilised to manage and deal with environmental problems. Large cities have also been places of innovation and the development of technologies that could conceivably contribute to enhanced sustainability in the future. The question of optimal city size for sustainability is a complex and contested one. Of greater importance than urban size is the 'internal organisation' of cities relating to their form, processes, management and the nature of urban regeneration activities (Hall 2006).

It is evident that a significant debate has developed over the relationship between urban form and sustainability (see Williams et al. 2000). This follows widespread advocacy of various high-density 'compact city' models (see Pacione 2005, p. 618; Hall 2006, pp. 160–164), which are deemed to offer a number of advantages:

- conservation: farmland, rural ecosystems, biodiversity;
- reduced need for cars;
- public transport, walking and cycling;
- efficient use of energy for heating – reduced urban heat island;
- access to services and facilities;
- efficient utility and infrastructure provision;
- inner city regeneration;
- 'community';
- greater social equity;
- high density urban living.

The idea of urban densification has been incorporated into government urban policy documents in a variety of countries since the mid-1990s, the most recent in the UK being the Urban White Paper (DETR 2000a), which was related to the work of the Urban Task Force (1999; see also 2005). This is also linked with the brownfield (previously developed land) versus greenfield (rural countryside) development debate (see Chapter 11). Densification has also been promoted by the European Union (CEC 1990). The compact city is represented in policy as a 'liveable city' connected to the ideals of urban regeneration and gentrification, which witnesses the social and economic upgrading of run-down areas of housing, often in the inner city (Lees 2002; see Chapter 11). The supporters of the compact city concept argue that the redevelopment of brownfield sites instead of suburban greenfields provides a way not just to protect the natural environment, but also to promote urban regeneration. However, the model of city that has dominated post-war urban growth, especially in the UK, North America and Australia, has been the model of the 'dispersed city'.

Problems of dispersed or decentralised urban forms include (Hall 2006, pp. 161–162):

- extensive consumption of land;
- high rates of storm water pollution;
- high rates of water consumption (for example, watering large gardens);
- increased fuel consumption (higher car use and associated problems);
- high rates of energy consumption (poor thermal qualities of single-storey and detached dwellings);
- low rates of waste water recycling;
- a failure to accommodate extensive or efficient public transport provision.

However, compact city and urban densification policies have proved highly controversial, especially in the US, and have been subject to academic critique. In the US, Lees (2002) pointed out that densification policies have been coupled with growth management policies to discourage suburbanisation through zoning restrictions on land use and taxes on driving vehicles. The pursuit of such policies in the UK would also be highly controversial.

Lees (2002) illustrated that the idea of the compact city has been criticised for being impractical, as it is impossible to halt the suburbanisation process; unrealistic in that planning is limited in its power and influence; and undesirable as people do not want to live in ever denser neighbourhoods (see Senior et al. 2004; 2006). Critics of the compact city point to little hard evidence to back up the claimed advantages of the model, and suggest that they rely on some false assumptions (see Haughton and Hunter 1994; Breheny 1995; Blowers and Pain 1999). Breheny (1995) put forward three related critiques of the compact city model.

The first was the lack of hard evidence and false assumptions about compact and decentralised forms. Significant savings in fuel consumption following compact city development are highly unlikely. Even following extreme reurbanisation, rates of energy saving are only likely to be around 10–15 per cent. Improved technology in vehicle design or raising fuel costs would be more effective and easier.

Second, there has been a failure to take into account social costs of the measures and dominant preferences for suburban living (Senior et al. 2006). There would have to be a fundamental reversal in the urban development process, resulting in increased urban cramming and density, perhaps resulting in overcrowding and homelessness. The lure of the rural environment would be in stark contrast to the increased population and densities of the compact city.

Third, Breheny (1995) argued that it is simply unrealistic to expect compact city models to be implemented in the UK given the current nature of urban development fuelled by the movement of people and employment to the suburbs and beyond. Urban containment policies have been ineffective, and restricting edge of town housing development results in house price increases and supply problems (see Barker 2004; CLG 2007a; SMF 2007; Dixon and Adams 2008).

It appears that the most practical response to the apparent irreconcilable demands of increasing urban development and sustainable development is not to try and reverse decentralisation, but to focus growth around strong sub-centres rather than allowing low-density sprawl by planning for 'sustainable city-regions' (see Chapter 7). Alternative concepts to the compact city have been put forward by urban geographers. One is the

'MultipliCity', the approach of which to urban sustainability proposes clusters of development that include urban infilling and the creation of new ex-urban (edge-of-urban) settlements, resulting in a sustainable city-region (Lees 2002; Hall 1998, Chapter 9; Pacione 2005, Chapter 30). Such ex-urban development is becoming common practice, with the emergence of edge cities, exopolis, technoburbs and metropoles (see Chapter 13). It is argued that sustainability should be encouraged at the regional rather than simply the city scale (see Boddy 2003d).

SUSTAINABLE URBAN DESIGN, HOUSING AND REGENERATION

To achieve sustainable urban regeneration and development, principles of sustainability must be incorporated in design, particularly of new housing, preferably in compact cities. CLG (2008) defined the qualities of sustainable urban design as comprising six criteria:

- Sustainable urban design delivers value for money, and design costs are a small percentage of construction costs.
- It is through the design process that the greatest impact can be made on the quality, efficiency and overall sustainability of buildings.
- Encourages local community identity over big-brand 'anywhere, anyplace' anonymity.
- Ensures attention to creating places that respond to people's needs, and helps community cohesion.
- Improves longer-term 'liveability', management and maintenance of the built environment, including public places such as streets and parks, to ensure spaces can stay clean, safe and green.
- Contributes to the achievement of sustainable development by respecting historical context, making best use of resources, and being able to respond to change.

A related concept is that of 'sustainable architecture', also known as 'environmental architecture', 'green building' or 'green architecture'. It is a general term that describes environmentally conscious design techniques in the field of architecture. Sustainable architecture is framed by the larger discussion of sustainability, and seeks to minimise the negative environmental impact of buildings by enhancing efficiency and moderation in the use of materials, energy, and development space. It is especially relevant to urban apartments.

There are a number of key features of sustainable or environmental architecture:

- energy-efficiency;
- natural heating, cooling and lighting;
- renewable energy sources;
- construction: local materials; renewable materials; low processing materials; sensitive extraction;
- waste management systems;
- insulation against noise pollution;
- living space with workspace;
- importance of contextual factors.

Each of these principles of sustainable urban design and architecture clearly links to aspects of contemporary urban regeneration in transforming cities. An example of an innovative sustainable housing developer is the social housing landlord Tai Cartrefi Cyf in Swansea, which is a member of Gwalia Housing Group and a member of the European Housing Ecology Network. The organisation has been recognised for its award-winning approaches and designs. An example of one of the environmentally and socially sustainable developments undertaken by this developer is the Swansea Foyer in the city centre, which provides 34 study bedrooms for young homeless people. It was constructed along sustainable building and housing methods and opened in 1997. Key features are:

- provides accommodation for 34 young people;
- allows them somewhere to live and learn new skills to assist them with secure permanent employment;
- provides extensive training, office and recreational facilities;
- has redeveloped the brownfield site of a former workingmen's club;
- has incorporated listed nineteenth-century façade;
- is an energy-efficient, environment-friendly building;
- design converts the internal circulation space into a covered street, bordered by a terrace of houses; each houses 4–5 residents;
- urban-style community space that allows natural light and passive solar gain;
- cost per resident/unit including communal facilities was £41,848;
- funded by local authority and Welsh Development Agency.

Once young residents gain skills and employment they move out of the foyer. Such a scheme represents sustainable social, economic and environmental urban regeneration. By the late 2000s six 'foyers' had been developed in South Wales. Another example of a sustainable housing initiative in Swansea is the City Living project run by Swansea Housing Association since the late 1990s. One of their developments is the Wind Street mixed-use development. Here, an old post office was regenerated to provide affordable housing for rent, which exists near to employment opportunities and close to all of the facilities and services of the city centre (Tallon 2003).

A recent £500 million urban regeneration project called Cabot Circus in Bristol city centre has attempted to incorporate sustainable practices into the design, building and regeneration process. These include the integration of the development with the existing built environment; the development of a mixed-use environment comprising retail, leisure and entertainment, offices, and housing; the recycling of 90 per cent of waste materials removed from the site; local sourcing of raw materials as far as possible; the incorporation of natural ventilation and solar design systems to reduce energy use required for heating and cooling in the design; city-wide consultation and community engagement; employment opportunities for neighbouring communities and the provision of some new affordable housing; and the local procurement of construction labour. However, there are criticisms regarding who benefits from this type of mega-project environment and the focus on the physical–economic–property dimension (see Clement 2007; Tallon 2008).

CASE STUDY: BEDZED – ENVIRONMENTAL ARCHITECTURE, SUSTAINABLE LIVING AND URBAN REGENERATION

Cities in the UK face dual issues relating to housing; first, meeting the obligation to provide affordable housing, for example, for key workers such as nurses and teachers; and second, minimising the environmental impacts of new housing development. One innovative scheme in the South London borough of Sutton offers some clues as to how these issues might be addressed and incorporated in the future, especially in terms of housing development (see Peabody Trust 2001; Hardy 2004).

The Beddington Zero Energy Development (BedZED) is one of the most famous examples of urban regeneration combined with sustainable living in the UK. BedZed is a development of 82 houses, maisonettes and apartments, along with business space, workspace units, and a range of public open spaces and facilities, that offers a model for large-scale, affordable, sustainable housing. It is built on a brownfield site of a former sewage works. There are three key elements of the BedZED development that allow it to achieve its aims. First, ecological design minimises energy use; second there is a high housing density; and third its location provides access to good public transport links (www.bedzed.org.uk).

Each home uses sunlight for heating and electricity generation, and 80 per cent of the building materials used in the construction were reclaimed timber and steel to create high-density live/work communities clustered around transport hubs. The BedZED development model has been hailed as a possible solution to London's housing crisis. It is viewed as the type of mixed tenure, mixed use, renewable-energy community that should be developed further.

One significant hurdle that developments such as BedZED face is the high building costs compared to traditional developments. This has made some developers wary of compromising profit margins. However, local authorities are increasingly being encouraged to account for the environmental and social benefits of schemes, as well as just the economic costs, in reaching decisions about housing developments. Another issue is that many of the residents of BedZED are middle-class professionals who live high-impact lifestyles and who like to be associated with such a 'trendy' development.

ECO-TOWNS AND SUSTAINABLE URBAN REGENERATION

Inspired by the BedZED model, eco-towns are proposed as commercial, private sector versions of eco-villages, and as such should not be confused with the values and lifestyles espoused by eco-village residents and communities. Eco-towns strive to incorporate elements of environmentally, socially and economically sustainable design and are essentially modern-day New Towns (see Chapter 2) for 5–20,000 homes (CLG 2008). In the UK, the government announced the approach in summer 2007, and 57 sites were identified by developers by early 2008 following an invitation for proposals from local authorities and other interest groups. Proposed sites ranged from a former petrochemical plant near the Manchester United training ground in Trafford to a greenfield site at Micheldever in Hampshire. Other sites included brownfield excess public sector land such as old army barracks and airports. These were reduced to a shortlist of 15 in April 2008, and subsequently, 10 are expected to go through to the full planning process (CLG 2008). It is anticipated that there will be up to five eco-towns by 2016 and 10 by 2020.

The driving forces for this policy are, first, the 3 million new homes forecast to be required by 2020 (CLG 2007a), and, second, the fact that around a quarter of carbon emissions that contribute to climate change come from housing. They are to be 'carbon-free sustainable communities'. At least 30 per cent of homes will be affordable and provided by housing associations, and a secondary school, shops, business space and leisure facilities are also required (CLG 2008; www.communities.gov.uk). Eco-town proposals are judged against the criteria of zero carbon and environmental standards; sustainable travel; design quality, including through design competitions; community involvement; and use of land which should be brownfield and public sector if possible. They aim to complement city centre regeneration, urban extensions and the redevelopment of major sites in existing urban areas.

Residents in towns and villages near several of the proposed eco-towns were set against the plans from their early stages. Critics argue that existing communities are strongly opposed to many eco-town plans; some are proposed for greenbelt and greenfield sites (see Chapter 11); the focus is on housing at the expense of wider infrastructure, public services and employment; and there will be an increase in car pollution as they will not be big or diverse enough to sustain viable public transport networks. In late 2008, it remains to be seen whether eco-towns will make a significant contribution to sustainable urban housing development, especially in the context of a housing market downturn.

SUSTAINABLE URBAN PLANNING AND REGENERATION: URBAN ENVIRONMENTAL MANAGEMENT SYSTEMS

In addition to policies such as urban compaction and eco-towns, environmental controls have emerged to address urban sustainability in the post-industrial city. Externalities stemming from urban growth are addressed by regulatory systems (rules and laws) and fiscal systems (taxing and spending) using the 'polluter pays principle'. However, the application of regulatory and fiscal systems in urban areas gives rise to several problems that can impact on city competitiveness, economic development and sustainability.

As successful cities expand, negative environmental effects are experienced in these increasingly competitive cities. First, population growth and increased demand lead to urban sprawl and pressure on green space and green belts, and higher house prices. Second, increased car ownership contributes to pollution and traffic congestion which acts as a cost to business. Third, growing and successful cities place more pressures on waste disposal from industrial and domestic property. There are some problems with regulations that seek to address these three major problems stemming from growing cities. There is a cost in implementing them at the local authority level within planning departments, they place increased costs on firms, which can cost jobs, and they can result in planning blight.

Regulations add to business costs due to the time delays in implementing them, the costs of application and compliance, and uncertainty. If 'compliance costs' are too high resulting from the cost in loss of value of a site due to planning restrictions, sites can remain derelict. Regulations can cost jobs, because firms locate where compliance costs are lowest. This is reflected in the de-industrialisation process, which has seen the loss of heavy manufacturing industry to emerging economies such as China and India. These countries demonstrate advantages to firms in terms of lower wages, looser planning regulations and weaker environmental laws. This competition leads to UK firms moving abroad, and makes it difficult to attract foreign investment to UK cities. Planning blight is caused by a slow planning system; the system presents uncertainty for firms who find it difficult to plan ahead. No new investment in an area is forthcoming; it is difficult to sell or relocate from an area; a drop in house prices occurs; and there is no investment in maintenance. Areas become run down and blighted because of a slow and unresponsive planning system that restricts urban regeneration.

In terms of fiscal policy, several advantages of taxing as an environmental management system can be identified. The first relates to the polluter pays principle. Whoever causes an environmental problem pays for it through policies such as congestion charging in cities. This reduces external costs and allows choice; regulations simply forbid. Taxes on environmentally problematic behaviour also raise revenue and are therefore a hypothecated tax (ring-fenced and spent on the same area); regulations are simply a cost. Taxes also have the advantage of flexibility as they can be changed quickly whereas regulations take time to change due to their complexity and possible legal challenges. Taxes allow business certainty as paying will result in what a business wants whereas regulations are subject to interpretation, appeal and delay. In the US, pollution permits give people the right to create a set amount of pollution and the money pays to clean up the environment. Those who pollute less pay less, and unused credits can be sold on the open market.

ENVIRONMENTAL REMEDIATION OF BROWNFIELD SITES

An example of an environmental issue in cities strongly linked to sustainable urban regeneration is the 'remediation' of brownfield sites (see Raco and Henderson 2006; Dixon et al. 2007; Henderson et al. 2007; Dixon and Adams 2008). Regulatory or fiscal policy can be mobilised to address this urban issue. A significant debate revolving around urban regeneration and sustainability since the mid-1990s has been the brownfield versus greenfield development debate (see Chapter 11), especially regarding new housing development, originally reflecting the forecast of 4.4 million new homes needed by 2016 (DoE 1995a; 1995b; 1996b; Allinson 1999) and more recently the 3 million more homes needed by 2020 (CLG 2007a). To encourage brownfield remediation and regeneration and to restrict greenfield development, regulatory policies can be instigated. These could include the fast-tracking of planning permission for brownfield sites, no Section 106 (planning gain) obligations, and allowing more floors on buildings. Fiscal policies to encourage brownfield regeneration rather than greenfield development can include tax breaks; government paying a part of remediation costs; government insuring the developer against all future claims; and government investing in office space on site.

CASE STUDY: THE THAMES GATEWAY

An example of a 'sustainable community' reflecting ecological modernisation and experiencing growth is the Thames Gateway east of London (Raco 2005a; Raco and Henderson 2006; www.thamesgateway.gov.uk; see Chapter 5). The Thames Gateway is a major regeneration programme which aims to create what can be termed an 'enviroregion'. The Thames Gateway stretches for 40 miles along the Thames Estuary from the London Docklands to Southend in Essex and Sheerness in Kent. The area was badly affected by the decline of industries serving London and the South-East. The vision for the Thames Gateway (www.thamesgateway.gov.uk) is to create:

- 180,000 new jobs, through business and economic investment;
- 160,000 new, well-designed homes which will include affordable homes which can be rented or bought by first-time buyers;
- a high quality transport system;
- better education facilities;
- high quality healthcare;
- attractive open spaces with access to the river.

New homes are to be built to meet the high demand in the South-East of England, with plans in place to build 160,000 between 2001 and 2016 focused on brownfield sites. The new homes will be set in attractive, green, well-designed public spaces, situated within the high-quality environment that will be known as the Thames Gateway Parklands. In Thames Gateway, high design standards for homes will be the norm, with energy and water efficiency built in. The developers aspire to see Thames Gateway housing leave a low carbon footprint, to contribute to environmental sustainability. There is an emphasis on providing enough affordable homes for those who have difficulty getting onto the housing ladder, or finding good quality rented accommodation (see Raco and Henderson 2006).

SUMMARY

There are clear interrelationships between urbanisation, development, regeneration and sustainability issues, which have been increasingly incorporated within urban regeneration policies since the 1990s.

It has been argued that economic growth, often exhibited through urban regeneration projects, is incompatible with sustainability; however, the ecological modernisation perspective suggests that economic development is compatible with sustainable development.

Sustainable urban regeneration can be delivered through development, planning, housing and design policies such as compact city development, urban intensification, mixed-use development, brownfield regeneration and eco-town principles, in the context of fiscal and regulatory policies.

The multi-use compact city is the promoted model of sustainable urban planning, although it can be critiqued as being impractical, undesirable and unrealistic.

Examples from across the UK illustrate contemporary sustainable urban regeneration, from the architecture of individual buildings to the design and development of whole urban areas.

STUDY QUESTIONS

1. Discuss the concept of 'sustainable urban regeneration' with reference to contemporary urban regeneration.
2. To what extent is it possible to reconcile economic growth with sustainable development in cities?
3. Compare and contrast the 'compact' and 'dispersed' models of the city in terms of their sustainability.
4. In what ways can planners and architects contribute to the creation of more sustainable urban landscapes?

KEY READINGS ON URBAN REGENERATION AND SUSTAINABILITY

Barton (ed.) (2000) *Sustainable Communities: The Potential for Eco-Neighbourhoods* [presents a global review of eco-villages and sustainable neighbourhoods with case studies]

Dixon, Raco, Catney and Lerner (eds.) (2007) *Sustainable Brownfield Regeneration: Liveable Places from Problem Spaces* [account of UK policies, processes and practices in brownfield regeneration with case studies]

Raco (2007a) *Building Sustainable Communities: Spatial Development, Citizenship, and Labour Market Engineering in Post-War Britain* [documents the ways in which policy-makers since 1945 have sought to use state powers and regulations to create better, more balanced, and sustainable communities and citizens]

Rudlin and Falk (1999) *Building the 21st Century Home: The Sustainable Urban Neighbourhood* [includes critiques of social, economic and environmental sustainability linked to neighbourhoods and urban regeneration]

Sustainable Communities Special Issue: *Local Economy* (2008, vol. 23, no. 3) [covers debates surrounding sustainable communities as a conceited metaphor or achievable aim globally]

KEY WEBSITES ON URBAN REGENERATION AND SUSTAINABILITY

BedZED: www.bedzed.org.uk [flagship example of sustainable living and urban regeneration]

Communities and Local Government: www.communities.gov.uk [government department with responsibility for sustainable regeneration and communities]

Sustainable Development Commission: www.sd-commission.org.uk [government's independent watchdog on sustainable development]
UK Government Sustainable Development: www.sustainable-development.gov.uk [government guidance on sustainable futures]

10 City centre retail-led regeneration

INTRODUCTION

This chapter deals with the rise and widespread adoption of commercial retail developments across urban areas in the UK since the 1950s, from refurbished high streets, through to small-scale city centre malls, and more recently to large-scale mega-projects incorporating a variety of uses. This chapter will draw upon historic and contemporary examples from across the UK and throughout the urban hierarchy.

There have been clear changes in the structure of the urban retailing system over the last 50 years, especially linked to the growth of out-of-town shopping centres associated with decentralisation. There are clear implications for urban regeneration in the UK as the changing daytime economy of central areas is a continuing major policy issue. Retail regeneration has formed an important part of many central government UK regeneration policies including Enterprise Zones and Urban Development Corporations (see Chapter 3), and retailing has an important indirect impact on recent initiatives such as New Deal for Communities (see Chapter 5).

In the post-war period in the UK, retail schemes have been widely used as a regeneration strategy in declining urban centres (Bromley and Thomas 1993a; Evans 1997) as have a wide array of other thematic approaches to city centre regeneration. These have been undertaken by city authorities in order to attempt to reverse their fortunes in the face of competition from out-of-town locations, and from home entertainment and the internet. The 1950s and 1960s witnessed the modernist reconstruction of war-damaged centres and the refurbishment of decaying central landscapes (see Chapter 2). By the early 1980s, city centre regeneration focused on the development of small-scale indoor malls and the enhancement of public space, which later expanded to include town centre management (Page and Hardyman 1996) and Business Improvement Districts (BIDs) (Symes and Steel 2003; Steel and Symes 1995). Associated with this period were attempts to attract office development back to the city centre (see Evans 1997, Chapter 4). During the late 1980s and early 1990s it was also considered essential to introduce more housing development and repopulate the city centre to establish an internal market and contribute to wider urban regeneration (Tallon and Bromley 2004; see Chapter 11). The focus was further widened to encompass the development of the evening and night-time economy and the twenty-four-hour city (Heath and Stickland 1997; see Chapter 12). The attempt to extend the daytime activity of the city centre has embraced café culture, late-night shopping and other leisure activities focused on the evening (Ravenscroft et al. 2000).

This chapter gives an overview of what has predominantly been retail-led regeneration in UK town and city centre since the 1950s. It first explores wider processes of city-centre change linked to retailing, incorporating the causes of decline and policy responses. It then moves on to illustrate the array of approaches to city centre regeneration, many of which incorporate retail. The chapter then focuses on the most important retail-led strategies pursued by urban authorities and exemplified by small-scale city centre malls, regeneration of deprived communities through superstore development, mega-retail projects, town and city centre management schemes, and Business Improvement Districts. The chapter finally offers a critique of retail-led regeneration responses that seek to facilitate urban regeneration in the UK.

CITY CENTRE RETAIL CHANGE AND REGENERATION

City centre development has been an important feature of urban change in the UK. The contemporary problems and future of the city centre are major issues within society and urban studies. There has been a general and continuing relative if not absolute decline in many UK inner cities and city centres since the 1970s as their traditional characteristic uses increasingly decentralise (Johnston 2000a). Rowley (1993) argued that literature on the inner city had generally overlooked the considerable importance of the city centre to the inner city, to its functions and to related employment and space requirements. The inner city can be viewed as an ill-defined area close to the city centre and central business district in Western cities, and more recently it has been a term associated with economic and social regeneration policies (Johnston 2000b). The central business district is an 'integral component of the overall inner city space economy' (Rowley 1993, p. 111), and is the nucleus of an urban area containing the main concentration of commercial land uses. The city centre can be viewed as synonymous with the central business district, although in the post-industrial city its role is more than merely a business function, as highlighted by Chapters 11 and 12.

The initial advantages of the city centre, as identified by Rowley (1993), emerged from the agglomeration and transportation advantages as the optimal location for serving the population of the whole city region. However, over time, the failure of the city centre to attract and retain certain functions became cumulative, and this initiated decline. Since the mid-1950s, the character and internal structure of the UK city centre have been transformed by several modifying influences associated with the competitive impact of functional decentralisation, particularly of retail and office activities. These problems relate to difficulties in accessibility, and to social and economic change resulting from de-industrialisation and ethnic ghettoisation (Herbert and Thomas 1997). As a consequence, city centres have presented fewer attractions for shoppers, employers and residents, and the segregation and fragmentation of functions have tended to increase significantly. The inner city until the mid-1990s was usually associated with 'dilapidation, poor housing and economic and social deprivation' (Johnston 2000b, p. 396). The UK inner city became synonymous with urban problems of economic and social malaise and with those areas requiring regeneration and renewal (Lawless 1989). Therefore, as Rowley (1993) argued, 'traditional' ideas of city centre dominance and structure required re-evaluation. With the growing problems of the inner cities and city centres in the UK, a move from short-term to longer-term strategies became necessary along with effective and concerted planning controls to revive the city centre. Approaches to urban regeneration encompassing retail, housing, leisure and entertainment functions escalated in status from the late 1990s.

Commercial functions including large retailers continue to invest in city centres, in the context of support from central government policies, and their presence is essential for the success of city centre schemes. Local authorities have also improved city centre environments through environmental upgrading (see Bromley et al. 2003b), and many office functions that have a high level of direct contact with consumers continue to locate centrally (Barlow and Gann 1993; 1994; Coupland 1997; Evans 1997; Heath 2001; see Chapter 11). City centre policies have tended to understate the job creation potential of retailing despite its labour-intensive nature. The long-term decentralisation of jobs and people from the city centre clearly affected retailing and an intra-urban hierarchy has emerged (Thomas 1989). In the inter-war period a specialised central city shopping area extended outwards along main traffic arteries, with local clusters and individual shops in the surrounding residential areas. Over time a clearer set of shopping areas has developed, comprising a central area serving a population of at least 150,000; regional shopping centres that developed from the smaller central areas in conurbations; district centres serving local catchments of around 30,000; neighbourhood centres selling convenience goods in catchments of 10,000; and local or sub-centres with small clusters of stores serving 500 to 5000 people (Herbert 2000). Much of this framework already existed but was consolidated in the 1950s when planned shopping centres were added to several layers of the hierarchy.

Massive reconstruction of city centre environments took place in the early post-war period with housing, industry and infrastructure tackled as priority redevelopment before attention turned to retail provision. This restructuring of space in post-war cities was characterised by striking modernist urban landscapes (Larkham 1997). Many city centres witnessed large-scale retail developments initiated during the 1950s, particularly in those cities that had suffered extensive wartime damage. However, early city centre redevelopment strategies also exacerbated the situation by introducing significant spatial fragmentation between functions.

The planned shopping centre began to have a profound effect on urban morphology from the late 1950s (O'Brien and Harris 1991), and subsequently proliferated (Reynolds 1993). Schiller (1985) outlined the growth and evolution of planned shopping centres as elements of the city centre environment and showed that planned centres first appeared in two different types of setting. One group comprised cities such as Bristol, Coventry and Plymouth, which suffered severely bomb-damaged centres where the traditional retail cores were completely demolished and redeveloped. The others were made up of the first group of New Towns, where planners designed from scratch with contemporary principles and ideals (see Chapter 2). The early schemes introduced a number of recurrent design features that now characterise the spatial structure of UK city centres. Inner ring roads with associated multi-storey car parking facilities frequently surrounded the central business district. Principal shopping areas were usually pedestrianised, and office districts, civic centres and entertainment areas emerged as distinct distinguishable 'zones' within the city centre (Davies 1984). Housing was almost totally absent as a significant component of city centre space. In this early phase of development, the public sector was the major investor in the construction of shopping precincts, as local authorities were dedicated to rebuilding their city centres in an imaginative and efficient way, acknowledging that a vibrant retail core was an essential element of vital and viable city centres (O'Brien and Harris 1991).

The early phase of shopping centre development was followed by the construction of covered shopping malls from the mid-1960s, when schemes of various sizes were incorporated into the intra-urban retail hierarchy. This second phase of shopping centre development lasted until the early 1970s (Schiller 1985), and these centres were enclosed and therefore weatherproof, representing a shift from the open-air precinct format. However, many of these early enclosed centres possessed architectural and stylistic features that later became outmoded and inadequate. Schiller (1985) identified the period between 1972 and 1980 as the 'golden age' of shopping centre development in the UK. More attention was given to design and location and this was the peak period in terms of the numbers and size of city centre shopping schemes. By the 1980s, design styles were increasingly influenced by post-modernism, and the monolithic brutalism of earlier post-war city centre buildings was progressively replaced by higher quality and more varied constructions (O'Brien and Harris 1991). Between 1965 and 1989, 8.9 million square metres of retail floor-space was provided in 604 town centre shopping schemes of 4650 square metres or larger (DoE 1992). However, from the 1980s, city centres started to witness a significant downturn in their fortunes for a variety of reasons (Table 10.1).

Dawson (1988) identified four major transformations that characterised city centre change during the mid-1970s and 1980s. One way in which city centres changed was in the overall design and appearance of shop buildings, along with the development of infill shopping centres, refurbishment and pedestrianisation schemes. Second, city centre streets changed in land use terms with other functions such as offices and leisure and entertainment facilities changing the city centre land use pattern. Third, there was a change in city centre land values, which increased considerably as retailing performed significantly better than office or industrial property. Finally there was a change in the spatial structure of the retail system. Although the city centre had traditionally been at the highest level of the intra-urban hierarchy, it was argued that a hierarchical model was no longer an appropriate form of urban structure.

Dawson (1988) argued that despite these transformations in the city centre, and the expectation that the high street might become a less dominant force in the city, other new uses would enter the landscape including leisure shopping and heritage tourism (see Chapter 12). Overall, despite the negative forecasts for the future of the city centre in the UK, the high street has survived but changed because consumers, retailers and society have changed. The city centre continues to exploit its strengths and address new market opportunities with intervention through innovative city centre planning and management approaches and energetic policies to revitalise city centres (Davies 1987; Whysall 1995; Evans 1997). These responses have included city centre management (Warnaby et al. 1998), place marketing (Page and Hardyman 1996), loyalty initiatives (Hallsworth 1997; Potts 2007a) and a wide range of associated policies and programmes to enhance attractions, accessibility and the environment (URBED 1994). These have been assisted by the intermittent constraint upon decentralisation prompted by successive governments.

Within the context of post-war city centre transformation and decline, continuous and rapid change has been the most characteristic feature of the retailing environment in the UK over the last 50 years. Increasingly mobile, affluent and discerning consumers have

TABLE 10.1 CAUSES OF THE DECLINE OF CITY CENTRE RETAILING

New competition in retailing
- regional malls
- smaller shopping centres
- ribbon developments
- 'edge cities'

Demographic shifts
- suburbanisation of population and spending power
- depopulation of inner city areas
- change in the socio-economic composition of the inner city population

Characteristics of city centre areas
- traffic and parking problems
- negative image
- obsolete physical landscape
- fragmentation of land use

offered retailers the opportunity to provide wider ranges of increasingly specialised products and services in new types of shopping outlets in a variety of locations outside traditional shopping centres such as the city centre. Despite successive central and local government policy aimed at preserving the 'vitality and viability' of traditional shopping opportunities in city, town, district and neighbourhood centres as the commercial and social foci of community life, the combination of consumer demand and commercial pressure from the supply-side has revolutionised the organisational and spatial structure of the retailing environment (Thomas et al. 2004). Retailers have responded to market opportunities and led consumers into new modes of shopping, usually at the expense of traditional shopping centres (Wrigley and Lowe 2002; Guy 2000; 2007). The larger multiple retailers were the first to recognise the commercial opportunities offered by changing consumer demands, and they responded with the provision of new shopping facilities offering the advantages of easily accessible out-of-centre sites, adequate car parking, larger premises to provide wider ranges of products and associated services, and a more attractive shopping environment in more secure surroundings (Thomas and Bromley 2002).

The process of retail change impacting on city centres in the UK has been characterised as a series of 'waves' (Schiller 1986). The 'first wave' was initiated in the 1960s and focused primarily on the development of out-of-town superstores, which have grown to dominate convenience goods shopping. By 1993, 868 superstores transacted 50 per cent of grocery shopping trade (Pacione 2005). By 1999, there were 1100 superstores, transacting 70 per cent of grocery shopping (Guy 2000).

This was followed by a 'second wave' during the late 1970s involving the sequential development of free-standing retail warehouses and retail parks, initially trading in bulky do-it-yourself goods. Retail parks subsequently developed to offer household furnishings and a wide range of domestic appliances sold in 'crinkly sheds' (Guy 2000; Thomas et al. 2004; 2006). By 1992, there were 2000 retail warehouses and 250 retail warehouse parks transacting 25 per cent of non-food trade. The steady growth of retail parks has been difficult to stem, and they have recently developed to resemble 'high streets in the suburbs' or 'fashion retail parks' (Guy 2000; Thomas et al. 2006). By 1999, there were 3000 retail warehouses in 485 peripheral retail warehouse parks, accounting for 30 per cent of non-food retailing (Guy 2000).

By the mid-1980s, demonstration of the commercial advantages of out-of-centre locations resulted in a 'third wave' of planned regional and sub-regional malls offering a full range of comparison and specialist goods which had previously been confined to the traditional 'high street'. Successive governments demonstrated ambiguity towards this form of decentralisation, acknowledging both the potential damage that could be done to city centres, and the commercial and social importance of the new centres. A strong element of planning gain through regeneration was attached to these centres as most were developed in areas of industrial dereliction (Walker 1996). Out-of-town areas have been regenerated (see Chapter 13) but arguably at the expense of central city areas.

In contrast to the previous two waves, central government has severely restricted the development of a 'third wave' of new planned regional malls due to their potentially adverse impact on nearby town and city centres (Guy 1994a; Hallsworth 1994). There were proposals for 50 regional shopping centres between 1987 and 1992 but only four had been completed by 1998 and in total just eight have been developed (Thomas et al. 2006; Lowe 2000b; 2005; Pacione 2005, Chapter 12). The first wave of regional shopping centres opened between 1985 and 1990 and is located at Merry Hill, Dudley; MetroCentre, Gateshead; Meadowhall, Sheffield; and Lakeside at Thurrock, east of London (Lowe 2000b; 2005). The second wave was developed between 1997 and 1999 and comprises White Rose, Leeds; Cribbs Causeway, Bristol (Plate 10.1); Trafford Centre, Manchester; Bluewater, Dartford; and Braehead, Glasgow (Lowe 2005; Thomas et al. 2006). Permission for these reflects complex planning considerations, associated with environmental improvement and

Plate 10.1 Out-of-town regional shopping centre developed in the late 1990s
Photograph by author.

economic regeneration in their vicinity, rather than simply their potential commercial impact on existing centres (Herbert and Thomas 1997; Lowe 1998).

An additional innovation from the early 1990s might also be considered to constitute a variant of the retail park phenomenon. This has been defined as a 'fourth wave' of decentralisation (Fernie 1995; 1998), and has been variously termed the factory outlet mall or outlet-shopping village, at a range of scales similar to retail park developments. The first, Clarks Village, opened in 1992 at Street in Somerset and by 2000, 29 factory outlet malls existed (Jones and Vignali 1994; Jones 1995; Norris 1999; Hillier Parker 2001; Thomas et al. 2004). These have been defined as 'groups of manufacturers or retailers who trade from separate shop units, or outlet stores, on a single site, often possessing associated leisure facilities' (Hignell 1996, p. 14). Functionally, they differ from sub-regional centres in that they comprise smaller individual shops, trading at discount prices in 'seconds' or 'surplus stock' of quality branded products. Also, they tend to focus on fashion clothing and footwear, although their range has been widened to include a full range of household goods, furniture and furnishings, jewellery and gifts. They also offer catering and associated leisure facilities, and were initially seen as oriented towards the leisure and tourism market segment rather than the conventional retail market. They have often displayed an element of planning gain as they contribute to the regeneration of former industrial sites and declining town centres (Jones and Vignali 1993; 1994; Hignell 1996). However, the recognition of their possible impact on traditional shopping centres has resulted in applications being subject since 1996 to similar constraints to the other out-of-centre retail innovations (see Table 10.4). It has been suggested that it is likely to be difficult to distinguish retail parks functionally from factory outlet centres, since it is unlikely that planners will be unable to restrict the latter to 'discount trading' (Guy 1998b; Thomas et al. 2006).

A 'fifth wave' comprising so-called limited line discounters has proliferated since the early 1990s (Baron et al. 2001). Retailers such as Aldi and Lidl entered the market to fill a gap at the lower end of the market, sometimes colonising stores abandoned by larger retailers and contributing to city centre regeneration. These are often located close to deprived inner city communities with a limited choice of food stores (Potts 2007a). There has more recently been a reverse reaction by the biggest retailers such as Tesco and Sainsbury's who have targeted city centres with small-store formats aimed at emerging city centre populations (see Chapter 11). This sector is currently the subject of much competition for central city markets and offers potential city centre regeneration opportunities. This development has implications for the continued future of independent food and grocery retailers in the UK (see Smith and Sparks 2000; Baron et al. 2001).

A 'sixth wave' since the late 1990s has witnessed the spectacular emergence of internet retailing which represents the ultimate level of decentralisation (Pacione 2005). The level of future demand is somewhat unknown (Jones and Biasiotto 1999).

The addition to the array of shopping opportunities has served to divert the shopping behaviour of car-owning households increasingly towards accessible decentralised retail environments (Bromley and Thomas 1993b). Such changes have resulted in negative impacts on the older, often more congested, traditional shopping centres, which are unable to compete adequately with the ease of access, car parking facilities and general environmental attractions of the innovative retail forms. Consequently, many have experienced a relative, and sometimes absolute, decline in their retail status and associated environmental attraction (see Tables 10.2 and 10.3; Plate 10.2). This resulted in a widespread concern for the

TABLE 10.2 IMPACTS OF RETAIL CHANGE

Economic

- sales transference
- investment levels

Social

- equality – disadvantaged consumers
- restrictions on choice
- poor, carless, disabled, elderly
- social malaise related to crime and vandalism
- accentuation of social polarisation

Environmental

- maintain access to old facilities
- maintain attractions of the city centre
- minimise congestion of new centres

Indicators of centres at risk

- environmental deterioration
- vacancy increases
- falling real rental levels
- low levels of investment and re-investment
- failure of proposals to reach completion
- poor maintenance
- poor car parking and acute congestion
- lack of sites for improvement
- expenditure leakage
- relocation losses

TABLE 10.3 NEGATIVE IMPACT OF RETAIL CHANGE: GENERAL MODEL

Diversion/transfer of sales from one centre to another →

Affects space and employee productivity →

Resulting responses

- lower levels of employment
- more part-time employment
- contraction of floor-space
- decline in functional status of outlets (more discount and charity stores)
- environmental deterioration →

Resulting loss of investor confidence →

Spiral of decline sets in

***Source:* DoE (1992).**

There is a problem of disentangling the effect of changing retail facilities from wider economic, social and environmental changes

Plate 10.2 Evidence of retail decline in the urban landscape
Photograph by author.

future economic health of older centres from the early 1990s and the necessity for urban regeneration initiatives (DoE 1992; URBED 1994; British Council of Shopping Centres 1996; Thomas and Bromley 1996; 2000).

The impact of government policy on retailing in the UK dates back to the Town and

Country Planning Act (1947), which took land into central and local government control and stated that the neatness of the traditional hierarchy should be maintained due to its value in terms of efficiency and equitability (see Guy 2007). There was an associated constrained approach to the car, to accommodate rather than promote, for environmental reasons. However, commercial pressures catered towards a decentralised, car-oriented retail system that eroded the traditional system.

However, despite waves of retail change and their impacts, traditional shopping centres continue to function due to a combination of restrictive regulatory planning controls exerted over new retail formats, and the positive promotional activities supported by successive central governments. Throughout the whole period of change, government has to varying degrees instituted regulatory controls and incentives designed to minimise the negative impact of the process of retail change, and to facilitate regeneration and revitalisation (URBED 1994; 1997; see Table 10.4). Despite the revolutionary changes since the 1960s, traditional centres co-exist with new forms in an often uneasy complementarity. In a context of continuing retail decentralisation, concern continues to be expressed for the future economic health of traditional shopping centres of all types (Thomas et al. 2004).

The main options in terms of retail planning policy appear to be the contrasting positions of a laissez-faire approach that leads to city centre decline, or investment and intervention, which means that city centres can be retained and grow.

Much attention has been focused on the impact of superstores on district centres and country towns, which have been largely detrimental to existing retailers (Thomas and Bromley 1995; 2002; DETR 1998c; Collis et al. 2000; Potts 2007a). In addition, research has been primarily concerned with the commercial impact of the relatively small number of regional shopping malls (Thomas and Bromley 1993; Lowe 1998; 2000b). Robertson and Fennell (2007) examined the economic effects of third wave regional shopping centres in the UK, and presented a generally favourable review. Because of their size and quality, Guy (2008) argued that they exemplify the potential for retail development to attract further retail use. Robertson and Fennell (2007) found that the MetroCentre and Merry Hill have additionally attracted mixed-use office, leisure and residential development (see Williams 1997; Lowe 1998). However, Lakeside and Cribbs Causeway have only been able to attract further consumer services to the locality (Robertson and Fennell 2007).

With the exception of a study of an unplanned retail park by Thomas et al. (2004), there

TABLE 10.4 IMPACTS OF RETAIL CHANGE: GOVERNMENT POLICY RESPONSES

1988–1994: Official attitudes to the development of out-of-town centres in the UK were characterised by caution and ambiguity

1988: DoE Planning Policy Guidance (PPG) 6 – the future of large off-centre shopping centres – 'Planners should not consider the competitive effects of large new centres on the existing system' except where 'they are on a scale and of a kind that would seriously affect the vitality and viability of a nearby town centre as a whole' (DoE 1988)

1988: PPG13 – reverse car culture

1993: Stronger views re-asserted in updated PPG6

1994: URBED report for the DoE – *Vital and Viable Town Centres: Meeting the Challenge* – stimulated the revitalisation of city centres (attractions, accessibility, amenity, action)

1994: House of Commons 4th Report: Shopping Centres and their Future

1994: PPG13 – reverse car culture

1996: PPG6 update – same message as before

2005: Planning Policy Statement (PPS) 6 emphasised the role of town centres and widened the focus to include leisure and entertainment (see Guy 2004; 2007)

is little evidence on the impact of the large number and variety of retail parks distributed widely throughout towns and cities of the UK. The early view that retail parks were essentially competitively 'benign' and functionally 'complementary' to traditional shopping centres remains largely unchallenged (DoE 1992), despite the expression of intermittent disquiet (Schiller 1986; Guy 1998a; 1998b; 1998c; Thomas et al. 2004; 2006). This deficiency is particularly important, considering the fact that many retail parks are currently undergoing the process of functional evolution highlighted by Guy (2000). Many of the early 'crinkly sheds' are being transformed into modern stores, providing new 'homes' for conventional 'high street' products. In effect, the UK has seen the emergence of new 'high streets' in the suburbs, without recognising the full competitive potential of the new facilities on existing middle and higher order shopping centres (Thomas et al. 2006). This is particularly important in view of the widespread observation that the great majority of substantial central regeneration schemes have been confined to the 50 largest city and town centres (Guy 2004). In fact, it is possible that changes within the retail park sector are capable of re-asserting the challenge to the primacy of town and city centres formerly associated with the development of regional shopping malls prior to the revision of PPG6 in 1996 (DoE 1996a; see Table 10.4).

Clearly there has been a complex set of changes to the retail scene that has resulted in problems of compaction of the city centre in the UK. However, retail decentralisation and change do not necessarily mean the end of the city centre; the effect has been for the city centre to re-image, regenerate and fight back. Urban regeneration of traditional centres has taken place, and the whole process of retail change and regeneration remains in a state of flux.

Commercial regenerative measures have been undertaken as a consequence of decentralisation and decline, in the context of the government's policy pronouncements since the 1990s. A number of principal approaches that UK city centres have undertaken are highlighted by Table 10.5. Each has potential, but each also displays problematic elements, many of which are further discussed in Chapters 11 and 12.

A major problem is integrating the full array of regeneration measures in a subtle way to address complex city centre problems and to minimise any negative outcomes. Public–private collaboration is needed along with effective management and promotion.

RETAIL-LED REGENERATION

Retail-led regeneration in town and city centres of the UK can be traced back to the 1950s. Lowe (2000a; 2005) placed the origins of the concept of planned shopping centres in 1950s North America. The world's first fully covered mall opened in Southdale, Minneapolis in 1956 and the model was subsequently replicated throughout the US (see Wrigley and Lowe 2002; Pacione 2005, Chapter 12). Retail development in the UK adopted the model, but used malls as a method of enlarging and modernising existing central shopping areas, rather than developing free-standing suburban entities as witnessed in the US (see Guy 1994b). During the 1960s and 1970s, many towns and cities in the UK gained one or two central shopping malls exemplified by the Arndale Centre in Manchester, Eldon Square in Newcastle, and Bull Ring in Birmingham (Lowe 2005). The trend towards centrally located shopping mall development was sustained until the move out-of-town in the mid-to-late 1980s, which was facilitated by a combination of supply, demand and policy motivations (see Chapter 3). This sparked an unprecedented period of retail change in the UK (see Bromley and Thomas 1993a), identified by the succession of 'waves' of retail innovation. Table 10.6 shows that city centres exhibited an array of characteristics that would facilitate successful retail-led regeneration.

TABLE 10.5 PRINCIPAL CITY CENTRE REGENERATION APPROACHES

- Pedestrianisation
  Returning people to the streets; accessibility issues related to less mobile groups (see Bromley and Matthews 2007; Bromley et al. 2007a)
- City centre shopping malls
  Mixed-use indoor centres; festival market places; commercial gentrification forcing out independent stores; privatisation of public space (see O'Brien and Harris 1991; Francis 1991)
- Historic preservation (see Chapters 11 and 12)
  Adaptive use of historic buildings; conservation; promotes gentrification and exclusivity (see Ashworth and Tunbridge 1990; 2000; Larkham 1996)
- Waterfront redevelopment (see Chapter 11)
  Restaurants; cafés; hotels; housing; promotes gentrification (see Hoyle et al. 1988; Tallon 2006; McNeill 2008)
- Office development
  Corporate activity; can lead to dead space out of working hours (see Evans 1997; Heath 2001)
- Transport enhancement
  Integration; public transport improvement; improved car parking; costly to implement (see Bromley and Thomas 1997)
- City centre managers and Business Improvement Districts (see Chapter 5)
  Aesthetics; promotion; city centre managers are under-financed and under-powered; limited impact (see Morphet 1991; Crouch 1992; Page and Hardyman 1996; Symes and Steel 2003; Steel and Symes 2005)
- Expansion of the 'out-of-hours' economies (see Chapter 12)
  Regenerative potential; negative consequences of the night-time economy (see Montgomery 1994; Bianchini 1995; Thomas and Bromley 2000)
- Diversification of activities and special events (see Chapter 12)
  Conventions; festivals; carnivals; sports events; cultural events; episodic in nature (see Bianchini and Parkinson 1993; Waterman 1998; Hughes 1999; Tallon 2007b; Smith 2007)

City centres have been able to maintain their status as the primary retail area in the city through a combination of new development, environmental improvements and restrictions on other major retail developments since the 1990s. This has been the experience of many Western European countries (Guy and Lord 1993; Davies 1995). During the 1980s and 1990s, Conservative government urban regeneration usually took the form of physical property-led regeneration with the intention of targeting run-down inner city areas, particularly those that suffered from the effects of de-industrialisation (Healey et al. 1992; Hall 2006; see Chapter 3). The aim was to encourage the regeneration of depressed urban areas by regenerating economic infrastructure and introducing private sector mechanisms, often through commercial property development.

In line with policies of place marketing (see Chapter 6), clear strategies have been formulated to take a positive, proactive approach to attracting investment back to city centres. Evans (1997) noted that ingredients for a successful town or city centre include a mixture of land uses; housing uses; social functions; economic vitality; good transport and accessibility;

TABLE 10.6 REGENERATION OF RETAILING IN THE UK

Historic advantages
- access to the city-region
- government administration – local and central
- office concentration
- historic buildings – tourism

Development opportunities
- gentrification of older enclaves
- redevelopment of former industrial areas – waterfronts, dockyards, railyards

Reorientation to 'internal markets'
- high-status residents of gentrified areas
- white-collar office workers
- festival shopping – leisure shoppers, visitors, tourists
- conference/convention visitors

Persistence of social divisions
- regenerated city centre not oriented to the inner city population
- divides between modern and post-modern architecture

amenities; a secure, safe environment; the promotion of arts, culture and heritage; and environmental sustainability. Jones (1989) stated that the high street was 'fighting back' and the challenge has since been taken up in a number of ways in UK city centres as illustrated by Table 10.5. New and refurbished shopping centres have had an impact on physically regenerating parts of city centres, as have larger-scale redevelopment schemes, which have provided high-quality, modern and attractive shopping environments. City centre developments of this type demonstrate high degrees of interior design and theming with juxtaposed food courts and leisure facilities. City centres have also aimed to improve access with traffic management schemes and improved car parking provision in addition to expanded and modernised public transport systems. Response to concerns regarding the vitality and viability of city centres has taken a number of forms, ranging from central government action through to various localised initiatives, exemplified by place marketing, city centre management and loyalty card schemes (see Ravenscroft 2000). Central within the array of approaches has been retail as a catalyst of urban regeneration.

Lowe (2005) noted that the relationship between retailing and local economic regeneration is not new. From the 1980s onwards in response to de-industrialisation, the role of services as alternative sources of employment and as catalysts of economic development became more important (Lowe 2005; see Chapter 1). Local authorities increasingly became supportive of retail expansion and shifts in retail structure in their areas. Retail-led strategies began to be allied with place promotion and image transformation strategies (Kearns and Philo 1993; Lowe 1993; Paddison 1993; see Chapter 6). Central government urban policies of the 1980s such as Enterprise Zones and Urban Development Corporations (see Chapter 3) encouraged the development of retail centres on brownfield sites. Examples are the Merry Hill regional shopping centre in the West Midlands (Lowe 1993; 1998) and Swansea's Enterprise Zone (Thomas and Bromley 1987; Thomas et al. 2004). Retail development was considered a way of kick-starting urban regeneration and developers were prepared to either buy sites outright or to go into partnership with agencies and inject funds into improving sites and providing infrastructure (Lowe 2005). In return for the investment, they were able to build stores that might not be allowed by the planning system in places such as Cardiff Bay (Wrigley et al. 2002), and in retail warehouse parks across the UK, which became inextricably linked with brownfield regeneration and light industrial sites through the 1990s (Lowe 2005; Guy 1998a; 1998b; 1998c; Thomas et al. 2006).

Lowe (2005) argued that links between retail development and urban regeneration became increasingly prominent as reflected in government planning policy. In the context of progressive regulatory tightening and the recognition by retailers that there were centrally located markets to serve, a retail-led regeneration agenda was elevated to prominence in the late 1990s. Guy (2002; 2008) and Wrigley et al. (2002) looked at the role of large store development in the UK where major superstore retailers presented a strong social inclusion and neighbourhood renewal rationale to support large-store redevelopment of long-term degenerating district shopping centres. This led to a noticeable shift in retail development away from the second wave of regional shopping centres, which had opened in the late 1990s after gaining permission in the very different regulatory environment of the 1980s and early 1990s, towards an exploration of the potential of city centre sites (Lowe 2005).

Retailing is one of the UK's largest industries (Guy 2008), employing around one-ninth of the workforce. In contrast to many other sectors of the economy, it is found in all urban areas, serving the local population. Retailing is therefore usually one of the main land uses under consideration where new development is concerned to facilitate the regeneration of the built environment. The relationship between retail development and urban regeneration in the UK has usually focused on employment creation (see Dixon 2005). Retail-led regeneration is also an important category of property-led regeneration (Turok 1992; Guy 2008). Here, initial property development attracts further development, not necessarily of the same land use. This process takes place partly because of synergies between types of development such as retail and leisure, and partly because the initial development indicates to other developers that the location is viable (Guy 2008).

The employment-related benefits of retail development in areas of deprivation have also been emphasised in studies of retail-led projects such as the Tesco Regeneration Partnership schemes (McQuaid et al. 2005; Potts 2007a). Research has also examined the ways in which existing shopping areas can be regenerated to the benefit of traders and the local population (Carley et al. 2001). Guy (2008) examined the relationship between retail development and urban regeneration from the perspective of the property-related benefits of incorporating retail development at an early stage of a project. Public sector regeneration agencies and private sector developers argue that retail development can 'kick-start' a programme of mixed-use development. Retail developers can provide finance at an early stage in the programme, allowing decontamination of land or provision of services to take place, as well as construction of the retail stores. This then encourages other types of development, including leisure and entertainment, sports (see Chapter 12), offices or housing (see Chapter 11) to take place at later stages. However, retail development has been encouraged for mainly pragmatic reasons, and it shows little synergy with other non-retail development.

Major retailers such as Asda promote the regenerative benefits of new stores, particularly in deprived areas and declining shopping areas (Potts 2007a; Guy 2008). The stores act as a catalyst for other forms of investment and regeneration. However, many large food stores exist in isolation and have attracted only further retail use. Guy's (2008) study of retail development in three off-centre locations in Cardiff showed that it can 'kick-start' further regeneration. The research in Penarth Haven, Cardiff Bay and the International Sports Village indicated that retail-led regeneration is problematic. First, the scale of retail development required financially to kick-start reclamation is likely to be much larger than that strictly needed to serve the new residential population, as a much wider catchment area is required. Second, Guy (2008) showed that there is no guarantee that the development that follows the retail initiative will turn out to be non-retail uses. Third, piecemeal regeneration of successive sites within urban brownfield areas might lead to broader negative outcomes such as competition with existing city, town and district centres, and congestion surrounding new retail park and mixed-use schemes.

Guy (2008) argued that retail-led regeneration of out-of-town sites might become much less common in future years, because planning policy is generally opposed to large-scale, out-of-town development, and that land with planning consent for housing is as valuable as that with retail consent. Retail is not always needed to initiate development and urban regeneration, particularly as it is much easier to obtain planning permission for housing on brownfield sites (see Chapters 9 and 11).

Potts (2007a) further examined the way in which superstores affect urban regeneration, particularly in deprived communities and those areas under-served by food retailers that are arguably conceptualised as 'food deserts' (see Wrigley 2002). Food deserts are defined as areas of inner cities where cheap and nutritious food is virtually unobtainable, impacting upon the disadvantaged and carless residents who are unable, or find it difficult, to access out-of-town retail opportunities. They also depend on independent stores with poorer choice and higher prices, as evidenced by Wrigley et al. (2003; 2004) in Leeds, and Clarke et al. (2002) in Cardiff, Leeds and Bradford. Potts (2007a) illustrated mixed evidence of the effects of superstores in terms of urban regeneration linked to the local economy, community life, and physical and environmental impacts.

In terms of the local economy, attractive features of deprived areas for investment include little or no competition, large supplies of available labour and potential access to profitable markets. 'Regeneration stores' that are economically viable and physically transformative of brownfield sites can contribute to previously underserved markets. Employment is created and innovative schemes such as the Leeds Seacroft Partnership encourage locally unemployment residents to work in the Tesco superstore. However, superstores can cause independent stores to close; in 1945 there were 500,000 independent food retailers, falling to 30,000 by 2006 (Potts 2007a). This transformation in the retail landscape has had severe economic and social repercussions.

The decline of local shops linked to superstore developments since the 1960s has contributed to the decline of the social fabric of local communities, and has resulted in poorly served markets and food deserts. However, superstores can have positive social impacts by acting as a neighbourhood hub and performing some of the functions previously provided by independents (Potts 2007a).

Local physical and environmental impacts of superstores are similarly contested. Some argue that their proliferation has heralded a loss of local character and distinctiveness. Town and city centres that have undergone substantial urban regeneration have often become 'clones' (Simms et al. 2005). Superstores have also been criticised for the move out-of-town, often to greenfield sites. Many town centre superstores have been closed and been subject to restrictive covenants preventing another firm operating a superstore from the premises. On the positive side, superstores can significantly contribute to the regeneration of centrally located brownfield sites, and can provide community facilities and social housing through Section 106 agreements. There are also recent examples of superstores being sympathetically designed to fit into the local environment (Potts 2007a).

CASE STUDY: SUPERSTORE DEVELOPMENT AND REGENERATION IN THE TOWN CENTRE OF LLANELLI, SOUTH WALES

Extensive and longitudinal research in South Wales has examined several dimensions of retail change (Tallon et al. 2005). Dating back to the 1980s, research looked at the functioning and effects of the Enterprise Zone retail park north of Swansea (Bromley and Morgan 1985; Bromley and Rees 1988; Thomas and Bromley 1987; Bromley and Thomas 1988; 1989a; 1989b; see Chapter 3), subsequently revisited in the late 1990s and early 2000s (Thomas et al. 2004).

During the early 1990s, research looked at retail change and the opportunities for the retail revitalisation of small shopping centres in South Wales (Bromley and Thomas 1995; Thomas and Bromley 1995). A more recently examined aspect has been the emergence and impacts of out-of-centre retail parks, which have morphed into 'suburban high streets' (Thomas et al. 2004; 2006).

Another strand of research linked to retail-led regeneration has explored the social, economic and environmental regeneration of a declining town centre through environmental improvements to the public space (Bromley et al. 2003a) and a major food store development in the centre of Llanelli (Bromley and Thomas 2002a; 2002b; Thomas and Bromley 2002; 2003).

Llanelli is a small town in South Wales with a population of around 40,000, placing it in the second tier of the regional hierarchy. With a strong industrial past, the town suffered decline in the second half of the twentieth century, resulting in coastal dereliction and town centre decay (Bromley et al. 2003a). The centre has further experienced social and economic decline associated with the development of an out-of-centre retail park. Policy responses at the local and regional level from the early 1990s comprised two principal elements – the improvement of the central town environment, and the purchase and preparation of land for a new central shopping centre.

Bromley et al. (2003a) examined the environmental improvements made to the existing town centre of Llanelli to facilitate regeneration. These were carried out from the early-to-mid 1990s at a cost of around £2 million. Evidence revealed that environmental improvements create short-term problems during and immediately following the improvement work, with a time lag before benefits are felt. They can also accentuate and create spatial inequality with divisions between the improved and unimproved areas. Environmental improvements to the built landscapes of town and city centres are only a first step towards change. Bromley et al. (2003a) showed that environmental improvements encourage private investment, and it is the investment in new and improved attractions that is the second critical step towards town centre regeneration. In this case, the additional boost was the construction of a new small-scale edge-of-centre mall and the attraction of an associated superstore (Thomas and Bromley 2002; Bromley and Thomas 2002a; 2002b). Evidence suggested that attractions for traders and residents were the range of shops available rather than the nature of the environment. It is expected that environmental improvements encourage further investment in shopping and other facilities that render the strategy viable. Public sector investment in environmental improvements is undoubtedly worthwhile if there is a reasonable expectation of subsequent private sector investment (Bromley et al. 2003a).

In the late 1990s, following the public sector environmental enhancements, a closely integrated new shopping precinct was developed incorporating an edge-of-centre Asda superstore. Surveys in the late 1990s and early 2000s found that there was a dramatic change in patterns of food shopping favouring the town centre. However, no real change in purchases of non-food items such as clothing or do-it-yourself goods was revealed (Bromley and Thomas 2002a). Disadvantaged consumers, the carless and the elderly (see Bromley and Thomas 1993b) were already heavily reliant on town centre shops before the opening of the new superstore, and changes in their shopping behaviour were no different from those of other shoppers. The increased proportion of shoppers was drawn from the full array of shoppers. Bromley and Thomas (2002a) suggested at least four strong positive effects of the new food store development. First, the food store represented a marked social advantage to the town, in particular for disadvantaged groups. Second, there was an increase in the satisfaction of shoppers in the town centre. Third, the social vitality of the town centre increased. Fourth, the addition of a new food store contributed to town centre vitality and viability stemming the spiral of decline. However, major spin-off shopping or economic vitality had not occurred, highlighting the limits of a strategy based purely on a new food superstore. Economic advantages of a new town centre food store should not be exaggerated, as non-food shopping behaviour is not significantly altered, at least in the short term (Bromley and Thomas 2002a).

Thomas and Bromley (2003) found that in Llanelli town centre, redevelopment that retained a compact structure based upon spatial proximity was found to encourage high levels of linkage between the component parts of the town centre, and generated favourable attitudes to the shopping environment. However, the successful spatial integration of the superstore with the centre needed a site that approximated to an in-centre or edge-of-centre site rather than a more peripheral edge-of-centre or out-of-centre site. Clearly, considerable care is required to define edge-of-centre locations for new developments if they are to assist in regenerating a declining centre (Thomas and Bromley 2003). A new central food store is no immediate panacea for regenerating the town centre economy, and should not be viewed as such in any town. However, Bromley and Thomas (2002a) made the case that a food store is a critical factor in town centre revitalisation, particularly for disadvantaged groups.

MEGA-RETAIL-LED REGENERATION

Since the late 1990s many cities in the UK have centred their regeneration strategies on developing high profile central retail developments. Case studies have now started to emerge on this recent wave of retail-led urban regeneration. Lowe's (2005) study of Southampton's West Quay development showed how a large-scale shopping centre could pump-prime the economic and urban regeneration of the city. West Quay can be visualised as a regional shopping centre in the inner city, which challenges the trend of 1980s-style out-of-town development. Lowe (2005) reported that it sought to retain and concentrate key city centre uses in the retail core, and act as a catalyst for wider urban regeneration. Such an approach has now been replicated across the UK with a variety of cities pursuing 'mega-retail-led regeneration' (Tallon 2008).

The term 'mega regeneration' reflects the iconic image projected of new centrally located large-scale developments, which have substantial impacts on the economy, society and environment of a locality. The term also echoes the complexity and finance involved, which set a relatively small number of schemes apart from smaller-scale projects. They can be classified as regional shopping centres, although situated in-centre rather than out-of-town (Lowe 2005; Guy 1994a). In the entrepreneurial era of inter-city competition, such approaches to city centre regeneration have often been funded by local public–private partnerships, although more recently many mega-projects have been entirely privately financed prestige developments (Loftman and Nevin 1995; Hall and Hubbard 1996; 1998) of extremely large scale, expense and public profile. Since the late 1990s there has been a shift in orientation of UK retail development and planning back towards city centre shopping malls, following a sustained period of out-of-town developments. This has resulted in a strongly urban-regeneration-led focus, which has established and confirmed the link between retail development and urban regeneration (Lowe 2005).

In the context of a buoyant economic and property market in the post-recession period between 1995 and 2008, examples of numerous post-2000 flagship city centre retail-led schemes include the £500 million redevelopment of Birmingham's Bullring (Emery 2006; Arlidge 2006); the completion of Manchester's redeveloped city centre following the 1996 IRA bombing, incorporating the £150 million Arndale Centre redevelopment (the UK's largest city centre shopping mall at 1.4 million square feet) (Pilkington 2006); and the Liverpool One Paradise city centre redevelopment, when under construction was the biggest development of its kind in Europe (Parker and Garnell 2006; Tallon 2008). Added to this list of centrally located retail-led schemes completed since the late 1990s are The Oracle in Reading, Buchanan Galleries in Glasgow, Overgate Centre in Dundee, Touchwood in Solihull, The Chimes in Uxbridge, and Festival Place in Basingstoke (Lowe 2005). More recent examples include Bath's £350 million SouthGate mixed-use redevelopment in the historic centre, and Exeter's £225 million Princesshay mixed-use scheme (Tallon 2008). Arlidge (2006) reported on some of the biggest urban malls ever built in the UK, forecast to rise to around 700 in number by 2010, a trend also being experienced across continental Europe, as well as in China, India and the Middle East. In the UK, 93 per cent of new malls developed between 2006 and 2010 are situated in central areas (Arlidge 2006).

These developments have emerged partly out of recognition of a supply gap in the destination market of city centres following the move of high-profile retailers and associated facilities out of town. Increased demand from growing city centre populations since the mid-1990s, including substantial young professionals and students, has also fuelled the development of city centre retail economies (Bromley et al. 2007b). These retail-led spaces have usually been developed in combination with leisure, entertainment, housing and office space (see Dixon 2005; Guy 2008). Supply and demand drivers emerged at a time when there was also a realisation by central government that city centre economies had suffered and must be favoured before peripheral retail and leisure development (URBED 1994; Urban Task Force 1999; ODPM 2005c).

Liverpool is an example of one of the UK's major shopping centres that had declined in the late twentieth century in the face of competition from its regional rivals (Couch 2003). Liverpool responded to this situation by developing a premier retail destination, the £900 million Liverpool One Paradise Project, which blends retail, leisure and residential landscapes. The development provides 1.5 million square feet of retail space in addition to hotels, apartments and a multi-screen cinema. Similarly, Cardiff's £675 million St. David's 2 (a retail-led, mixed-use scheme) comprises a 967,500 square feet regeneration and expansion of the city centre. A further example of mega-retail-led regeneration is Cabot Circus developed in Bristol city centre in the late 2000s (Plate 10.3). This £500 million retail-led scheme combined 1 million square feet of retail space with leisure, office and housing functions (Tallon 2008).

Alongside the visible and striking retail-led mega schemes have been two other major advances in the UK's city centre shopping areas. Town and city centre management schemes and Business Improvement Districts (BIDs) have sought to improve the vitality and viability of central shopping areas through a variety of interventions, which encompass the array of strategies listed in Table 10.5.

TOWN AND CITY CENTRE MANAGEMENT SCHEMES

There has been a growing interest in the concept of city centre management since the 1990s (see Morphet 1991; Crouch 1992; Page and Hardyman 1996; Reeve 2004). This has been recognised as one of a number of possible responses to city centre change, and has emerged as a component of the wider place promotion and marketing of city centres (see Chapter 6). Central and local government as well as the private sector acknowledge the concept as a valid mechanism in the city centre revitalisation process. Oc and Tiesdell (1998) commented that the initial concerns and justification for city centre managers were primarily focused on the economic viability and physical environment of city centres. Increasingly, however, they have become concerned with social vitality and ambience, and more importantly issues of safety. As Jones (1990) summarised, the concept of city centre management involves a carefully planned and co-ordinated programme of policies and activities to establish, maintain and enhance city centre shopping environments. City centre management involves a team approach to a comprehensive range and combination of activities and improvements, and demands close planning, working and financial co-operation. A strategy typically includes a diversity of projects covering the economic, environmental, social and cultural needs of the city centre (Hollins et al. 1997). City centre management can embrace a

Plate 10.3 Central mega-retail-led regeneration built in the late 2000s
Photograph by author.

number of functions and activities including routine maintenance, the co-ordination of local authority services and of public and private sector initiatives, the creation and management of a safe shopping environment, the integration of retail interests with leisure and tourist facilities, and the promotion of events and publicity within the city centre (Jones 1989). Where a commitment from both the local authority and the retail sector exists, city centre management schemes can provide a focus for the co-ordination and control of a wide range of city centre activities, and for the pursuit of programmes designed to stimulate environmental improvements and to make city centres more attractive and consumer friendly (Jones 1990).

Only a handful of town and city centre managers were in place in 1991 (Jacoby 1996), and by 1995, 125 managers were in post (Hollins et al. 1997). The practice has continued to escalate and few major cities are without their own appointed individual or team, with the total rising to more than 500 by 2008 throughout the UK (see www.atcm.org). As Warnaby et al. (1998) stated, there has been much comment on the rapid growth of city centre managers and much of this has focused on descriptive case studies of specific schemes. Overall, city centre management is seen as part of a new solution to old problems faced by city centres (Morphet 1991), but there are constraints encountered by city centre managers, with lack of funding and their position in middle management being imponderables. City centre management schemes are commercially oriented, and other uses such as residential schemes are only considered of importance when these are related to commercial uses.

Associated with city centre management schemes have been other innovative reactions against out-of-town retailing to aid revitalisation. These are exemplified by localised loyalty schemes such as that employed in Leominster (Hallsworth 1997; 1998; Potts 2007a).

The 'Loyal to Leominster' initiative involved a loyalty card scheme that was launched in 1996. Over 100 retailers participated and around 8000 cards were issued (Hallsworth 1997). Although the Leominster initiative has had some success in retail revitalisation of the centre, other similar schemes that have followed this lead have had mixed fortunes (Hallsworth 1998). Worthington (1998) noted that over 50 town centres in the UK had researched, developed or implemented a town centre loyalty card programme by 1998. Such cards can have a role to play in the revitalisation of an urban centre and they can provide an icon around which city centres can build a programme to emphasise their virtues to potential consumers, although recent evidence suggests that such initiatives are declining in importance.

BUSINESS IMPROVEMENT DISTRICTS (BIDs)

Business Improvement Districts have emerged in England and Wales since 2004, although not in Scotland (Peel and Lloyd 2005) or Northern Ireland, encouraged by central government regeneration policy (see Chapter 5). BIDs originated in Canada in the 1970s and have most commonly been established in the US since the 1980s (Ashworth 2003; Symes and Steel 2003; Brown 2008; Cook 2008). BIDs represent a voluntary tax that local businesses impose on, administer and spend themselves on strategies such as crime prevention measures, improving the built environment, and place marketing and promotion, all aimed at boosting the local economy (Peel and Lloyd 2008). BIDs are usually formed in central retail environments and totalled almost 80 by late 2008, including 17 in London (Grail and Dawkins 2008), with most major cities following the model (see www.ukbids.org).

Examples of BIDs in North America, from where the concept was imported (Ward 2006; Cook 2008), have shown the main benefits to be the emergence of productive public–private partnerships, the development of a safer and more attractive environment, and the successful revitalisation of declining city centre retail economies at little expense to the public sector (Symes and Steel 2003; Steel and Symes 2005). However, the BID model has been the subject of critical scrutiny in the North American context (Steel and Symes 2005; Cook 2008). Criticisms encompass concern about the privatisation of what should properly be public services and public space; the exclusion of certain groups from BID areas; the emphasis on promotion and marketing rather than solving the causes of local economic and social problems; the displacement of crime; the risk of failure; separation of interests between owners and tenants of property; and the promotion of sameness, replication and banality through the commodification of space. Symes and Steel (2003, p. 312) made this point by suggesting that 'BIDs represent a type of Disneyfied urban space aimed only at the affluent' where 'the messy vitality' of cities is 'diluted into the carefully designed and homogenised blandness found in malls and business parks'. BIDs cannot be seen as a panacea for urban regeneration (Symes and Steel 2003; Steel and Symes 2005; Cook 2008). However, the experience so far among business and local authorities in England and Wales has been largely positive (Walburn 2008).

RETAIL-LED CITY CENTRE REGENERATION: CRITICAL ISSUES

A number of tensions, controversies and issues are brought about by pursuing regeneration based on retail-led schemes. Although striking post-modern architecture characteristic of new developments looks impressive, these developments should not be uncritically accepted as positive in nature. The critical contemporary issues brought about can be divided into issues of exclusion (Jackson 1998; MacLeod and Ward 2002), gentrification (Smith 2002) and the privatisation of central city space (Steel and Symes 2005); the involvement of local communities in retail-led regeneration projects (Clement 2007); and the lessons for future economic and physical regeneration of town and city centres.

Exclusion, gentrification and privatisation are to an extent linked to wider processes of globalisation and entrepreneurialism (see Chapters 3 and 6). Both out-of-town and in-centre shopping malls have traditionally been characterised as highly privatised, enclosed spaces subject to high levels of security and surveillance (see Goss 1993; Jackson 1998; Norris and Armstrong 1999). Particular groups in society, such as teenagers, the homeless and other 'undesirable' groups, are disproportionately subject to targeting and ejection from these environments. Such tactics reinforce academic criticisms of the purification and privatisation of what was formerly often part of urban public space. Retail-led regeneration of city centre space has often resulted in a subtle privatisation of public space and the exclusion of those who detract from the consumption experience (Bannister et al. 1998). Goss (1993), Christopherson (1994), Norris and Armstrong (1999), Williams and Johnstone (2000) and Atkinson and Helms (2007), for example, have illustrated the exclusionary nature of new city centre shopping spaces and malls linked to the rise of fear and the fortress city (see also Chapter 11).

Exclusion is also likely to be experienced by retailers, owing to higher rents in new developments. This could lead to the dominance of national and multi-national brands in new retail-led developments, resulting in sameness and serial replication across the UK's city centre retail landscape as reflected in the 'clone towns' discussion (Simms et al. 2005). This research revealed how retail spaces once filled with a thriving mix of independent butchers, newsagents, tobacconists, pubs, bookshops, greengrocers and family-owned general stores are rapidly being replaced by faceless superstore retailers, fast-food chains, mobile phone shops and global fashion outlets (Simms et al. 2005; Plate 10.4). Shopping centres, malls and high streets exhibit the same shops whether in Southampton or Stornoway, and malls have replaced 'once vibrant city centres with sanitised wastelands of jerry-built nowheres' (Arlidge 2006).

Globalisation has been the driving force behind this trend, as the rise of multi-national retailers has been a significant element of the regeneration of the built environment of town and city centres. Buildings and spaces in town and city centres are increasingly carrying the signs of global firms. Most places now have their Benetton store, Gap store, Starbucks coffee shop and McDonald's restaurant. This has also prompted initiatives to try to regenerate a sense of local identity. Economic globalisation has also had profound impacts on the power relations between those who control investment and those who are responsible for looking after the interests of particular places. Towns and cities now find themselves in competition with other places to attract the 'mobile investment' controlled by global firms. This has made it more difficult to turn down applications to build on sensitive sites, or to ignore requests to provide facilities that inward investors are looking for (Short and Kim 1999).

However, the homogenisation of high streets is not benign or inevitable and can be addressed through policy changes. Promoting local shops can enhance diversity, and increase the vitality and stability of local economies. Simms et al. (2005) argued that this might begin to reverse the trend in towns that have already been overtaken by clones (see also Simms et al. 2002; Oram et al. 2003).

A further criticism of retail-led regeneration has revolved around the limited social housing provision, public space and community facilities within many new city centre mixed-use projects (Clement 2007). This reinforces gentrification, and exclusionary trends as explored in Chapter 11. The lack of involvement of local communities has been a frequent criticism of city centre property-led regeneration schemes since the 1980s (see Chapter 8). The extent to which new developments such as Liverpool One or Bristol's Cabot Circus are integrated with surrounding deprived inner city areas, and the extent to which local groups benefit from opportunities presented by new mega-schemes, remains an issue of debate. Local

Plate 10.4 'Making sameness': the cloned urban landscape
Photograph by author.

communities must be involved and integrated with retail-led plans if they are to benefit the wider social and economic regeneration of a locality (Potts 2007a).

Re-centralisation of retail space and capital continues into the late 2000s (Lowe 2005; Guy 2007; Tallon 2008). This has taken place in a climate of continued competition from out-of-town and Internet retailing, a buoyant property market until the late 2000s, policy support for in-centre development, and demand from affluent consumer groups. Within this context, some lessons for future retail-led town and city centre regeneration can be drawn from the experience of recent approaches. The development of these environments relies on a thriving property and consumer market. This means that there are issues associated with potential saturation levels for regional-scale malls in city centres. Additionally, any significant downturn in the property market and consumer spending, such as that being witnessed in the late 2000s, quickly and directly affects the retail sector of the economy. Such a trend would dampen and affect regeneration based on flagship property-led schemes consistent with the experience of the late 1980s and early 1990s.

Place competition, with its roots in the desire to capture the attention of those controlling investment, has also encouraged towns and cities to channel resources into high-profile 'flagship' projects such as concert halls and museums (see Chapter 12). It has encouraged city authorities to create leisure zones that project a high quality of life. It has encouraged places to devote efforts to staging major sporting events and cultural festivals, such as the Olympic Games and the European Capital of Culture.

In the face of significant exurban retail and leisure competition, investment in centrally located retail-led projects is a more recent and continuing trend, and presents an apparent panacea for town and city centres previously considered problematic, depressing and ruined places. Concurrently, superstore investment in deprived neighbourhoods is seen as a way of regenerating inner cities and peripheral estates (Bromley and Thomas 2002a; Potts 2007a). However, concerns over the exclusion of particular social groups and businesses, and over the extent of real and meaningful involvement of local, often deprived communities in the development of new central retail and housing environments, should be given greater precedence.

SUMMARY

Successive 'waves' of retail innovation comprising superstores, retail warehouse parks, regional shopping centres, factory outlet malls, limited line discounters and internet retailing have had negative impacts on traditional city, town, district and neighbourhood shopping centres.

Since the 1980s, town and city centres have reacted to the decline of the traditional high street through a range of measures such as pedestrianisation, development of city centre shopping malls, historic preservation, waterfront redevelopment, office development, transport enhancement, employment of city centre managers, Business Improvement Districts, expansion of the evening and night-time economies, and diversification of activities and special events.

Superstores have been developed in declining central areas and deprived neighbourhoods as a regenerative measure, with both positive and negative effects.

Mega-retail-led regeneration in the form of mixed-use mall developments has re-emerged since the early 2000s across UK city centres.

Critical issues related to retail-led regeneration encompass exclusion of particular social groups, the promotion of gentrification, the privatisation of public space, the development of clone landscapes, the dependence on a buoyant property market, and the lack of involvement of local communities.

STUDY QUESTIONS

1. In what ways has the regeneration of the central business districts of US and UK cities changed their form and function since the 1980s?
2. Examine the positive and negative effects of superstores in regenerating the local economy, society and the environment.
3. Explain how selected streets in the UK's towns and cities have been 'malled' or 'cloned', and examine the implications of this for the continuing heterogeneity of urban public space.
4. What considerations should be taken into account when using retail investment as a way of regenerating town and city centres?

KEY READINGS ON CITY CENTRE RETAIL-LED REGENERATION

Bromley and Thomas (eds.) (1993a) *Retail Change: Contemporary Issues* [covers major changes in the retail landscape from the 1970s until the early 1990s including the organisation of retailing, urban transformation and retailing, planning implications of retail change, and social issues of retail change]

Evans (1997) *Regenerating Town Centres* [examines town centre change and decline, and covers thematic approaches to regenerating town and city centres, including office development, public space enhancement, as well as public policy initiatives]

Guy (2007) *Planning for Retail Development: A Critical View of the British Experience* [documents retail planning policy in the UK from the 1960s, within its broader economic and social context; includes a chapter on urban regeneration and retail policy]

KEY WEBSITES ON CITY CENTRE RETAIL-LED REGENERATION

Association of Town Centre Managers: www.atcm.org [organisation dedicated to helping town and city centre regeneration]

Business Improvement Districts: www.ukbids.org [National BIDs Advisory Service contains latest news on BIDs and case study examples of retail-focused BIDs]

CLG (Communities and Local Government): www.communities.gov.uk [government department with responsibility for urban regeneration in the UK; contains examples and practice of city centre regeneration]

New Economics Foundation: www.neweconomics.org/gen [independent think tank; research 'hot topics' includes 'ghost town Britain']

Online Planning Resources: http://planning.rudi.net [contains reading lists on town and city centre regeneration, office development, town centre management, and Business Improvement Districts]

11

Housing-led regeneration and gentrification

INTRODUCTION

This chapter looks at the use of housing as a driver of wider urban regeneration objectives in the UK. The chapter traces the changing nature and scale of public and private investment in housing. It also highlights the controversies associated with housing-led regeneration such as unsustainability, exclusion and affordability within the housing market. The chapter also deals with the much-debated issue of the gentrification of central cities, and notes some international examples of this process.

The chapter first briefly looks at approaches to housing-led regeneration in the post-war period. It then looks at the trend towards mixed-use development, adaptation and reuse in the urban environment. The chapter next moves on to a discussion of the gentrification debate. It also illustrates the emergence of common interest developments such as master planned or gated communities. The chapter concludes with a discussion of the contemporary concern of housing supply in the UK and the issue of brownfield redevelopment. The focus is very much on the central city as a place that has experienced the most striking transformation of the housing landscape.

URBAN REGENERATION AND HOUSING DEVELOPMENT IN THE CITY CENTRE

In many parts of the world, the loss of population from the city centre has been an issue of concern, and many housing regeneration efforts have focused on city centre space. Although some European city centres have managed to retain a substantial residential population, this has not been the case in most North American cities, and even in parts of the developing world city centre population losses have been noted (Bromley and Jones 1996; Champion 2001; Pacione 2005, p. 322). The population losses have prompted policies to revitalise and repopulate city centres as the areas in the greatest need of urban regeneration. Cities such as Denver and San Diego in the United States have successfully combined retail regeneration with historic conservation and repopulation in the central area.

After decades of disinvestment, concerted attempts to regenerate and revitalise the city centre have been undertaken since the late twentieth century. This has been facilitated through boosting retail and other service functions (see Chapters 10 and 12), but also by encouraging more people to live in the area. Policy has been extremely influential in generating some considerable repopulating of the UK city centre since the mid-1990s (see Nathan and Urwin 2006; Unsworth and Nathan 2006; CLG 2006a). The arrival of new residents constitutes a key feature of the transforming and regenerating city centre and is a major contemporary process (Bromley et al. 2007b). Similar trends and policy effects are likely to be occurring in other parts of the world.

In the UK since the 1980s there has been evidence of a reversal of the trend of population loss, at least in limited areas of the central city. Initially, population growth was restricted to isolated new developments and small areas of converted warehousing in cities such as London, Manchester and Liverpool (Oc and Tiesdell 1997c). Central accommodation became fashionable, particularly with young professionals, because of its proximity to central office areas, and arts and cultural facilities, so avoiding lengthy commuting trips (Evans 1997). In cities such as Manchester overall population growth in the city centre increased from 200 in 1993 to over 6000 by 2000 (DETR 2000b) and to over 15,000 in 2003 (Barber 2007), and is forecast to reach 25,000 by 2010 (Garrett 2008). By the second half of the 1990s the pattern of city centre population growth had spread to many regional cities in the UK. Evidence in the literature has revealed significant trends towards centre city living since the mid-1990s as a visible sign of urban renaissance. Between 1991 and 2001, city centre populations increased by 100 per cent in Dundee to 2900, around 40 per cent in Liverpool to 13,500 and by almost 300 per cent in Manchester to 10,000 (Nathan and Urwin 2006, p. 1). Over the same period, Bromley et al. (2007b, p. 145) reported increases in city centre residents of 66 per cent in Bristol (to 8642 residents), 60 per cent in Cardiff (11,070), 13 per cent in Swansea (1274) and 8 per cent in Birmingham (14,867) (see also Barber 2007). Leeds, Sheffield, Nottingham, Newcastle and Glasgow have all experienced similar, rapid growth (Nathan and Urwin 2006; Unsworth and Nathan 2006; Garrett 2008).

Until the 2000s, there was little empirical research exploring the repopulating of the city centre due to the extent and novelty of city centre policy initiatives and residential changes (Tallon 2002). Much research on residential change has focused on the unique example of London (Butler 1997; 2003; 2007a; 2007b; Lees 2000; Butler and Robson 2001; 2003; Butler with Robson 2003; Hamnett 2003a; 2003b; Davidson and Lees 2005; Hamnett and Whitelegg 2007). However, an emerging research base has examined the process in other UK cities such as Birmingham (Porter and Barber 2006; Barber 2007; Bromley et al. 2007b), Bristol (Lambert and Boddy 2002; Tallon and Bromley 2004; Bromley et al. 2005; 2007b; Boddy 2007), Cardiff (Senior et al. 2004; 2006; Bromley et al. 2007b; Hooper and Punter 2007; Punter 2007), Dundee (Nathan and Urwin 2006), Glasgow (Seo 2002), Leeds (Dutton 2003; 2005; Unsworth and Nathan 2006), Liverpool (Tang and Batey 1996; Couch 1999; Nathan and Urwin 2006; Unsworth and Nathan 2006; Allen 2008), Manchester (Couch 1999; Seo 2002; Nathan and Urwin 2006; Unsworth and Nathan 2006; Young et al. 2006; Allen 2007), NewcastleGateshead (Cameron 1992; 2003; 2006; Townshend 2006), and Swansea (Tallon 2003; Tallon and Bromley 2004; Bromley et al. 2005; 2007b).

The growth in city centre population has been encouraged by policies that promote city centre living, in a context of rising demand for small urban dwellings associated with changes in household structure and significant lifestyle shifts which favour central city living (Couch 1999; Hamnett 2000; Tallon and Bromley 2004; Bromley et al. 2005; 2007b). The process has been facilitated by a ready supply of vacant city centre sites and premises for housing, for which more profitable uses did not exist. A buoyant housing and wider property market since the mid-1990s has made a significant contribution to the trend. Policy statements from central government throughout the 1990s and 2000s (DoE 1995a; 1995b; DETR 2000a; 2000b; 2000c; CLG 2006a; 2007a; see Table 11.1) demonstrate the emerging and growing emphasis on housing development and repopulation of the city centre.

TABLE 11.1 CENTRAL GOVERNMENT POLICY ENCOURAGING CENTRAL CITY HOUSING DEVELOPMENT SINCE THE MID-1990s

DoE (1994a) *Planning Policy Guidance 13: Transport*
DoE (1994b) *Planning Out Crime*
DoE (1995a) *Projection of Households in England to 2016*
DoE (1995b) *Our Future Homes: Opportunity, Choice and Responsibility – The Government's Housing Policies for England and Wales*
DoE (1996a) *Planning Policy Guidance 6: Town Centres and Retail Development*
DoE (1996b) *Household Growth: Where Shall We Live?*
DoE (1997) *Planning Policy Guidance 1: General Policy and Principles*
DETR (1998a) *A New Deal for Transport: Better for Everyone*
DETR (2000a) *Our Towns and Cities: The Future – Delivering an Urban Renaissance*
DETR (2000b) *The State of English Cities*
DETR (2000c) *Planning Policy Guidance 3: Housing*
DETR (2000d) *Living in Urban England: Attitudes and Aspirations*
DETR (2001) *Planning Policy Guidance 13: Transport*
ODPM (2003a) *Sustainable Communities: Building for the Future*
ODPM (2005b) *Planning Policy Statement 1: Delivering Sustainable Development*
ODPM (2005c) *Planning Policy Statement 6 (PPS6): Planning for Town Centres*
ODPM (2005f) *Housing Land Availability Assessments: Identifying Appropriate Land for Housing Development – Draft Practice Guidance*
ODPM (2006a) *State of the English Cities*
CLG (2006a) *Planning Policy Statement 3 (PPS3): Housing*
CLG (2007a) *Homes for the Future: More Affordable, More Sustainable*

The vitality and viability of city centres, which had suffered from the earlier decentralisation of retail and office facilities were seen as enhanced by an enlarged residential population (URBED 1994). The DoE (1996a) report and subsequent statement (ODPM 2005c) advocated a mixture of small businesses, houses and offices in or near to city centres and the occupation of flats above shops. New residents, as well as workers, were seen as stimulating retail activity, restaurants and cafés and thus contributing to the vitality of the centre, and improving safety by introducing more eyes on the street (see Newman 1972; Oc and Tiesdell 1997c). This new residential policy for the city centre came at a time of rising housing need resulting from increasing family fragmentation and declining household size (Oc and Tiesdell 1997c), and coincided with the debates about where the additional houses might be provided. An additional 4.4 million households were expected to be formed between 1991 and 2016, and it was hoped that many might be accommodated in existing urban areas, which already possessed buildings and land suitable for conversion (DoE 1995a; Allinson 1999). Better use of these urban resources would not only revitalise city centres but would also reduce the pressure for new housing on greenfield sites.

The most recent forecast suggests the need for 3 million new homes between 2007 and 2020 equating to around 240,000 per year (CLG 2007a). The continued increase in the number of households has been attributed to increasing immigration (particularly since the mid-2000s); an ageing population (in 2001 the number of people aged over 60 exceeded the number of children aged under 16 for the first time); more young people leaving home at earlier ages (in 2006, 42 per cent of men and 61 per cent of women aged 20 to 24 did not live with their parents) and a rise in university students (to around 40 per cent in the late 2000s); and a rise in divorce and separation (to over 40 per cent).

ADAPTATION AND REUSE OF BUILDINGS

Adaptation and reuse of derelict buildings have been part of the drive to mix uses and repopulate the central city. Within this agenda, it has been recognised that one of the major opportunities for more residential units in the city centre is through the conversion of vacant space above shops. Every city centre has a significant amount of vacant or under-used space above shops and the potential additional housing available nationally from unused space above shops was estimated at between 0.5 million and 1.5 million units which equates to at least half a percent of the population of any given town (LOTS 1998). Flats and upper floors have become empty as retailers now tend to live away from their premises and modern retail outlets are rarely used for storage since retailers operate on a 'just in time' system relying on regional distribution centres. The tens of thousands of empty flats above shops, if occupied, would bring benefits both to the retailers and the city centre community. Many city centre office blocks, which were the result of post-war urban planning, are becoming redundant and outmoded and many of these could be converted to residential use, which is seen as a major factor in urban renewal. In addition to increasing the resident population of city centres and the greater vitality that results, further benefits accrue from the development of this space into flats such as physical improvements to buildings externally and the increased surveillance of retail properties out of normal retail hours. Although Planning Policy Guidance (DoE 1996a; ODPM 2005c) gives general support for such conversions, local authorities seem unable to secure the implementation of these policies and the development of these schemes (Oc and Tiesdell 1997c).

In the UK, a commitment to utilising space above shops was formalised with the setting up of the Living Over The Shop (LOTS) Project in 1989, which ran until the late 2000s. As Petherick and Fraser (1992) outlined, the central feature of the approach was the involvement of a housing association and the use of a commercial lease. Until 1988 there was no simple solution to the problem of the long-standing reluctance of commercial property owners to grant residential leases due to a combination of genuine fears and other misunderstandings. Around 80 per cent of all retail property is controlled by national companies (LOTS 1998), and therefore decisions about the use of these properties must be made at head office rather than in the city centre concerned. LOTS (1998) found that national companies are willing to co-operate in the initiative once familiar with its workings. Funding programmes were available in the early 1990s from the Housing Corporation and a government-sponsored LOTS scheme, which gave more than £45 million of public funding from the DoE (Beck and Willis 1995, p. 214). The Flats Over Shops pilot scheme ran from 1992 to 1995, but the results of this were disappointing due in part to a lack of co-ordination and poor targeting.

Adaptive reuse of offices for residential use in city centres also offers part of the solution to increase the number of homes in city centres as a tool of urban regeneration and sustainability policies, despite a number of problems that must be overcome (Barlow and Gann 1993; 1994; Coupland 1997; Heath 2001). In London, for example, conversions of obsolete post-war office buildings to residential use have registered a significant impact in terms of new homes created (see Hamnett and Whitelegg 2007). Although the challenge of encouraging the conversion of offices is far greater in other cities, a housing-led revitalisation strategy that works in one city may not be appropriate in another and it is therefore essential that strategies be customised to the specific context (Heath 2001). In a critique of mixed-use development, Rowley (1996) argued that mixed-use development is not an automatic panacea and there are obstacles to promoting and maintaining more integrated environments, in particular regarding perceived links between the physical environment and socio-economic processes.

Barlow and Gann (1994) argued that the conversion of redundant office buildings into flats can help create new mixed-use 'urban villages'. As Oc and Tiesdell (1997c) suggested, city centre living should generally be encouraged on an area basis with measures to improve the locality planned on a comprehensive scale rather than as a piecemeal initiative. The conversion of individual buildings or construction of individual residential developments may make little contribution to the vitality and safety of an area. Oc and Tiesdell (1997c) advocated that parts of the city centre can be promoted and developed as residential quarters with positive support from planning authorities in facilitating, developing and building confidence in the city centre as a potential residential area.

The Urban Villages Group (1992) and the Urban Task Force (1999; 2005) have promoted concepts of 'urban villages' and 'urban quarters'. However, as Oc and Tiesdell (1997c) pointed out, creating a vibrant 'urban village' in an area without a tradition of residential population needs a critical mass of development and therefore inevitably takes a number of years to achieve. Since the mid-1990s, new residential quarters have emerged within many UK city centres. Such developments tend to be open, integrated developments with an element of segregation, and many have arisen for conservation reasons as a way of using old buildings. Tiesdell (1995) quoted the example of the functional diversification of the Nottingham Lace Market, which was developed as a mixed function urban quarter. The area demonstrated the potential for housing development incorporating an 'urban village' quarter amongst alternative viable uses evoking the notion of a European style of urban living. Tiesdell (1995) added that the centres of regenerated and 'liveable' cities of Europe might hold important lessons for the historic centres of UK cities (see Tiesdell et al. 1996).

Linked to the promotion of mixed-use areas in cities has been the strategy to create mixed-tenure housing environments (Bailey et al. 2006; Roberts 2007). Social mixing and the mixing of tenures within housing schemes are seen as reducing segregation and as more sustainable, and have been promoted by central government since the early 2000s (see Chapter 9; DETR 2000a; ODPM 2003a; CLG 2006a; 2007a). Mixing in this sense relates to a mix of demographic characteristics such as income, age, family structure and ethnicity, set within wider mixed-use zones of housing, retail, office, leisure and entertainment uses. Social mixing has been criticised for promoting gentrification of communities and the approach requires critical attention relating to its role in an inclusive urban renaissance (see Lees 2008).

THE GENTRIFICATION DEBATE

In theoretical terms, much of the research on residential change and related urban regeneration has focused on gentrification. City centre regeneration has been linked specifically with a new, post-recession gentrification (Lees 2000; 2003a; 2003b; Hackworth and Smith 2001; Hackworth 2002; Boddy 2007). The urban renaissance envisaged by the Urban Task Force (1999), which fed into the Urban White Paper (DETR 2000a), is viewed by some researchers as remarkably similar to visions of gentrification (Wyly and Hammel 1999; Lees 2000; 2003a; 2003b; Smith, N. 2002; Lees et al. 2008). Through its policy initiatives, the government can be seen as not only promoting residential development, but as effectively promoting gentrification for the regeneration of the city centre. Cameron (2003, p. 2373) termed this 'positive gentrification' or 'gentrification as a positive public policy tool'. Hall (2006, p. 144) cautioned that gentrification is by no means caused exclusively by urban regeneration, but it has served to exacerbate the process.

However, it appears that the recent period of housing-led urban regeneration is closely associated with a gentrification process. Gentrification represents a transformation in central urban characteristics and spatial patterns. Gentrification 'is the process by which poor and working class neighbourhoods in the inner city are refurbished by an influx of private capital and middle class homebuyers' (Smith, N. 1996, p. 2). Within the concentric rings, spatial model of the city, gentrification has traditionally affected the zone of transition and stagnation that was typically the inner and cheaper suburban zone (see Hall 2006, Chapter 2). Gentrification in its traditional form involves the reuse of the existing building stock which usually focuses on Victorian and Georgian housing in the UK, or even warehousing. Affluent groups displace lower-income residents, which results in a gain for landlords and a problem for tenants. Neighbourhoods subject to the gentrification process experience a change in their social, economic and environmental character (see Pacione 2005, pp. 212–218; Plate 11.1). It can be described as a 'process of social upgrading and/or social displacement' involving 'the movement of people to realize their consumption or lifestyle choices' (Phillips 2002, pp. 282–283), and as a process involving movement of capital into 'rent-gaps', the refurbishment of property, and the development of retail, financial and consumption spaces.

Gentrification is a multi-faceted and heterogeneous process, subject to a range of different emphases in its definition, and it has been described as a 'chaotic concept' (Beauregard 1986, p. 40). The process has been fuelled by a variety of much debated explanations (see, for example, Hamnett 1984; 1991; 2000; Ley 1986; 1996; Lees 1996; 2000; Smith, N. 1996; 2000; Lees et al. 2008). The process embraces physical, economic, social and cultural transformations (Hamnett 1984; Warde 1991; see Table 11.2).

Plate 11.1 'Traditional' gentrification, Southville, Bristol
Photograph by author.

TABLE 11.2 CHARACTERISTICS OF GENTRIFICATION

Physical transformation
A transformation of the built and unbuilt environment, via revitalisation activities, that exhibits multiple 'pockets' of distinctive, aesthetic features and the emergence of new types of local service provision.

Economic transformation
An economic reordering of renovated and unrenovated property values, a commercial opportunity for the conversion industry, and generally, an extension of the property system of the private ownership of domestic property.

Social transformation
A process of resettlement and social concentration, a process involving the displacement, replacement and/or marginalisation of a variety of indigenous residents (with diverse characteristics) by invading 'outsiders' (also with diverse characteristics), entailing territorial patterns of social clustering.

Cultural transformation
The emergence of a mosaic of distinctive enclaves of individuals, who have a putatively shared cultural lifestyle and consumer preference.

***Source:* Warde (1991, p. 225).**

The term 'gentrification' was coined by Glass (1964), who studied the movement of middle-class households into working-class neighbourhoods in London. Since the 1960s the process of gentrification has evolved to the extent that three distinct waves have been recognised by Hackworth and Smith (2001), largely based on the process in the US (see Lees et al. 2008, Chapter 5).

According to Lees et al. (2008, p. 174), a first wave of 'sporadic gentrification' dated prior to 1973 when the process was confined to small neighbourhoods in the north-eastern US and Western Europe. A transition period during the early 1970s witnessed 'gentrifiers buying property' in New York and other cities; here developers and investors used the downturn in property values to consume large sections of devalorised neighbourhoods, which set the scene for 1980s second-wave gentrification.

The second wave experienced 'the anchoring of gentrification' within which the process became implanted in previously disinvested central city neighbourhoods. In contrast to the first wave, the process became increasingly important in smaller, non-global cities. Second-wave gentrification was associated with the return of investment to the central city and intensified during the 1980s with major redevelopments, funded by public–private partnerships, including the transformation of former docklands into residential enclaves (Robinson 1987; Short 1989; Smith, A. 1989; Robinson and Williams 1990; Plate 11.2). A transitional period during the late 1980s and early 1990s saw a 'slowing of gentrification' coinciding with economic recession and the restriction of the flow of capital into gentrifying and gentrified neighbourhoods, arguably representing a 'de-gentrification' process (Lees and Bondi 1995).

A third wave can be traced from the post-recession era since the mid-1990s, which has witnessed 'the return of gentrification' in which the role of large-scale capital has appeared greater than ever as large developers redeveloped whole urban neighbourhoods, often with state intervention and support (Wyly and Hammel 1999; Hackworth and Smith 2001; Hackworth 2002; Lees et al. 2008). It has been characterised as a 'highly integrated conquest of urban space' (Smith, N. 2002, p. 9). In addition to residential rehabilitation and refurbishment, gentrification is now considered to involve urban reinvestment in commercial developments such as shopping centres, cultural complexes, restaurants and new

Plate 11.2 1980s Maritime Quarter regeneration in Swansea
Photograph by author.

offices (see Smith, N. 2000; 2002; Slater 2006). The central city, in particular the inner city, remains the classic place of gentrification.

Lees et al. (2008, pp. 173–187) contended that a fourth wave of gentrification in the US has emerged since the early 2000s, with 'intensified financialization of housing combined with the consolidation of pro-gentrification politics and polarized urban policies' (Lees et al. 2008, p. 179). Further encouragement of the process by central government and local authorities has been exhibited in the US and the UK. The US and the UK appear to be entering another transitional period, with emerging evidence suggesting a slowing of gentrification linked to a decline in the housing market. This has resulted from the global effects of the 'credit crunch' that originated in the US from late 2007, which forced developers to retrench in a similar pattern to the late 1980s and early 1990s (Gardiner 2007; Garrett 2008; Hamnett 2008). A continuing global economic downturn is serving to restrict the gentrification process, and there is evidence that developers are not building schemes that have planning permission, which is clearly having an impact on urban regeneration initiatives.

Gentrification is now recognised to be a global urban strategy (Smith, N. 2002). Research has investigated the process in UK cities (for example, Butler and Robson 2001; 2003; Butler 2003; Butler with Robson 2003; Cameron 2003; Dutton 2003; Butler 2007b; Hamnett and Whitelegg 2007), North America (for example, Filion 1991; Carpenter and Lees 1995; Hackworth and Smith 2001; Hackworth 2002; Hamnett 2003b; Lees 2003c; Slater 2002; 2004), Australia (for example, Badcock 1991; 1995; 2001), and other countries in Western Europe (for example, Van Kempen and Van Weesep 1994; Uitermark et al. 2007).

A number of wider processes have facilitated and driven the gentrification process globally. First, changing demographics have resulted in more single-person households, linked to the need for more housing. Second, house prices in the suburbs have become increasing expensive. Third, travel time to the suburbs has become greater. Fourth, life in the suburbs has been seen as increasingly dull and repressive (see Ley 1996). Fifth, there is a growing section of the population who do not wish to live conventional lifestyles in suburban three-bedroom semi-detached houses (see Florida 2002; 2005).

By contrast, the city centre can offer cheaper space, faster access to leisure, entertainment and cultural facilities, a better lifestyle, and a shorter commute for growing sections of the population, particularly in the post-modern city. The causes of the gentrification process are linked both to demand and supply, and it can be facilitated by public policy (see Table 11.3). On the demand side, de-industrialisation resulted in a large stock of Victorian housing being located close to where docks, warehouses, mills and factories had closed. This left underused housing, which became relatively cheap to buy or rent in the zone of transition between the city centre and the suburban fringe as depicted by Burgess's model of the city (see Chapter 1). The growth of the service economy comprising the tertiary and quaternary sectors has led to a demand from an increasing proportion of workers in the new economy. A growing city centre close to the cheaper space in the zone of transition was increasingly seen as an ideal location for urban professionals and graduates working at close proximity.

Demographic and lifestyle changes have contributed to the demand-side causes of gentrification as graduates seek to delay or avoid marriage, and aspire to a single lifestyle, city centre living, and living in neighbourhoods with 'character' by contrast to the perceived sterility and suppressive nature of the suburbs (Ley 1996; Florida 2002; Tallon and Bromley 2004). This group also wishes to be in proximity to cultural and entertainment offerings in city centre environments. A growing section of the population demonstrates the characteristics of being suitable for city centre living. For example, 15 per cent of the UK population lived alone in 1961 and this figure had grown to 30 per cent by 2001. Nathan and Urwin (2006) found that city centre residents were mostly young, single and not expecting to live in the area for many years. Consistent with the findings of Tallon and Bromley (2004),

TABLE 11.3 SUMMARISING THE GENTRIFICATION DEBATE

The 'rent gap' theory

- Production-supply-driven; profit motive of capitalism; concentrates on producers: developers, landlords, mortgage lenders, etc.
- Logical aspect of industrial society; capital flows where rate of return is highest.

Mechanism

- Suburbanisation/decentralisation → fall in land values in city centre.
- Opportunity for profit in city centre: buy up properties, renovate, redevelop.

→ Gentrification: producers of housing drive the process when the rent gap is large enough for them to secure satisfactory profit.

Consumer choice

- Consumption-demand-driven; cultural expression.
- Post-industrial/post-modern.

Mechanism

- Growth of the 'service class'.
- State actively involved.
- Assertion of identity through consumption by service class.

→ Gentrification: service class reproduction through residential choice.

Source: **Based on Smith, N. (1996; 2000); Ley (1996); Hamnett (2000); Lees et al. (2008).**

students and young people were well represented, and attracted by convenience and buzz. It has also been argued by Bromley et al. (2005) and Nathan and Urwin (2005) that city centre living is not appropriate for families who prefer the space and better services in the suburbs (Karsten 2003; Silverman et al. 2005). Demand-side explanations are consistent with the 'emancipatory city' thesis, which argues that gentrification is a liberating experience and a process that unites people in the city centre, and creates opportunities for social interaction, tolerance and cultural diversity (see Caulfield 1989; Lees 2000; 2004; Lees et al. 2008, pp. 209–212; Florida 2002; 2005).

Gentrification has also been driven by supply-side motivations, as advocated by Neil Smith's rent gap theory (Smith, N. 1996; 2000). This asserts that gentrification occurs when the gap between what current rents are and what they could be grows sufficiently to entice developers to move into a neighbourhood and buy up housing, improve the streetscape, convert former industrial buildings, and market and re-brand the neighbourhood. The rent gap covers the costs and risks borne by the developer. On the supply side, Smith, N. (1996) argued that the process in the US was synonymous with the 'revanchist city' thesis, whereby the middle classes sought revenge against low-income groups who had 'stolen' the city from them (see also Lees et al. 2008, pp. 222–234).

Along with supply and demand drivers, it has been argued that the repopulation and associated gentrification of city centres since the mid-1990s have been a policy-led process (Lees 2003a; 2003b; Tallon and Bromley 2004; Boddy 2007; Bromley et al. 2007b). National and local policy has specifically encouraged the repopulation of the city centre exemplified by urban renaissance policies, brownfield redevelopment and mixed-use development.

The winners in the gentrification process are initial owners of property, 'first wave' or pioneer gentrifiers who buy in before prices rise, local government, which gains greater tax revenues, developers who derive profits, speculators who make money from holding on to property as the market improves, and citizens who benefit from a regenerated area (Pacione 2005, pp. 215–216; Herbert 2000). Losers in the process are renters of property who are displaced by the upgrading process, the indigenous communities that are broken up, and those who stay in gentrifying areas and experience increases in rent, food prices and local services (see Table 11.4). Gentrification is generally recognised to have an array of both positive and negative effects (Atkinson 2004b; Atkinson and Bridge 2005, p. 5; Lees et al. 2008, Chapter 6; see Table 11.5). However, Slater (2006, p. 752) argued that 'gentrification is not . . . the saviour of our cities', and that 'the only positive to gentrification is being able to find a good cup of coffee'. Research often reveals contradictory messages in terms of the valuable and destructive dimensions of the process (Atkinson 2004b; Table 11.5).

Gentrification has risks associated with it, and the reverse process, de-gentrification, can be initiated by a slump in the service sector (Lees et al. 2008, p. 174). The significant increase in buy-to-let in the 2000s has resulted in an inflated demand in the short term from speculative investors; when prices start to fall the market is likely to witness a surplus of city centre

TABLE 11.4 WINNERS AND LOSERS IN THE GENTRIFICATION PROCESS

Winners	*Losers*
Local government	Displaced households and businesses
Initial owners of property	Less well-off
Incoming owner-occupiers	
Developers	
Speculators	

TABLE 11.5 POSITIVE AND NEGATIVE EFFECTS OF GENTRIFICATION

Positive	*Negative*
Stabilisation of declining areas	Community resentment and conflict
Increased property values	Loss of affordable housing
Reduced vacancy rates	Unsustainable speculative property price increases
Increased local fiscal revenues	Homelessness
Encouragement and increased viability of further development	Greater draw on local spending through lobbying by middle-class groups
Reduction of suburban sprawl	Commercial/industrial displacement
Increased social mix	Increased cost and changes to local services
Decreased crime	Loss of social diversity (from socially disparate to affluent ghettos)
Rehabilitation of property both with and without state sponsorship	Increased crime
	Under-occupancy and population loss to gentrified areas
	Displacement through rent/price increases
	Displacement and housing demand pressures on surrounding poor areas
	Secondary psychological costs of displacement

Source: **Atkinson (2004b, p. 112); see also Lees et al. (2008, p. 196).**

apartments on the market (Norwood 2005). Gardiner (2007) reported on the rise of buy-to-let associated with new city living apartments emerging in UK city centres, offering the prospect of home ownership for young professionals and first-time buyers. However, a high proportion has been mothballed by distant property investors and stand empty; in Leeds city centre around 20 per cent of 5653 flats were empty. The practice of 'buy-to-leave' involves property investors remortgaging or using bonuses to purchase new apartments to watch their value appreciate, leaving them lying empty until sold on for a significant profit. Gardiner (2007, p. 19) called this 'the ugly flip-side of city living: the dirty underbelly of the much-hyped urban renaissance'.

However, by mid-2007 it had become apparent that the market might have seen the development of flats reaching saturation point. Flats have been at the vanguard of contemporary gentrification. In 1997, 16 per cent of all new build was in the form of flats, and this had increased to 23 per cent by 2002 and 47 per cent by 2007 (www.communities.gov.uk, 2008). Garrett (2008) provided evidence that new city centre apartment schemes in cities such as Manchester, Liverpool and Leeds were being mothballed because of uncertain market conditions and oversupply of city centre apartments aimed at investors. For example, in Leeds city centre, 7184 apartments were completed between 1998 and 2008, with a further 3640 under construction and 7216 having received planning consent. Of the apartments sold, around 75 to 80 per cent have been sold to investors (Garrett 2008). Saturation, inflated prices, unrealistic rental expectations and rising interest rates have conspired to turn 'buy-to-let' into 'buy-to-lose' (Hamnett 2008). The nature of recent city centre housing development has made it extremely susceptible to a serious downturn in the property market.

Other risks have been borne by first-wave gentrifiers, who take a chance, as there are no guarantees for incomers that the area will experience an upward trend in prices. The revanchist city thesis (Smith, N. 1996) argued that active hostility to incomers can be provoked. Gentrified areas can be poorly provided with police presence, green space and public facilities such as libraries.

CITY CENTRE HOUSING REGENERATION AND GENTRIFICATION

City centre regeneration and repopulation are strongly linked with gentrification, and the process is the single most important theory of residential change relating to the city centre (Tallon and Bromley 2004; Bromley et al. 2005; 2007b). Therefore, recent research has explored whether the characteristics of the new residents suggest that this process is occurring (Bromley et al. 2007b). In many UK city centres, the process of gentrification is strikingly evident, and research has supported the notion that city centre regeneration involving a substantial repopulating is associated with such a trend. Empirical work by Bromley et al. (2007b) and Nathan and Urwin (2006), for example, showed that gentrifier types are moving to live in city centres.

Although the gentrification literature does not offer a consistent message on the characteristics of the gentrifier, and some have argued caution in the use of the term because of diversity (Bondi 1999, p. 277), the emphasis on young adults and professional managerial occupations is a recurrent theme. Although Smith, N. (1996, p. 104) cautioned against strongly associating yuppies and gentrification, this is a frequent assumption, and holding a professional managerial occupation was one of Ley's (1996) key criterion for a gentrifier. Using evidence from Tyneside, Cameron (1992, p. 7) showed a far higher proportion of young single adults in city centre housing and emphasised that new housing brings a new population that is 'affluent, skilled, dynamic' to inner city areas (Cameron 1992, p. 11). Mills (1988) pointed to the presence of many young professionals in the gentrifying areas of Vancouver. Gentrified households are seen as consisting of young professional partners or single person households (Bridge 2003), and the importance of the dual career household is often stressed (Bondi 1991; Hamnett 2000). Some gentrification researchers have associated the process with an increasingly professionalised workforce and with expanding financial and business service sectors (Hamnett 2000; Dutton 2003; Hamnett and Whitelegg 2007).

Evidence from Bristol city centre, a location where the financial and business service sectors have experienced growth (see Boddy 2003b; Boddy et al. 2004), revealed that 62 per cent of new residents were classified as professional or semi-professional compared with 48 per cent of the longer-term residents, emphasising the upward social trend of the area (Bromley et al. 2007b). However, the proportions evident in provincial cities do not match the 'solidly professional middle class' character of gentrifying areas of North London (Butler 2003; Butler and Robson 2003; Butler 2007b).

Work on Fairview Slopes in Vancouver (Mills 1988) emphasised how those living in gentrifying areas valued proximity to work and an array of leisure pursuits. In terms of the more lively social life of new city centre residents, much has been written about the cultural habits of gentrifiers, their urban ethos and the desire for a social capital-rich environment. Butler and Robson's (2003) study of gentrification in North London found that residents were motivated by such aspects as minimising the journey time to work and by gaining the opportunity to sample the cultural assets of the cosmopolitan city of London. Gentrifiers also value flexibility rather than commitment (Mills 1988, p. 181), and the short-term predictions of the new residents for their continued stay in the UK city centres of Bristol and Swansea match this profile (Tallon and Bromley 2004; Bromley et al. 2007b).

Allen (2007) cautioned against characterising city centre dwellers as young, single professionals as if this were a more-or-less undifferentiated social group. Instead it is argued that new city centre populations are a complex and differentiated group that occupy a number of distinct positions within the city centre housing market. Allen (2007) put forward three typologies of city centre residents based on the Manchester experience. First, 'counter-culturalists' tend to originate from within the 'new' middle class and were the first main group to move back into the city centre. They had an authentic interest in the city centre and tended to buy converted warehouse properties for reasons of cultural distinction. Second, 'city-centre tourists' tend to originate from the service class and have a passing rather than authentic interest in the city centre. They are attracted to the city centre to sample the vibrant life and tend to rent bland apartment developments before returning to suburban living. Third, 'successful agers' tend to be over the age of 50, and have an authentic interest in the city centre. They use their high-volume economic capital to facilitate their move into the city centre and immerse themselves in its cultural scene. This classification would appear to resonate with other major UK city centres such as Bristol and Cardiff.

The availability of particular types of housing is a key factor associated with the characteristics of gentrifiers. Recent housing change and regeneration in UK city centres usually fall into the two main categories of new building and conversions in non-residential areas typically on the edge of the commercial core or on nearby, previously industrial, riverside or dockside land; and new infill in existing residential areas. However, a further and much less important category of housing change in the city centre is the major upgrading of existing residential property. This is associated with classic gentrification, involving the restoration of urban property in working-class neighbourhoods by the upper and middle classes, resulting in the displacement of the former inhabitants (Hamnett 1984; Smith and Williams 1986). This conventional gentrification process has been a slight feature in the contemporary regeneration of many UK city centres (Bromley et al. 2007b).

The majority of city centre housing development in provincial cities belongs to the main categories of new buildings on non-residential land, the conversion of industrial or commercial buildings into residential use, and residential infill (Plate 11.3). Many cities experienced new residential development in their central areas during the 1970s and 1980s, before the housing market collapse of the late 1980s and early 1990s. Development took off again in the second half of the 1990s, following the policy initiatives focused on the city centre and in the context of a more favourable economic climate. Bourne (1993), Cameron (1992), Van Wessep (1994), and more recently Lambert and Boddy (2002), Tallon and Bromley (2004), Bromley et al. (2005) and Boddy (2007) have argued that new housing development on mainly derelict industrial sites does not constitute gentrification, because the key element of displacement is not involved. It is perhaps more appropriately termed 'residentialisation' whereby housing replaces other land uses (Tallon and Bromley 2004; Bromley et al. 2005). This process can be seen as a 'response to the new spaces and opportunities created by de-industrialisation, decentralisation and suburbanisation' (Bromley et al. 2005, p. 2423).

However, the majority view among gentrification researchers seems otherwise, as forcibly argued by Davidson and Lees (2005), Slater (2006), Butler (2007a) and Lees et al. (2008). The 'conditions of systematic disinvestment in inner-city locations' are fundamental to the gentrification process (Smith, N. 1996; Slater 2006, p. 745). There might also be overspill effects on nearby residential areas (Lees 2003a). New development infill has been seen as part of the gentrification process (Smith, N. 1996, p. 115), as is large-scale new residential building on former industrial or commercial land (Slater 2006). Punter (1992) suggested that Bristol's dockland redevelopment of the late 1980s and early 1990s led to large-scale gentrification, even though this process did not involve direct displacement. Other studies of waterfront redevelopment have followed a similar argument (for example, Hoyle et al. 1988; Tang and Batey 1996). New developments on vacant land are a key part of the current wave of post-recession gentrification, which may involve ambitious planning and mega-regeneration projects (see Smith, N. 2002; Lees 2003a; Davidson and Lees 2005; Slater 2006).

Evidence on the character of new city centre housing development in the UK helps to explain why new city centre residents are showing characteristics synonymous with those of

Plate 11.3 City centre regeneration and new-build gentrification on a brownfield site
Photograph by author.

gentrifiers. If the new developments are directed at higher status residents, akin to a new middle class (Bridge 1995; Hamnett 2000; Hoskins and Tallon 2004; Boddy 2007), then the argument for the occurrence of gentrification is clearly strengthened. In Bristol city centre, the construction of many exclusive, high-status developments is a feature of the city centre, with high quality and expensive conversions and new build reflecting a vibrant property market since the mid-1990s (Boddy 2007).

Lambert and Boddy (2002) and Boddy (2007) explained that Bristol has a number of characteristics that support the targeting of new housing development at the affluent consumer (Plate 11.4). This is consistent with evidence from across the UK's city centres (Nathan and Urwin 2006). Developers can be seen as creating a 'new product' in the form of city centre apartments with marketing that concentrates on a fairly narrow section of people who tend to be young, professional, affluent and childless. Marketing aims to construct the image of an 'urban idyll' (MacLeod and Ward 2002; Hoskins and Tallon 2004; Colomb 2007). The urban idyll represents the idea of a vibrant city neighbourhood that has emerged as central to the commodification of urban space, which is implicit in New Labour's urban renaissance agenda (see Chapter 5; Colomb 2007).

GENTRIFICATION AND THE 'URBAN IDYLL'

Many of the selling and marketing practices promoting city centre living mobilise notions of an idyllic city environment to attract middle-class gentrifiers. Many people moving to the contemporary city centre may feel that they are moving or escaping to an urban culture

Plate 11.4 Developing 'new product': extreme high-end city centre housing
Photograph by author.

(Urban Task Force 1999; Holloway and Hubbard 2001) contained within some kind of utopian 'urban idyll' (Hoskins and Tallon 2004). This operates in a similar way to the more academically recognised and scrutinised mental construct of the traditional 'rural idyll' of the UK countryside (for example, Mingay 1989; Short 1991; Bunce 1994; Boyle and Halfacree 1998). The pro-rural myth of the rural idyll was constructed in opposition to the urban and suburban.

The experience of urban places can often be one of rapid change and as a result there has arisen a very powerful set of social myths concerning the stability of the countryside in relation to the dynamism of the city. The idea of the rural idyll is an important consequence of the way the contemporary countryside is represented (Mingay 1989). Some elements of this idyll include an impression of timelessness; an emphasis on traditional 'family' and community values; harmonious relations between 'nature' and 'culture'; an absence of social problems; and the fostering of good physical, spiritual and moral health (Holloway and Hubbard 2001). The contemporary inner city, in contrast, has usually been seen through the lens of anti-urban mythologies where 'the inner city is the dark underside of the city, a place of crime and disorder' (Short 1991, p. 50). Such views are created and sustained by the suburban middle classes. In sum, the countryside is imagined to be a more pleasant place to be and live than the city. However, this representation is selective and is imagined mainly from an urban perspective and has been strongly associated with particular social groups composed of wealthy professionals linked with the process of counter-urbanisation.

More recently, however, it can be argued that a contrasting argument applies to new populations returning to the central city and particularly to areas of urban regeneration in places such as London Docklands and Manchester city centre. This is associated with a process of reurbanisation, in line with an urban idyll consisting of pro-urban myths (Short 1991; Crilley 1992a; Holcomb 1994). In this context, new urban professionals view cities and their centres in particular as attractive, vibrant and cultured places to live in the early twenty-first century. Here, pro-urban myths exist in opposition to an imagined countryside stereotype of the 'ignorant and brutish yokel or "country bumpkin", and against visions of a monotonous and dreary suburbia. The city can be viewed as a cultured and vibrant "melting pot", . . . providing opportunities for variety, social mixing and vibrant encounters between very different social groups' (Holloway and Hubbard 2001, pp. 120–121). Thus, the contemporary city centre can, like contemporary rural villages, be said to be 'ideological' in that it selectively embodies the interests of people of specific social classes, ages, genders, ethnicities and sexualities. This is captured within the idea of the 'urban idyll', which has been set up as an alternative to the rural and suburban as another set of false, narrow, selective and exclusive images. It must be asked what sort of city centre is being created and what groups are being excluded from this 'idyll' of new city centre housing (Hoskins and Tallon 2004; Colomb 2007).

The city centre in the UK has undergone somewhat of a renaissance since the mid-1990s serving certain kinds of people and not others with a shift from production to consumption uses and the emergence of new populations and arguably a gentrification process. Single professionals, those with double incomes and no kids ('dinkies'), and those benefiting from new wealth created by financial and service sector industries during the late 1990s and early 2000s, have a strong presence in the youthful neighbourhoods emerging in many city centres. Urban regeneration has facilitated this shift and its design is partly informed by a bohemianesque 'urban idyll' (Hoskins and Tallon 2004). The term 'urban idyll' can be used to represent the vibrant city neighbourhood so often portrayed in contemporary policy and urban regeneration promotions, and is central to the commercialisation and marketing of urban space (Plate 11.5).

The urban idyll is exemplified by the development of the Walt Disney Corporation-backed town of Celebration near the company's theme park in Orlando (see Frantz and Collins 1999; Ross 2000). Celebration's planners avoided the cul-de-sacs typical of modern suburbia and opted for a layout more closely resembling the grids of old. Narrow streets reduce traffic speed and houses have front porches where residents are meant to while away their evenings in conversation. However, here as with any other emerging city landscape comprehensively designed as such, the capacity residents have to engage with their location and to shape their surroundings is very limited. To achieve a pleasant architectural profile, Celebration houses must follow a 'pattern book' that mandates everything from tree placement to window size, grass length to curtain colour. The frequency of attempts to construct and materialise the perfect neighbourhood is such that Celebration is neither particularly innovative nor unique. More and more, image, brand and vision have inspired architects to design city centre housing within themes that maintain the integrity of a selected narrative as has become widespread in city centres and, in particular, gated communities as discussed later.

Concentrating on the affluent new resident through marketing the urban idyll has inevitably led to the neglect of the poor, and has invited criticisms similar to those directed at the social exclusivity of the London Docklands redevelopment (for example, Goodwin 1991; Brownill 1999; Foster 1999). However, experience of housing regeneration and change in less affluent cities with a strong industrial legacy such as Swansea in South Wales has been rather different. Social and affordable housing has been a more significant element of housing regeneration. Swansea's Maritime Quarter redevelopment, ongoing since the 1970s, includes a significant social housing component in addition to the more extensive higher-status accommodation (Robinson and Williams 1990). During the late 1990s and early 2000s, housing developments in Swansea city centre were largely confined to those

Plate 11.5 Marketing the 'urban idyll'
Photograph by author.

by housing associations (Tallon 2003; Bromley et al. 2007b). Given its orientation to those of middle and lower socio-economic status, these developments are not a boost to gentrification.

Although supply and demand factors inevitably operate in the style of new central-city housing developments that seek to contribute to urban regeneration (see Zukin 1989, pp. 13–17; Maclennan 1997), the producers of housing, such as planners, developers and architects, have played a strong role. In many cities, city centre regeneration policies have been centred on waterside development. Large-scale cultural flagship schemes (see Chapter 12) have been used to initiate private sector investment in waterfront and other residential developments (Griffiths et al. 1999). In countless contemporary city centres throughout the UK, exclusive and private sector developments are dominating housing supply, which underpins the strong trend towards gentrifying characteristics (Nathan and Urwin 2006; Allen 2007; Boddy 2007; Bromley et al. 2007b).

Gentrification as a key urban process and method of regeneration is inextricably linked with several other transformations associated with the shift to the post-modern city (see Chapter 1). De-industrialisation gave rise to a supply of cheap terraced housing and unused warehousing close to the city centre. These areas are often near a 'cappuccino box', including new arts centres, which seek to accomplish a 'Guggenheim effect' to 'brand' city image (see Plaza 2006). The Guggenheim Museum, designed by architect Frank Gehry, opened in Bilbao, Spain, in 1997. It attracted over 1 million visitors in its first year, 85 per cent of whom were from beyond the region, and 7 million by 2005 (Plaza 2006; see Chapter 12). This single flagship spectacular post-modern development stimulated a wider revival of the city centre. Local government has undertaken city branding to market the city centre as an attractive place to live (see Chapter 6). Increasing commuting times and changing commuting patterns have also favoured central city living. New retail patterns have witnessed a reinvestment in small superstores in central locations to serve a growing market of professionals without cars (see Chapter 10), and Business Improvement Districts have been established to improve the safety and appearance of central retail areas (Symes and Steel 2003).

The recent phenomenon of city centre living appears to be highly localised and benefits the more idyllic and central parts of cities. Allen (2007) argued that the inner-urbanisation of city centre living in Manchester is unlikely to have a regenerating impact on inner-urban areas, largely because the social, cultural and economic interests of inner-urban dwellers are located in the city centre. The urban renaissance as reflected by new city centre housing development is largely divorced from surrounding areas and communities, which should perhaps become the focus of urban regeneration beyond the city centre (see Unsworth and Nathan 2006).

'STUDENTIFICATION' AND CITIES

Since the late 1990s, the process of gentrification has widened to encompass a process defined by Smith, D. (2002; 2005) as 'studentification', representing the colonisation of parts of cities by a growing student population. The proportion of students living in rented accommodation in particular parts of city centres, inner cities and suburbs of university cities or towns results in the take-over of the area, typically leading to increasing numbers of student-oriented services including DVD rental shops, pizza deliveries, off licences, and traditional pubs being converted to theme pubs (Ravetz 1996). 'Studentification' gives rise to positive and negative impacts within urban localities, which have resonance with the wider impacts of gentrification (see Tables 11.6 and 11.7).

Students are attracted to contemporary inner cities and city centres and parallel aspects of the gentrification process in university cities; indeed Smith and Holt (2007) conceptualised the group as 'apprentice gentrifiers' or 'a potential grouping of future gentrifiers' (Smith, D. 2005, p. 86). Attractions of city centre living for students revolve around proximity to university, city centre nightlife and cultural facilities; good public transport, shops, services and facilities; and concentrations of other students (Chatterton 1999; Chatterton and Hollands 2002; Smith, D. 2002; Tallon and Bromley 2004; Allinson 2006). 'Studentification' often involves the displacement of an established working-class population by predominantly middle-class students (Smith, D. 2002). Smith, D. (2002, p. 16) asserted that 'distinct student geographies are the product, in part, of well-defined tastes and preferences for particular types of location, housing and rental costs . . . to satisfy the "hedonistic" cultural orientations of students'.

TABLE 11.6 POSITIVE IMPACTS OF 'STUDENTIFICATION'

Positive impacts
Contributes to regeneration of declining central city areas
Contributes to repopulation of central cities
Contributes to revitalisation, sense of buzz and a thriving community
Contributes to social capital
Maintains viability of public infrastructure
Increases spending power in the local economy (shops, pubs, services)
Inflates property prices

Sources: **based on Allinson (2006) and Hubbard (2008).**

TABLE 11.7 NEGATIVE IMPACTS OF 'STUDENTIFICATION'

Conflicts and boundaries between 'students' and 'locals'
Student lifestyle (late nights, noise)
Adds to increase in incidence of burglaries
Displaces indigenous population
Seasonal occupation of housing
Transient population with no commitment to the area
Creates student ghetto monoculture
Results in loss of facilities oriented towards other social groups

Sources: **based on Allinson (2006) and Hubbard (2008).**

Allinson (2005) argued that much of the residential growth of central cities during the 1990s was the result of the expansion of higher education, which injected around 510,000 extra students into the housing market. This process has arguably revitalised many parts of decaying central cities to the degree that the challenge has been transformed from urban regeneration to the management of the impact of a homogeneous and lively group in 'studentified' districts (Smith, D. 2002; Allinson 2005; 2006). Patterns and trends of 'studentification' have been identified in inner urban areas of cities such as Bristol (Chatterton 1999), Leeds (Smith, D. 2002), Manchester (Allinson 2005), Birmingham (Allinson 2005; 2006) and Loughborough (Hubbard 2008). Hubbard (2008) revealed some of the most 'studentified' wards of UK cities with over one-fifth of student residents to be Cathays in Cardiff (a 27 per cent student population in 2001), Dunkirk and Lenton in Nottingham (24 per cent), Treforest in Glamorgan (24 per cent) and Headingley in Leeds (21 per cent). Smith and Holt (2007) presented evidence that showed 352 wards in the UK to be 'student communities' or 'enclaves'.

In cities across the UK, the rapid growth of students in higher educational institutions has had significant impacts on the social, economic and cultural geographies of central cities in particular. There is a clear regenerative potential in terms of the built environment, local economy and social revitalisation, but potentially degenerative impacts in terms of displacement of existing residents, and the social and cultural conflicts stemming from the concentration of a largely homogeneous social group. Hubbard (2008) concluded by arguing that 'studentification' can be interpreted as an ambivalent process whose cost and benefits must be assessed on a case-by-case basis with policy similarly tailored. 'Studentification' as an urban process remains a relatively under-researched topic, including its significance for urban regeneration in the UK.

GATED COMMUNITIES

Gentrification and studentification have resulted in the apparent regeneration of many formerly declining urban neighbourhoods since the mid-twentieth century. In addition, since the 1970s, master-planned, gated communities or common interest developments have emerged as middle-class residential environments, which are often established as a part of urban regeneration schemes. These have been developed particularly in the US, but increasingly so in the UK and elsewhere globally in places such as Australia, China and India (Blakely and Snyder 1997; Webster 2001; Webster et al. 2002; Grant and Mittelsteadt 2004; Atkinson and Blandy 2006). In the US, 4 million people lived in gated communities in 1995, and this figure had risen to 16 million by 1998 (Low 2003). In the UK, Atkinson and Flint

(2004) identified around 1000 gated communities, the majority of which were located in and around big cities such as London. Gated communities have primarily emerged in response to urban social problems as a way to protect their residents from crime and to create a sense of community among interest groups (Blakely and Snyder 1997; Low 2003).

In the US, privatised cities have been embodied in the development of master-planned communities whereby private sector corporations take over functions of the state (see Chapters 3 and 7). The arguments for this are that choice is increased and private sector efficiency is introduced. The arguments against master-planned communities are that choice is available to wealthier groups only, that they are undemocratic, and that the poor are no longer subsidised by the rich through redistributive tax (Blakely and Snyder 1997). Such philosophies have permeated into UK residential environments.

Gated communities, sometimes referred to as master-planned communities or common interest developments, vary in their geographies but exhibit a set of general characteristics. Gated communities are literally gated or walled urban environments. Property in gated communities has a high purchase price, on which a high additional annual management fee is payable, similar to a tax. The developer in effect acts as a government as it duplicates public sector provision of services such as security staff, schools, parks, churches and shops. The environments are highly regulated and residents must follow strict rules (see Blakely and Snyder 1997; Webster 2001; Atkinson et al. 2005; Manzi and Smith-Bowers 2005). They are in essence 'private cities' (Glasze et al. 2006).

A range of regulations in gated communities in the US includes dictating colour of house, restricting external changes, banning political posters, levying fines for untidy gardens or not putting rubbish out, and limiting the number of pets that can be kept. It can also be compulsory to own a firearm and not to drink alcohol in the public environment. These are all supplemental regulations and cannot be contrary to US law. Residents are therefore subject to three sets of laws and taxes from federal, state and community levels. Although the overall design idea of gated communities involves security and separation, they vary in terms of affluence, location, level of service provision and demographics (see Grant and Mittelsteadt 2004). The three main types of gated communities can be classified by characteristics of lifestyle, prestige and security (Blakely and Snyder 1997; Forrest and Kearns 2001). Although all gated communities demonstrate all three criteria, different developments display different levels of emphasis on these.

Lifestyle gated communities are top-end gated communities that exhibit extreme wealth and are dominated by older married residents. They are located outside the city on key transport routes and are in essence gated by distance, although often still walled. All services are provided on site, and even schools and churches are run by the private developer. 'Themed' architecture and lifestyle themes revolving around activities such as golf or watersports are common (Blakely and Snyder 1997). Prestige gated communities have been developed in suburban, edge of city locations. Big detached houses dominate the residential environment, and some services are provided on site such as crèches, tennis courts and jogging paths. These developments are in effect simply a wealthy suburb with gating and private security (Blakely and Snyder 1997). Gated communities dominated by security rather than lifestyle or prestige qualities can be found in inner city areas. They are heavily gated and high-rise in nature and can be termed security zone communities. They are characterised by a younger demographic profile comprising individuals or couples. Fewer on-site services are provided, as they are all available at close proximity to the city centre (Blakely and Snyder 1997).

A new variant of gated community has emerged to house low-income urban residents in inner city areas demonstrating high unemployment and high crime. Neighbourhoods are gated off into blocks of a few streets with closed circuit television installed at the gates. The irony is that these environments are provided by the state, and represent a merit good to groups who cannot afford to purchase property in private sector gated communities.

Increasingly common in the US, such developments are emerging on a smaller scale in the UK (Webster 2001; Atkinson and Flint 2004).

A number of factors have driven the growth and development of gated communities, which are often constructed to regenerate inner city and edge of city areas. These are linked to wider urban regeneration issues, including the fear of crime in urban areas (see Oc and Tiesdell 1997a); the hope that closed, private streets behind gates lead to a more open, friendly and cohesive community (Blakely and Snyder 1997; Low 2003); the rise of the creative class (Florida 2002; 2005); the urban problems of de-industrialisation, declining cities, unemployment and pollution; and the growth of car ownership and the motorway network, which have enabled gating by distance.

Whether gated communities are positive or negative developments and can contribute to urban regeneration is a subject of debate (see, for example, Atkinson and Flint 2004; Atkinson and Blandy 2006; Atkinson et al. 2005). It is difficult to turn the fortunes of failing cities around if affluent social groups retreat into gated communities. Gated communities represent club goods, which are public but excludable, rather than merit or public goods (Manzi and Smith-Bowers 2005). The private sector often provides services normally supplied by the public sector (Gordon 2004). Additionally, divides open up along social, political, racial and religious lines when those who are able to take flight to gated suburban developments leave behind a residual inner city population (see Blakely and Snyder 1997; Forrest and Kearns 2001; Webster 2001; 2002; Manzi and Smith-Bowers 2005; Glasze et al. 2006). It can be asserted that gated communities represent the 'ultimate private city'.

URBAN REGENERATION AND HOUSING SUPPLY IN THE UK

Providing enough housing, especially affordable housing, in the UK has been a problematic issue in the post-recession period since the mid-1990s. This issue is connected with gentrification, the emergence of gated communities and evolving patterns of urban space. Housing can be conceptualised as a merit good that is undersupplied at the current market price; as a consumer good that is bought for consumption; and as an investment good bought to make a profit. Changing tenure patterns over the twentieth century have seen a shift from a largely private rented market (90 per cent in 1901) to a market dominated by owner occupation (70 per cent) and the state (20 per cent) in the early twenty-first century (see Mullins and Murie 2006).

The housing shortage is a symbol of the successful, growing, competitive city. Housing becomes unaffordable for a significant section of the population and the price of rented accommodation increases. It can be difficult to attract new firms to cities with affordability problems, and this acts as a drag on labour mobility. The contrasting issue of limited demand for housing is a problem in declining cities that have experienced de-industrialisation, representing a mismatch between the location of housing and population.

Attributing blame for the housing shortage in the UK is problematic. Local authorities have been accused of being slow to respond and badly managed; housing developers are accused of hoarding land banks; and central government is accused of presiding over outmoded planning laws. The causes of the shortage of housing in growing cities are both short-term and long-term in nature.

The short-term causes closely reflect macroeconomic success. Successful management of the macroeconomic targets of economic growth, low unemployment and low inflation (with low interest rates) resulted in a boost in demand for housing between the mid-1990s and late 2000s. Long-term causes can be divided into demand side and supply side categories. On the demand side, long-term de-industrialisation resulted in a changing geography of demand, with less demand in the industrial north and more demand in the new economy of the south leading to pressures on house prices. Demographic change has seen an increase in single households reflecting fewer marriages, later marriages and a rise in divorces. There has been an increase in the UK population to almost 61 million but a more significant increase in the number of households to over 25 million by 2007 (www.statistics.gov.uk, 2008).

On the supply side, shortages in housing can be attributed to four key factors. First, construction time lags, which are a structural problem with all property, mean that slight changes in demand trigger much greater changes in price. Second, a slow and painstaking planning process adds to construction time lags. Third, a small number of suppliers of housing has resulted in a high concentration ratio confirmed by the fact that 85 per cent of housing is constructed by five firms. This concentration means that firms can restrict new supply of housing to keep prices high, and can assemble land banks of sites with planning permission to watch their value appreciate. Fourth, a shortage of sites in growing cities has fuelled the emergence of edge cities (Garreau 1991), and the need for housing development on the edges of urban areas has been hampered by green belts and other protected areas. There is a need for balance between more land for housing development and environmental protection.

To address the problems of housing supply, the New Labour government commissioned the economist Kate Barker to formulate some possible solutions (Barker 2004; see also HM Treasury 2005a; 2005b; CLG 2007a). The first main possible solution recommended by the Barker Report (2004) was to accelerate the planning process by speeding up minor applications, speeding up the appeals process, making more money available for planning departments, outsourcing some of their business to private sector consultancies, and fostering more co-ordination between local authorities on cross-boundary planning applications. A second possible solution to the supply problem was to modify the 60-year-old green belt system accounting for 13 per cent of land in England (see Chapter 2), as it can be seen to act as a 'corset' on urban expansion and as a drag on economic development, leading to higher house prices, and to longer commuting times as new housing development 'leapfrogs' the green belt. Barker (2004) recommended a decrease in the area of green belt and encouraged green corridors or wedges instead. A third key recommendation was to promote more mixed-use zoning comprising housing, retail, offices and leisure in the same area (see Coupland 1997). Mixed-use development seeks to cut commuting times and to restrict the development of mass volume constructed housing estates devoid of services and facilities. A fourth proposal by Barker (2004) was to encourage more brownfield housing development consistent with city centre living (see Chapter 9; Raco and Henderson 2006; Dixon et al. 2007), to recycle old industrial land and buildings, to reduce commuting, and to make more efficient use of existing infrastructure. Finally, the influential report argued for more 'shared equity' schemes whereby low-income households buy a certain percentage of a property and rent the remainder from a social landlord.

The government responded to the Barker (2004) report by proposing a package of measures to reform the planning system and to deliver increased investment in infrastructure (HM Treasury 2005a; 2005b; Cullingworth and Nadin 2006). The government commitment to achieve 60 per cent of new housing development on brownfield land was emphasised (HM Treasury 2005a), and guidance on identifying suitable brownfield sites was provided (ODPM 2005f). The weight of policy favoured housing development on brownfield sites, the re-use of existing buildings and sustainable development (see Chapter 9).

To encourage a greater supply of housing in the UK in the future, incentives need to be provided for volume house-builders to build more housing. The planning system might also be adjusted to grant time-limited planning permission to prevent developers from building up large land banks. Firms with unused sites might also be prevented from being granted further planning permissions. Stepped taxation on land with planning permission which is not used could be instituted as a way of increasing housing supply and contributing to urban regeneration in the UK.

BROWNFIELD SITE REGENERATION

Linked to the problems with housing supply and the regeneration of formerly developed urban areas has been the trend towards brownfield regeneration, particularly since the mid-1990s. Private sector developers have historically been more likely to avoid brownfield sites, usually in the central city, for a variety of reasons. These include the costs of assembling a site for development; difficulties of achieving economies of scale on relatively small sites; difficult access; expensive surveys; high site remediation costs; easements; and consumer suspicion of brownfield sites (see Dixon et al. 2007).

There have been three key trends since the 1970s that have driven the process of brownfield development and regeneration. The first is the long-standing process of a north to south drift of both population and economic activity in the UK. This trend fluctuates in response to economic cycles and the migration has been lower since the end of the 1980s. London has been the main attraction, although there has been a south to north net migration since 2000. Clearly there are social and economic issues relating to the continued movement of people and capital from the north to the south; a lack of demand in the north leads to housing decline and abandonment, and an over-demand and lack of supply of housing and services in the south.

An urban to rural shift associated with the counter-urbanisation process is a persistent and continuing second trend (see Chapter 1), which has seen a general population move away from metropolitan areas. Population gains have been in resorts, ports and accessible/remote rural areas, and in non-metropolitan cities (CLG 2007a). However, as discussed earlier, city centre living has witnessed a revival since the mid-1990s.

Housing market change has been a third stimulus for brownfield site regeneration in the UK. The number of households is increasing, and supply is responding to these changes with developers building more small homes, especially apartments in city centres. However, the supply of new housing has not kept up with demand (Barker 2004; CLG 2007a) and there is especially a lack of affordable housing. CLG (2007a) put forward a target to construct 3 million new homes by 2020. The controversial issue is where these homes are to be built; there is a debate as to whether they should be constructed on brownfield sites or on greenfield sites. The government has stipulated that 60 per cent must be built on brownfield sites, a target originally set by DoE (1996b) that has since been exceeded (CLG 2007a).

A brownfield site can be defined as previously developed land, or any land that has previously been used for any purpose and is no longer in use for that purpose (see Dixon et al. 2007; CLG 2007a). Arguments for using brownfield sites are inextricably linked with urban regeneration and sustainability concerns (see Chapter 9). It prevents urban sprawl and saves precious countryside, and it is also argued that UK urban areas could be more densely populated as they are less so than others in Europe. There are attractions for business in city centres and zones of transition close to Victorian transport hubs on former industrial land. The attractive housing environments match the demand for city centre apartments from new single and double-professional households in particular (Tallon and Bromley 2004), and new central brownfield housing can also cut commuting (Bromley et al. 2005). These advantages have contributed to a rapid increase in brownfield development since the mid-1990s, within a favourable central government policy context (CLG 2007a; Lees 2003a; Bromley et al. 2007a). Regulatory and fiscal policies have sought to coerce developers into a greater take-up of brownfield development. Thus brownfield development revives and regenerates city centres on the one hand and preserves rural areas on the other. However, there are higher costs of clearing and cleaning sites, extra pressures are placed on existing urban infrastructure, and there can be a loss of remaining central green or open spaces in urban areas (Dixon et al. 2007).

Regulatory policy has sought to encourage brownfield development through fast-tracking planning permission; lower planning fees; the authorisation of more floors; exemption from planning gain obligations; the use of compulsory purchase orders to secure site assembly (ODPM 2004h); and the granting of planning permission to develop greenfield sites to developers who also regenerate brownfield sites. Fiscal policy has mobilised tax-and-spend incentives to facilitate brownfield development. These encompass tax breaks, grants to cover remediation costs (for example, some states in the US give 110 per cent of clean-up costs to developers, representing a 10 per cent profit), and government insurance against all future claims on the site. Further encouragement from central government could be in the form of an agreement for government to buy office space on a completed project, or to improve infrastructure to enhance access and land value. Greenfield developments could be taxed and the money raised then used to offer tax breaks for brownfield housing development in the form of a hypothecated tax.

The government argues that the housing development target of 3 million new homes by 2020 is possible by building 2 million homes by 2016, then another 1 million by 2020 (CLG 2007a). The government states that this can be achieved through policies such as Regional Spatial Strategies; Growth Areas and Growth Points; eco-towns; more use of public sector land; better use of disused land; and better use of existing buildings and land (CLG 2007a).

The Barker Report (2004) on land-use planning and housing argued that planning restrictions on green-belt land should be eased. Barker recommended maintaining 'wedges' of green space in urban areas, rather than retaining rigid green belts. However, according to the Social Market Foundation (SMF 2007), the target cannot be met with 60 per cent of new homes being built on brownfield sites. Even if density is high at 80 dwellings per hectare, 2.1 million houses can be built on brownfield land, which will run out in 2016. If housing density is low at 20 dwellings per hectare, 500,000 houses can be built on brownfield land, which will run out in 2012.

The upshot of this is that more green belt or greenfield sites will need to be developed to extend urban areas. Green belt land around cities designated for long-term protection from development was established 50 years ago to prevent urban sprawl and protect the countryside (see Chapter 2). Greenfield is any undeveloped land that accounts for 87 per cent of land in England, of which 31 per cent is designated Areas of Outstanding Natural Beauty, National Parks and Sites of Special Scientific Interest. Arguments for developing greenfield sites are that towns and cities are running out of development land and that town cramming is occurring, particularly in areas where the bigger household increases are focused. The very high densities that some central city housing schemes are achieving might well contribute to future regeneration problems in these areas (see Senior et al. 2004; 2006). Greenfield development also corresponds with aspirations as the majority of the population desire a house plus outdoor space and a better quality of life environment (DETR 2000d; Senior et al. 2004). There is also a need for low-cost and affordable housing in rural areas which further greenfield development could help supply, and it would also mean that towns and cities can retain their green spaces.

However, there are concerns related to allowing greenfield development. It reinforces the trend of urban sprawl; existing rural communities lose their identity; it is unpopular among locals and the population at large; and it encourages reliance on cars. The debate over greenfield development suggests that certain groups in society are likely to be gainers or losers as a result of policy decisions in this area. Gainers include landowners and developers who stand to benefit from a relaxation of controls over greenfield sites. If the increase in housing supply leads to lower prices, households at present excluded from the housing market due to high prices might benefit. Losers could include groups who wish to preserve existing rural landscapes and wildlife habitats, and those individuals who derive benefit from living within these areas. Rural communities might be swamped by commuters and lose small place identity.

Overall, brownfield site regeneration and the debate over greenfield development can be summarised under three key contemporary issues. The first is where new homes should be built: on greenfield sites, brownfield sites, in new 'growth areas', or in new eco-towns (see Chapter 9). The second is the type and size of housing to build, and in particular whether houses or apartments are required. The third issue is the provision of affordable housing both generally, and in London and the South-East where net international migration is highest, and in popular rural areas in the south.

SUMMARY

Housing has been a key driver of urban regeneration in the UK, particularly since the mid-1990s in inner cities and city centres.

Housing regeneration can often promote gentrification of the urban landscape, which can be seen either to breathe new life into derelict spaces, or to displace and exclude indigenous lower income populations.

Gated communities have emerged in the US and more recently in the UK to improve the safety of housing environments and to create a sense of community, although they reinforce segregation and exclude lower social groups.

Urban housing supply can be increased to match demand through developing central city brownfield sites, which can promote regeneration and preserve greenfield sites.

STUDY QUESTIONS

1. 'The only positive to gentrification is being able to find a good cup of coffee' (Slater 2006, p. 752). Critically evaluate this claim with reference to the housing-led urban renaissance of UK city centres since the mid-1990s.
2. Critically assess the differing views as to why gentrification has occurred in Western cities.
3. What factors have led to the rise of gated communities in the US, and to an extent in the UK? How and why do such communities vary, and what criticisms have been levelled at them by urban researchers?
4. Explain the mismatch between housing supply and demand in the UK, and assess what urban regeneration policy can do to try and correct it.
5. The growth of edge cities demonstrates a lack of confidence in the city and a movement of economic activity to the periphery; gentrification evidently demonstrates the reverse. Explain and account for this apparent paradox in US and UK cities.
6. Assess the greenfield versus brownfield development debate and its implications for urban regeneration.

KEY READINGS ON HOUSING-LED REGENERATION AND GENTRIFICATION

Atkinson and Blandy (eds.) (2006) *Gated Communities* [international perspectives on the rise of gated communities]

Atkinson and Bridge (eds.) (2005) *Gentrification in a Global Context: The New Urban Colonialism* [theoretical and empirical research on gentrification at a global scale]

Blakely and Snyder (1997) *Fortress America: Gating Communities in the United States* [examines the debate in US cities over whether any neighbourhood should be walled and gated, preventing intrusion or inspection by outsiders]

Gated Communities Special Issues: *Environment and Planning B: Planning and Design* (2002, vol. 29, no. 3) [theme issue on the global spread of gated communities]; *Housing Studies* (2005, vol. 20, no. 2) [special issue on the 'new enclavism' and the rise of gated communities]

Gentrification Special Issues: *Urban Studies* (2003, vol. 40, no. 12) [special issue on the meanings and problems of gentrification]; *Environment and Planning A* (2007, vol. 39, no. 1) [special issue on extending the meaning of gentrification]; *Urban Studies* (2008, vol. 45, no. 12) [special issue on gentrification and public policy]

Lees, Slater and Wyly (2008) *Gentrification* [key text on the history, explanations and critique of the gentrification process with global examples]

KEY WEBSITES ON HOUSING-LED REGENERATION AND GENTRIFICATION

Centre for Cities: www.centreforcities.org [part of the Institute for Public Policy Research; carries out independent research on urban policy and the economies of cities]

Gentrification Web: http://members.lycos.co.uk/gentrification [comprehensive resource for anyone interested in learning more about gentrification, maintained by gentrification researcher Tom Slater]

Joseph Rowntree Foundation: www.jrf.org.uk [independent research organisation includes research on housing and neighbourhoods]

12 Leisure and cultural regeneration

INTRODUCTION

Urban regeneration through leisure and cultural regeneration has emerged as a crucial feature of the post-modern city of consumption. One of the most dramatic elements of contemporary urban regeneration is the visual transformation of the cityscape associated with the rise of post-modern architecture and high-quality design. This has assisted the development of a range of architecturally innovative showpiece cultural projects that contributes to the cultural economy. 'Cities of spectacle' and 'fantasy cities' are labels that have been attached to places experiencing regeneration that is linked with leisure, culture and consumption. Examples of this style of urban regeneration are evident in cities such as Birmingham, Manchester, NewcastleGateshead and Liverpool. This trend is international in scope, as witnessed in cities across North America, Europe and Australasia, including Baltimore, Pittsburgh, Barcelona, Bilbao, Dublin and Brisbane. These cities have transformed their fortunes, strongly influenced by cultural regeneration.

This chapter covers the most significant of the increasingly wide range of styles of urban regeneration, undertaken in the UK and globally, that fall within the leisure, cultural and creative economies. The chapter first outlines the nature of the expanding leisure, entertainment and cultural economy and spaces in cities. It moves on to look at the character of the cultural city, the emergence of cultural quarters and the promotion of new urban festivals and carnivals. A significant component of the leisure economy driving urban regeneration has been the development and promotion of the evening and night-time economy in cities in the context of the shift to the 24-hour society.

The chapter then evaluates the role of urban tourism and sport in the entrepreneurial city. Measures to promote the creative city economy have also been instrumental in the regeneration of the post-industrial city, as has the selling of the city of spectacular, post-modern architecture. The chapter concludes by briefly noting some critical issues and policy implications for the leisure and cultural economy.

EXPANDING THE LEISURE AND CULTURAL ECONOMY IN CITIES

The way in which people spend their leisure time is changing and this has implications for the leisure and cultural economies of contemporary cities (see Casado-Diaz et al. 2007). In terms of cities, elements of the urban leisure economy encompass restaurants, fast food outlets and cafés; bars, pubs and clubs; museums, galleries, theatres and concert halls; festivals and other cultural events; sports stadia, fitness centres and health clubs; casinos and betting shops; and hotels, guest houses and short lets (Tallon et al. 2006). The diversity of the leisure economy in cities is illustrated by the groupings of types of leisure activities (Table 12.1). These activities share common characteristics of involving people leaving their homes, moving around the city, and purchasing the experience of leisure, fun and enjoyment. Questions surround the inclusion of shopping and retailing within the leisure experience, as the sectors are becoming increasingly conflated (Worpole 1991; Newby 1993; see Chapter 10), and the role of leisure in the home. Leisure can also be viewed as a public good in the city represented by activities such as walks in public parks. Leisure can be defined as 'what draws people out of their homes and workplaces to spend time and money going out and about' (GLA 2003, p. 12). Leisure is essentially a product of the post-modern world, where the city is constructed as a product to be consumed by visitors (see Urry 1995).

Post-modern cities are increasingly taking the leisure and cultural economies seriously for a variety of reasons (Landry et al. 1996; Short and Kim 1999, Chapters 5, 6 and 7; García 2004a). These include population growth in cities and city centres in particular; sustained economic growth that has led to rising disposable incomes; the potential of job creation in the leisure and cultural economy of cities; the regenerative and transformative benefits of the leisure economy across urban areas; the prominence of place marketing and positioning in a globally competitive world; the value of the leisure economy in purely monetary terms; and the perceived social benefits of encouraging interaction. Within these wider social changes, the development of leisure and cultural attractions has emerged as an element of wider policy for regenerating post-industrial cities in the UK, with culture recognised as a significant component of urban success in the post-industrial city (see Griffiths 1995a; Evans 2001; Aitchison et al. 2007; Aitchison and Pritchard 2007).

Despite these perceived benefits for the regeneration of urban areas in the post-industrial economy, there are some cautionary implications in terms of public policy. These include neighbourhood effects of noise, rubbish, displacement and loss of local character; public service effects with additional pressures on health care services and policing; and labour market effects regarding low-paid, seasonal and casual jobs in the urban leisure economy.

The leisure economy has transformed and expanded since the 1970s. First, there has been a huge growth in home-based leisure activities including television, home personal computers, the internet, game machines, eating and drinking at home, and entertaining guests. Second, there has been an enormous increase in attendance at sporting and cultural events,

TABLE 12.1 CLASSIFICATION OF LEISURE ACTIVITIES IN CITIES

Home-based leisure activities: hobbies, gardening
Health and sports related: stadia, pools, pitches, fitness centres, health clubs, parks, the sea
Dining: restaurants, fast food outlets, outdoor dining provision
Drinking (alcoholic and non-alcoholic): bars, pubs, nightclubs, cafés
Gambling: online at home, casinos, betting shops
Cultural-based activities: museums, galleries, theatres, outdoor facilities, cinemas, festivals, carnivals

and in visiting bars and restaurants. Third, there have been subtle changes in the ways in which people spend their money; consumers are becoming more discerning in their choices. On the one hand evidence suggests that people are participating more in the life of the city (see Worpole 1992), but on the other hand many stay at home partly due to perceptions about the groups towards which leisure and cultural attractions are aimed (see Bromley et al. 2003b).

Since the 1960s, the UK city centre has moved through a well-documented process of economic decline, losing manufacturing industry, residential population, shops and offices to suburban locations (Evans 1997; see Chapters 1, 10 and 11). In addition, recent decades have seen leisure and entertainment functions relocating to out-of-centre, edge-city locations (Hubbard 2002). At the same time the attraction of the city centre has declined in relation to out-of-town entertainment centres (Hubbard 2003) and entertainment in the home (Lovatt and O'Connor 1995). City authorities have reacted to these economic and social changes in a number of ways over the last 30 years, with a number of successive approaches to city centre regeneration (see Chapter 10). These successive phases have comprised a sequence of retail, office and leisure uses, while attempting to build upon the vestiges of 'high' and 'popular' culture facilities of the past (Tallon et al. 2006).

Since the 1970s, central government has recognised the need to regenerate city centres, especially those in old industrial areas, to address the economic and social problems that have developed over many decades. During the 1980s, central government-sponsored Enterprise Zones, Urban Development Corporations and Garden Festivals all sought to create new uses for central city sites, each incorporating some degree of leisure and entertainment function. Characteristically, waterfront developments were favoured incorporating mixed-use developments of retail, leisure, office and housing. The late 1980s initially saw a retail focus for regenerative policy, which later expanded to include town centre management. Associated with this period were attempts at attracting offices to the city centre. More recently, central government policies focused on the city centre have aimed to broaden the attractions base with the promotion of leisure and cultural facilities and the introduction of special events (Comedia 1991; URBED 1994; ODPM 2005c). There has also been an enhanced policy of encouraging city centre living with new residents constituting a pool of consumers for city centre leisure and cultural economies (Bromley et al. 2005; 2007b). During the 1990s, the focus was widened to encompass the development of the evening and night-time economy. This attempted to extend the daytime activity of the city centre and embraced a café culture concept, late night shopping and other leisure activities focused on the evening (Comedia 1991; Ravenscroft et al. 2000).

However, by the late 1990s this aspiration had in practice been overtaken by the market to become focused on 'popular culture' and dominated by leisure and entertainment uses of pubs and clubs (Chatterton and Hollands 2002). This trend is inextricably linked to the evening and night-time economy and the 24-hour city concept (Lovatt and O'Connor 1995; Heath and Stickland 1997), as discussed later in this chapter. An intensifying residential repopulation continued alongside this trend (Tallon 2003). Since the late 1990s there has been an interest in adding 'culture' as an additional element (Griffiths et al. 2003). This has been part of the aim to extend the leisure dimension of the city centre through a combination of 'high culture' and 'heritage tourism', highlighted by the proliferation of theatres, concert venues, galleries and museums. This has been extended since the late 1990s via a subset of crafts as arts, consisting of arts industries provided by artists, potters and glass-arts industries which focus on a social enterprise dimension (Worpole 1992; Tallon et al. 2006). An additional related element promoted by cities more recently is tourism, which straddles entertainment, sport, and high culture and heritage (Page 1995; Law 2000; Bradley et al. 2002), as illustrated later. Additionally since the late 1990s, events, festivals and carnivals have increasingly been staged in city centres, which illustrate the trend towards popular culture, although some of these may cross to high culture (Landry and Bianchini 1995; Hughes 1999; Tallon 2007b).

The elements of the entertainment, leisure, tourism and culture spectrum are diverse and what is included within this category for the purposes of research and policy is debated. Although there is a lack of clarity regarding the exact boundaries of each of these, it has been argued that distinctions can be made between traditional 'high culture' and 'arts', and 'leisure', and 'mass' or 'popular culture'. 'High culture' was historically linked with arts policies that were concerned with art collections, museums, theatre and classical music. 'The arts' is now seen to encompass visual art such as fine art, sculpture and photography; performance art such as music, dance and theatre; and literary art including poetry. Since the 1980s, it has been expected that culture and arts would be subjected to market forces like other aspects of social policy in the city (see Bianchini 1990). Rather than culture and arts being separate from the wider economy, they have become viewed as a fundamental part of it. Culture has become commodified as a means to an economic end, and has been increasingly associated with product development, marketing strategies, branding and image (Griffiths 1995a; Watkins and Herbert 2003).

The leisure and culture approach to urban regeneration is more deeply entrenched in Western European cities such as Barcelona, Cologne, Dublin, Hamburg and Rotterdam (see Bianchini and Parkinson 1993; Jensen-Butler et al. 1997; McCarthy 1998a; 1998b). Post-industrial cities globally have been following leisure and cultural strategies to revive declining central cities. For example, such approaches have been followed in Brisbane (Stimson and Taylor 1999), Pittsburgh (Holcomb 1993; 1994a) and other cities in the United States (see Short and Kim 1999). Following these European, Australian and North American cities, UK cities have followed the trend towards leisure and cultural regeneration. In fact, Miles and Paddison (2005) argued that the rise of culture within urban policy in the UK has been little short of extraordinary, and even more so than in other countries. Examples of leisure and culture-based regeneration have been highlighted in major cities such as Birmingham (Hubbard 1996a), Bristol (Griffiths 1995a; Griffiths et al. 1999; Tallon 2007a), Glasgow (Booth and Boyle 1993; Paddison 1993), Leeds (Harcup 2000), Liverpool (Madsen 1992; Parkinson and Bianchini 1993), Manchester (Williams 1996), NewcastleGateshead (Hollands and Chatterton 2002), Sheffield (Oatley 1996; Brown et al. 2000), and Swansea (Watkins and Herbert 2003; Tallon et al. 2006).

THE CULTURAL CITY

The cultural city has been engineered as a representation of city, regional and national identity. There has been supra-national interest in this idea with culture being mobilised as a means of branding the 'European project'. Culture has been defined as 'a lifestyle indulgence for urban elites' (Montgomery 1995, p. 136) and it is more complex than being a product for consumption. As a way of life, culture relates to the way people talk, eat, think, meet, engage in transaction, and spend free time. Culture in its broadest definition helps contribute to an understanding of place and what cities are all about (Montgomery 1990; 1995). The environmental and employment benefits that have been attributed to urban culture offer an irresistible cocktail for politicians responsible for urban areas, cultural producers and the private sector.

The shift towards leisure and cultural regeneration in the promotion of the 'cultural city' has been reflected in public and private sector policy initiatives, which have performed a key role in driving and shaping the changes taking place in post-industrial cities (Bailey et al. 2004; Evans 2005). Arts and cultural activities have been increasingly incorporated into wider policies for regenerating the post-industrial city at national, regional and local levels of government. Policies of the Department for Culture, Media and Sport in the UK since the late 1990s have sought to locate culture at the heart of regeneration and have argued that the cultural element can become the driving force for urban regeneration (DCMS 1998;

2000a; 2000b; 2004a; 2004b; Miles and Paddison 2005). Additionally, the Core Cities Group, which works in partnership with the government and other key agencies, has sought to promote a cultural-city-led approach to regenerating cities (Comedia 2002). These policies provide a context of established cultural provision and illustrate the current tensions that exist, particularly concerning conflicts between authentic cultural development and economic objectives (Mommaas 2004; DCMS 2004a).

Clusters of culture often emerge organically only to be snuffed out as culture-led regeneration leads to the gentrification and homogenisation of urban space (Harvey 1989; Griffiths 1998a; Brown et al. 2000; Tallon et al. 2006; McCarthy 2007b). There is a particular role for the design and purpose of contemporary art museums and cultural complexes in the cultural city. New sites must represent the universality of culture as the homogenisation of culture is to the detriment of difference and individuality. 'International landmarks' have been created as cultural sites, and there has been an international public experience of these products of consumption such as the famous example of the Guggenheim Museum in Bilbao, Spain (see Plaza 2000; 2006; Vicario and Manuel Martinez Monje 2003).

New Labour's urban policy since the late 1990s has also supported arts and culture as a method of addressing neighbourhood renewal in inner cities (Bassett et al. 2003; Bianchini 2004). At the European level, the European Capital of Culture programme, launched by the European Commission in 1985, has supported cultural city regeneration in a variety of industrial and post-industrial cities (see Table 12.2). Although involving no direct financial reward, competition for designation is fierce and can attract considerable inward investment to assist and promote cultural regeneration (see Balas 2004; Wilks-Heeg and North 2004; Jones and Wilks-Heeg 2004; Griffiths 2006).

Research into Glasgow's European Capital of Culture year in 1990 revealed that culture was successfully employed as a tool of city marketing and regeneration, and the demands of the local community were balanced with externally marketing the city's image. It was defined as an 'urban spectacle' with some 3000 events staged in the city during the year. However, criticism surrounded the designation, as there was little evidence of its contribution to local economic development, and there was a great deal of opposition as the local community felt their culture had been devalued when it came to represent a marketable commodity (see Booth and Boyle 1993; Paddison 1993; Boyle and Hughes 1994; Hall 2001; García 2004b).

The competition for the 2008 UK European Capital of Culture involved 14 cities including Newcastle, Bristol, Birmingham and Belfast, such is its perceived importance in transforming and improving city image to attract investment and visitors. The competition was won by Liverpool, which witnessed a significant upward trajectory in visitors and cultural activities during 2008. Liverpool had been expected to benefit from around £2 billion of investment over the five years from 2003 to 2008. It was expected that employment in the culture sector encompassing tourism, sport, heritage and the creative industries would be boosted by the creation of up to 14,000 jobs. Nearly 2 million extra visitors a year were anticipated to boost spending in the city by more than £50 million as a result of the designation as European cultural capital. Local community objectives were more directly addressed by Liverpool's approach, rather than solely the economy and business interests, and the bid attempted to counter some of the criticisms levelled at Glasgow's year as Capital of Culture (see Jones and Wilks-Heeg 2004; Griffiths 2006).

CULTURAL QUARTERS IN CITIES

Much leisure and cultural regeneration in the post-industrial or cultural city has focused on the notion of cultural 'clusters' or 'quarters' (Montgomery 2003; Mommaas 2004; Miles, M. 2005; Tallon et al. 2006; McCarthy 2007b). These are concentrations of leisure and

TABLE 12.2 EUROPEAN CAPITALS OF CULTURE, 1985–2008

Year	Capital(s)
1985	Athens (Greece)
1986	Florence (Italy)
1987	Amsterdam (the Netherlands)
1988	Berlin (West Germany)
1989	Paris (France)
1990	Glasgow (UK)
1991	Dublin (Ireland)
1992	Madrid (Spain)
1993	Antwerp (Belgium)
1994	Lisbon (Portugal)
1995	Luxembourg (Luxembourg)
1996	Copenhagen (Denmark)
1997	Thessaloniki (Greece)
1998	Stockholm (Sweden)
1999	Weimar (Germany)
2000	Avignon (France), Bergen (Norway), Bologna (Italy), Brussels (Belgium), Helsinki (Finland), Krakow (Poland), Reykjavik (Iceland), Prague (Czech Republic), Santiago de Compostela (Spain)
2001	Porto (Portugal), Rotterdam (the Netherlands)
2002	Bruges (Belgium), Salamanca (Spain)
2003	Graz (Austria)
2004	Geneva (Italy), Lille (France)
2005	Cork (Ireland)
2006	Patras (Greece)
2007	Luxembourg (Luxembourg), Sibiu (Romania)
2008	Liverpool (UK), Stavanger (Norway)

Source: **http://ec.europa.eu/culture/index_en.htm.**

cultural uses, often located in former industrial areas of the city centre, and usually combine authentic artistic and cultural production activities with a variety of leisure and entertainment consumption elements such as bars, restaurants and cultural retail spaces (Mommaas 2004; DCMS 2004a). Cultural quarter is a somewhat nebulous and slippery term (Mommaas 2004). However, Montgomery (2003) conceptualised 'cultural quarters' drawing upon theories of urban growth, economic development and urban design. Despite their wide variety of forms, a number of essential ingredients make 'a successful cultural quarter'. These consist of a mixture of activity (for example, diversity of uses, an evening economy and a small-firm economy); an appropriate built environment (for example, fine-grained morphology, a variety of building types, and a good quality public space); and cultural meaning (for example, a sense of history and progress, identity, and design appreciation and style). Further, Mommaas (2004, pp. 516–517) stated that 'ideal' cultural quarters 'cater for dense project-based intracluster transactions . . . as part of close face-to-face contacts between cultural professionals within an independent cultural community, accommodated within a self-managed and emotionally charged urban environment'. Urban authorities have attempted to preserve and encourage cultural production and consumption in 'successful' cultural quarters and as part of wider strategies for urban economic and cultural development (Montgomery 2004; DCMS 2004b). Montgomery (2003) further argued that the 'ideal' cultural quarter is a mixed-use public realm which combines strategies for increased consumption of the arts and culture with cultural production.

Cultural quarters appear to date from the 1980s in the US and have spread across the Western world. Most contemporary cities display identifiable quarters to which artists and cultural entrepreneurs are attracted (see Bell and Jayne 2004). Famous international examples include Soho and Lower East Side in New York City, Marais and the Left Bank in Paris, and researched examples in Dublin (Montgomery 1995; 2004); Rotterdam, Amsterdam, Utrecht and Tilburg (Mommaas 2004); and Adelaide (Montgomery 2004). Many cultural quarters within cities have developed in an 'accidental' fashion over a period of some time whereas some recent cultural quarters have been developed and marketed as purposeful models or policy instruments for urban regeneration (Montgomery 2003; Mommaas 2004). Examples in the UK include those in Manchester (Brown et al. 2000; Montgomery 2004; McCarthy 2007b), Glasgow (McCarthy 2007b), Liverpool (Couch and Farr 2000), Bristol (Griffiths 1995a), NewcastleGateshead (Hollands and Chatterton 2002; Miles, S. 2005), Sheffield (Oatley 1996; Brown et al. 2000), Dundee (McCarthy 2007b), Swansea (Tallon et al. 2006; Figure 12.1) and Wolverhampton (McCarthy 2007b). Montgomery (2004) and McCarthy (2007b) provided evidence from a range of case studies that showed that arts and culture-led urban regeneration works in a variety of contexts, and that strategic planning and policy can achieve successful cultural quarters. However, Miles and Paddison (2005) argued that a more critical appreciation of the aspirations of culture-led regeneration is required.

CASE STUDY: DUBLIN'S TEMPLE BAR

Dublin's Temple Bar emerged as the Irish capital's cultural quarter in the 1990s and acted as a model for similar developments in the UK (Montgomery 1995; 2004; Tiesdell et al. 1996; McCarthy 1998b). European Union funding kick-started the physical regeneration of the 218-acre urban quarter which is situated along the south of the River Liffey. Much of the built environment dates from the seventeenth and eighteenth centuries, however, there are many layers of buildings from the Danish settlement to Victorian, Georgian and more recent. The area was a backwater in the early 1990s even though it is strategically situated in the central city. Locally driven regeneration has transformed its fortunes from a place of dereliction into a place of discovery, vitality and exchange.

In Temple Bar, an urban place marketing strategy was employed to facilitate the emergence and development of the cultural quarter, which consisted of a number of components (Montgomery 1995):

- Going with the grain of the area
- Fostering greater diversity
- Encouraging development of culture-led activity
- Widening usage of local amenities
- Giving a sense of the public realm
- Enlivening the evening economy
- Programming of events designed to attract
- Animating the street
- Re-branding the city and the quarter to attract tourists and visitors

Montgomery (1995) argued that Temple Bar proves that culture is an economy, as it acts as

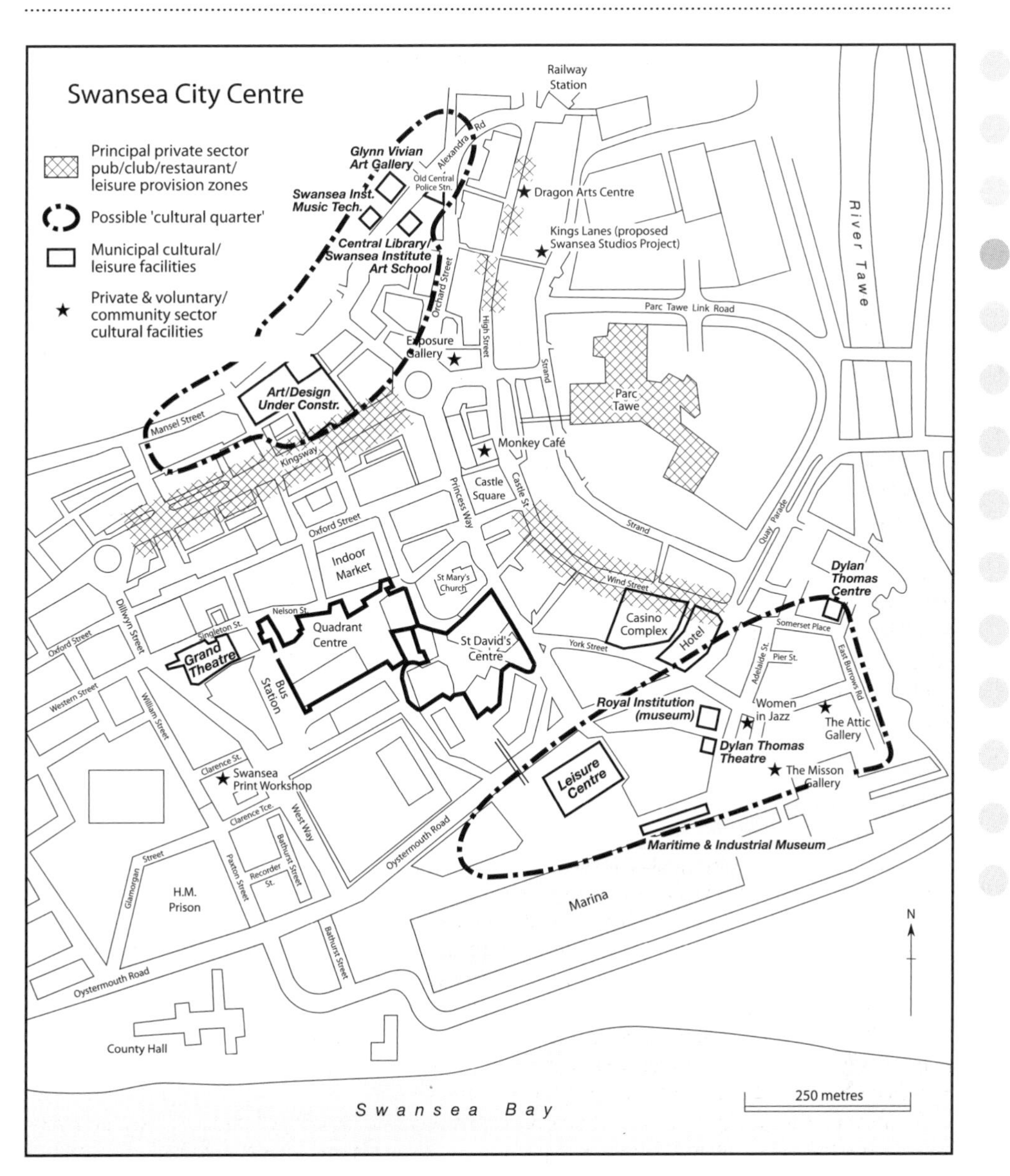

Figure 12.1 Swansea city centre's cultural quarters and cultural attractions

a generator of new work, new ideas, new wealth and new products. The city can be used as a showcase for human creativity. However, cultural quarters can become victims of their own success as evidenced by the proliferation of evening and night-time activities which have plagued many cultural quarters (see Tallon et al. 2006; Roberts 2006; Roberts et al. 2006).

URBAN FESTIVALS AND CARNIVALS

Many urban cultural quarters and wider city centre areas provide the location for the ever-proliferating number of urban festivals in the city. City festivals are analogous to flagship and post-modern developments in that their aim has been to improve city image and create jobs in the service sector. Festivals as a tool of urban regeneration emphasise the city of consumption and culture (see Short and Kim 1999, Chapter 7). Contemporary urban festivals and carnivals are closely linked to issues such as urban change and restructuring; tourism and place promotion; the leisure, arts and entertainment economy; the development of the night-time economy; and local community and culture. Urban festivals have long been recognised by city authorities as a fundamental element of the leisure and cultural regeneration of the post-industrial city (see Law 2000; Ravenscroft et al. 2000; Tallon et al. 2006; Tallon 2007b). Examples of high-profile urban festivals in the UK include the Edinburgh Festival, New Year's Eve celebrations across the UK, and festivals associated with city designations and major events (see Hughes 1999; García 2004b). Urban festivals seek to promote the carnivalesque and festive, often in response to economic, social and environmental problems that have emerged in city centres. They have frequently been developed as a component of wider 24-hour city strategies since the mid-1990s, often imposed top-down by city authorities such as in the case of the Leeds St Valentine's Fair (Harcup 2000). However, festivals have sometimes emerged from bottom-up community origins such as the illuminated carnivals that take place in smaller town settings in south-west England (Tallon 2007b; Plate 12.1).

The famous London Notting Hill Carnival, which takes place across the August Bank Holiday weekend and regularly attracts upwards of 1 million spectators, began in 1959 to celebrate the cultural diversity of the area in response to the race riots of the previous summer (see Chapter 1). Traditional carnivals have also evolved around celebrations of particular historical events, such as the Gunpowder Plot of 1605, or around religious events.

Plate 12.1 Somerset carnivals: economic and community regeneration
Photograph by author.

Many of these festivals or carnivals have now become synonymous with place marketing and urban tourism strategies.

In purely economic terms festivals and carnivals are drivers of urban regeneration. For example, £36 million was spent at the 2002 Notting Hill Carnival (DCMS 2004a), and estimates of the economic impact of Edinburgh's Hogmanay Festival have been put at over £10 million (Hughes 1999). Local festivals and carnivals can also have social and cultural benefits, and some places revolve around carnival activity throughout the year with activities such as fundraising, float building, meetings and social events (Tallon 2007b). Formal and informal learning can take place throughout somebody's life when involved in and actively participating in festivals and carnival, and such voluntary activity is a vibrant component of society. Local people use specialist skills and self-expression in the art and entertainment of carnival. Festivals and carnival have therefore been identified as being of social and cultural as well as economic potential, and can be considered as 'inclusive' in that people of any social class background, ethnic group, gender or sexuality can be, and are, represented in carnival circles (Tallon 2007b).

THE EVENING AND NIGHT-TIME ECONOMY AND THE '24-HOUR CITY'

Urban festivals and carnivals have been staged within the evening and night-time economy. This 'out of hours' economy has expanded rapidly since the 1990s as part of the shift to a post-Fordist, post-modern and entrepreneurial city (Comedia 1991; Lovatt 1996; Kreitzman 1999). The night-time economy has been trumpeted as being central to the rediscovery of the post-industrial Anglo-Saxon city (Lovatt and O'Connor 1995).

Cities also reflect, in spatial terms, changes in patterns of spending and consumption as reflected in the liberalisation of licensing hours through the Licensing Act 2003 (DCMS 2004c). Participation in the night-time economy has not been such a mass phenomenon as in other European countries (Lovatt and O'Connor 1995), and there has been a recognition that cities offer opportunities beyond the limited realm of the traditional pub in the 1980s (Heath and Stickland 1997). The development of the evening and night-time economy has been broadly supported by local authorities as a component of urban regeneration, but at the same time there has been increasing awareness of its downsides (Thomas and Bromley 2000; Hobbs et al. 2000; 2005; Hadfield et al. 2001; Roberts 2004; 2006).

The development of the 24-hour city is inextricably linked to safety issues, both in the daytime and night-time city (Oc and Tiesdell 1997a). The competitive impact of functional decentralisation has frequently resulted in either the loss of existing retail, office and entertainment facilities, or the failure to attract new investment. This has consequently reduced the vitality of the daytime city centre and generated safety concerns for shoppers, employees and residents (see Chapters 10 and 11). The safety of shoppers and residents in UK city centres has emerged as an important element in the strategies to retain their vitality. Safety considerations already have negative consequences for the daytime vitality of city centres with a fear of peripheral locations (Thomas and Bromley 1996). This has frequently been accompanied by the emergence of adverse environmental features, vacant premises and the proliferation of security shutters, particularly around the peripheries of city centres (Nelson 1997). Also, early redevelopment strategies contributed to anxiety by introducing significant degrees of spatial fragmentation between functions, and the loss of a substantial residential population (Thomas and Bromley 2000). Combined, these changes have negative implications for the perception of safety of all users of the daytime city centre. Fear of crime has impeded the ability of retailers to serve existing and potential customers (Oc and Trench 1993), and gained a place within the regeneration and planning agendas since the 1980s.

Bromley et al. (2000) noted that issues of city centre revitalisation at night are inextricably linked with safety concerns as many people avoid using the city centre through fear of crime. The initial focus was on women's fear of the city centre environment (Valentine 1989; Trench et al. 1992) and calls to 'reclaim the night' (Trench 1991). Research on urban space dating back to the work of Jacobs (1961) and Newman (1972) on people's perceptions of safety, long advocated the importance of a residential population in providing 24-hour activity, and a feeling of safety and security associated with natural surveillance. Housing has a major role in city centres aspiring to become 24-hour zones (Oc 1991, p. 238; Tallon and Bromley 2002; see Chapter 11). Fears for personal safety are a powerful barrier to the development of the evening and night-time economy (Bromley et al. 2000; Thomas and Bromley 2000). It was recognised at the end of the 1990s that opportunities existed to revitalise the evening and night-time economies but that progress in the direction of a 24- hour city centre was likely to be slow and incremental due to adverse perceptions of many safety and security issues, which remain apposite in the late 2000s (Roberts and Eldridge 2007a).

Crime and fear of crime in the daytime and night-time city are major contributors to the problems facing city centres and therefore many city authorities have implemented various measures to make city centres safer by day and night for leisure, shopping, business and residents (Pain and Townshend 2002). Oc and Trench (1993), Trench et al. (1992) and Oc and Tiesdell (1997d) explored measures to create safer city centres based on the US experience. Similarly, design and development solutions were seen by Paumier (1988) as essentials for reshaping the space–use composition and economic vitality of city centres, and instrumental in building safer environments in city centres. These principles were to promote diversity of use; emphasise compactness; foster intensity; provide for accessibility; build a positive identity; ensure balance; and create functional linkages. In order to overcome fear of crime in the city centre, it is often necessary to create a dense, compact and multifunctional core area. However, Oc and Tiesdell (1997b) argued that design can only create the preconditions for a safer environment.

Physical and functional fragmentation associated with the widespread development of new indoor shopping centres, office precincts and apartment complexes in US city centres raised concerns about the 'privatisation of public space' and the reduction of 'democratic space' (Brooks and Young 1993; Loukaitou-Sideris 1993; Robertson 1995). Feelings of anxiety are exacerbated by diffusing rather than integrating pedestrian activity by reducing levels of natural surveillance and self-policing through the introduction of features such as pedestrian subways and skywalks. Similar situations emerged in UK city centres (Thomas and Bromley 1996). The closure of many city-centre shopping precincts at the end of the trading day further fragments the structure of the city centre and creates a privatised space (Francis 1991). Thomas and Bromley (2000) suggested this curtails informal pedestrian browsing as well as presenting barriers to pedestrians, constraining them to use alternative less attractive and less safe routes. The shift towards the 'fortress' approach to solving safety problems in the city centre might 'design-out-crime' but 'design-in-anxiety' (Beck and Willis 1995).

In addition to physical fragmentation of the city centre, temporal functional fragmentation has always been a significant element of city centre life in the UK. Worpole (1998) noted that time as much as space separates urban functions, and this presents regeneration and planning problems. Activity during the day has focused on retail and business quarters, while evening and night-time functions have focused on concentrations of entertainment and cultural facilities such as public houses, nightclubs, restaurants, cinemas and theatres (Thomas and Bromley 2000). It has been widely argued that city centre revitalisation would be assisted by merging and integrating the working day into an expanded evening and night-time economy along the lines of the experience of many Western European cities (URBED 1994; Bianchini 1994; 1995). However, it has also been argued that several physical, social and cultural changes over the last 40 years have introduced temporal functional fragmentation into the city centre with a related series of safety considerations and concerns which detract from the revitalisation process (Thomas and Bromley 2000).

Social and cultural changes in the UK have altered the nature of the city centre out of daytime business hours. There has been a decline of older city centre entertainment facilities, such as smaller cinemas and dance halls, in favour of larger multi-leisure complexes combining multi-screen cinemas and bowling alleys, often in peripheral or decentralised locations (Hubbard 2000; 2002; 2003; Jones and Hillier 2000; 2001). This has narrowed the range and number of leisure opportunities available to people visiting the night-time city centre economy, as well as the kinds of people who visit. This has been accentuated by the extension of home-based entertainment activities for younger family groups, and the night-time 'withdrawal of people into privatised leisure' (Lovatt and O'Connor 1995, p. 132). These changes have been linked to the growth in the fear of crime in the city centre at night, particularly among women, the elderly and the suburban middle classes (Comedia 1991).

Thomas and Bromley (2000) indicated that strategies that have aimed to reduce the temporal divide associated with the 5.00pm 'flight' from the city centre had proved problematic as changes in the city centre have often created a night-time economy consisting of monocultural and exclusionary 'entertainment enclaves'. The night-time city centre is widely perceived as a threatening environment with high levels of anxiety with explicitly spatial dimensions which acts as a barrier to the regeneration and revitalisation of city centres (Jones et al. 1999; Thomas and Bromley 2000; Hadfield et al. 2001; Roberts 2006; Roberts and Eldridge 2007a).

In response to social change and the physical and temporal fragmentation obstructing the inclusive regeneration of the evening and night-time economy, a plethora of strategies have emerged to revitalise the night-time city centre economy. The evening and night-time economy is seen as essential in urban renaissance strategies and represents one of the most dynamic sectors of the city centre economy with considerable potential for development. There is scope for late-night shopping and a rolling programme of events through the day and night making use of the 'out-of-hours' period currently associated with the 'leisure and entertainment city'. While the daytime economy is dominated by retail and office activity, the evening economy revolves around entertainment and leisure activities (Montgomery 1994; Bromley et al. 2000; Roberts 2004; 2006).

However, the night-time economy currently thrives in most major city centres in the UK (Chatterton and Hollands 2003; Hobbs et al. 2003; Roberts and Turner 2005; Roberts 2006). For example, Jones et al. (1998) noted that central Newcastle had 154 public houses, 125 restaurants and 14 nightclubs and, similarly, Sheffield city centre contained 90 public houses, 2 cinema complexes and 17 nightclubs with a total capacity of 18,000 people in the late 1990s. Between 1997 and 1999 there was a 243 per cent increase in the capacity of licensed premises in Manchester city centre; and in Nottinghamshire the capacity rose from 61,500 in 1997 to 105,000 in 2004 (Roberts 2006). Across the UK as a whole, the nightclub industry was estimated to be worth around £2 billion a year (Malbon 1998). The economic consequences of nightclubbing offer potential regenerative possibilities that clubs and their associated cultural industries can contribute to declining city centres. 'Superpubs' have proliferated since the 1990s and are almost always established in primary shopping locations on the sites of former banks, post offices, cinemas, shops and hotels (Jones 1996), and fill the gaps left by retail decentralisation. Local authorities have generally expressed their support for such developments due to the regenerative economic and environmental benefits they bring. The evening activity extends the vitality of city centres beyond traditional working hours (Jones and Hillier 1997), and Ravenscroft et al. (2000) argued that leisure attractions and entertainment venues can have a major role to play in enhancing the vitality of city centres, both during the day and at night.

During the late 1990s, it became apparent that the growth and continuing development of the night-time economy of city centres raised a number of problem issues (Jones et al. 1998; Table 12.3). Reservations about the increase in number of pubs in city centres have been expressed due to resulting environmental impacts (Delafons 1996). There have also

TABLE 12.3 PROBLEMS OF THE NIGHT-TIME CITY CENTRE

Spatial and temporal fragmentation of functions
High levels of crime and disorder; perceived lack of safety
Loss of residential population
Lack of things to do; existence of pub and club culture
Single activity visits
Monofunctionality of city centres
Pedestrian-unfriendly environment
Poor public transport provision
Restrictive licensing laws (historically)
Night-time city is marginalised, controlled and regulated

been concerns regarding the detrimental effects of pubs on daytime users (Jones 1996) and on local residents due to the noise and disturbance that can be created (Tallon and Bromley 2004). The unfavourable image of the night-time city centre culture may in addition impact adversely on the daytime image (Thomas and Bromley 2000; Roberts 2006). This leads to a tendency for shoppers to leave the city centre after 5.00pm due to safety concerns associated with lower levels of activity on streets and in car parks as closing time approaches (Thomas and Bromley 1996; 2000). In addition, although night-time functions are a significant component of the city centre economy, their contribution to its continued 'vitality and viability' is restricted due to their exclusionary or 'mono-cultural' characteristics (Oc and Tiesdell 1997b; Roberts 2006; Roberts et al. 2006; Roberts and Eldridge 2007a). This is because the transformation of the night-time city centre has frequently resulted in a youth-dominated 'threatening environment' (Thomas and Bromley 2000), associated with a drinking culture (Jayne et al. 2006; 2008). This growth in 'youth culture' tends to be associated with males, heavy drinking, drugs and late-night violence, which have all contributed to a negative image (see Hobbs et al. 2003). This has reduced the attractions of the night-time city centre for a wider spectrum of the population as evidenced in cities such as Cardiff, Swansea and Worcester (Bromley et al. 2000; Thomas and Bromley 2000; Nelson et al. 2001; Bromley and Nelson 2002).

As shops and offices close after 5.00pm there is a virtual abandonment of the city centre and there is a hiatus of activity until after 7.00pm when early-evening diners and theatre and cinema-goers generally enter the evening economy, followed later by the late-night frequenters of the pub and club scene (Thomas and Bromley 2000). Consequently it has been recognised since the early 1990s that city centre revitalisation needs to address the problem of the 5.00pm 'flight' from the city centre and to ameliorate the negative image that has resulted from the emergence of an exclusionary night-time culture (Roberts and Eldridge 2007a). The infrequency with which the majority of people visit the night-time city centre due to safety concerns constitutes a major threat to the vitality and viability of the evening economy (Bromley et al. 2000; 2003b), and action is therefore required with a focus on attraction, access and amenity in the same way as recommended for city centre economies as a whole (URBED 1994), in order to extend the consumer base across the social and demographic spectrum.

The spatiality of fear in the night-time city centre emphasises places of poor design, transport termini and peripheral streets with a strong pub and club function in the late evening economy that result in very low levels of visits to the night-time city centre, particularly among women, older people and higher status groups (Bromley et al. 2000). City centres will only be able to develop successful and thriving evening economies if they are able to attract a substantial proportion of the population in an attractive, safe environment, with design solutions and quality in urban design also being promoted (Parfect and Power 1997; Chapman and Larkham 1999). However, as Bromley et al. (2000) indicated, the critical influence of function has received less attention than issues of environmental aesthetics and access in analysing the spatiality of fear. Although a degree of spatial segregation is often seen as an accepted practice for nightclubs in the city centre, the same approach might be applied to public houses as part of a zonal strategy of planning in the city centre (Bromley et al. 2000). Although there have long been conflicting views about the zoning approach to functions within the city centre (Oc and Tiesdell 1997b, p. 9), the creation of city quarters came into vogue in the 1990s and has been adopted as a loose form of functional zoning in many city centres in the UK. The endorsement of functional segregation in the short term should not detract from a long-term objective of achieving a more balanced evening city culture across the whole city centre which attracts people of both genders and of all ages and social groups, emulating the European model (Bianchini 1994; Montgomery 1994; Bromley et al. 2003).

The 24-hour city approach to revitalising city centres came to prominence in the 1990s and is associated with the revitalisation of the evening and night-time economies (Lovatt et al. 1994; Lovatt 1996; 1997; Heath 1997; Heath and Stickland 1997; Hughes 1999). The strategies associated with this aim to extend the activity period and social mix of city centre users by offering a wider range of evening and night-time functions, whilst, in the process, creating safer city centres with an image more likely to attract future inward investment. An actual 24-hour activity pattern is rarely the ultimate aim. The intention for most cities is to extend the economic, social and cultural life of city centres from early morning, around 8.00am, to the early hours of the next morning, between 2.00am and 4.00am.

The origins of the 24-hour city concept in the UK can be traced to Manchester's 'more hours in the day' initiative in 1993 (Lovatt et al. 1994). Since then the idea has rapidly spread and elements of the approach have been adopted by most UK cities for a variety of reasons (Table 12.4) and through a range of strategies (Table 12.5).

TABLE 12.4 REASONS FOR 24-HOUR CITY STRATEGIES

Reasons
Development of a safer city centre
Improve city image/attract inward investment (business, tourists, residents)
Economic regeneration
Enhance local service provision for the population
Specific local reason, festival or sporting event
Following another city's lead
Prompted by central government

Sources: **Heath (1997); Heath and Stickland (1997).**

TABLE 12.5 TWENTY-FOUR-HOUR CITY INITIATIVES

Initiatives
Licensing changes (overcome restrictions; special events)
Retail initiatives (late-night shopping)
Restaurant/café promotion initiatives (café culture; café quarters; corridors of activity)
Street entertainment and festivals
Lighting/CCTV schemes

Sources: **Heath (1997); Heath and Stickland (1997).**

The benefits of the development of the evening and night-time city centre have been mainly economic. The night-time economy is now thriving and big business with around 110,000 to 130,000 revellers visiting Manchester city centre at weekends and 80,000 going out in the centre of Newcastle (Roberts 2006). Even smaller sub-centres, such as Kingston upon Thames attract up to 12,000 people at night (Roberts 2006). In Manchester in the late 1990s the night-time economy employed over 5000 people and was worth around £300 million annually (Lovatt 1997). The evening and night-time economies are estimated to be worth between 5 and 15 per cent of the UK economy (Kreitzman 1999). Café culture has also proliferated in recent years along the lines of continental Europe. Nightclubs and associated 'cultural activities' have clearly offered regenerative possibilities to declining city centres such as Manchester, Leeds, Sheffield and Bristol (Plate 12.2).

However, a range of problems has emerged to blight the development of the city centre night-time environment (Thomas and Bromley 2000; Roberts 2006). Problems have been the compaction of the city centre due to decentralisation with dead space on the margins; a reduction of the overall attraction of the city centre in relation to the new decentralised and privatised indoor malls and entertainment complexes; increasingly privatised space and a reduction of public space due to the proliferation of indoor shopping malls; fragmentation of functions due to post-war planning ideologies; a loss of 'eyes on the street' (Jacobs 1961) and 'natural surveillance' (Newman 1972); the creation of 'hot spots' of crime, disorder and fear in the city centre (Nelson et al. 2001); the proliferation of physical incivilities including vacant premises, graffiti, litter, and shutters, and social incivilities such as the homeless, alcoholics, drug addicts, and the mentally ill; and a marked temporal–functional fragmentation (Thomas and Bromley 2000; Tallon 2001). A socio-temporal polarisation is also apparent in

Plate 12.2 Landscape of the night-time city
Photograph by author.

the city centre with certain groups visiting at certain times (Bromley et al. 2003b). These have all contributed to the 5.00pm 'flight' from the city centre; a limited emergence of late-night shopping or café culture; city centres remaining largely unpopulated out of working hours with little impact on their vitality and viability; and the dominance of an exclusionary male-dominated youth culture of 16–30 year olds (Bromley et al. 2000; Thomas and Bromley 2000). This alcohol-dominated landscape forms the stereotypical image of the night-time city in the media and among the population at large. Hadfield et al. (2001) reported that in Manchester the rise in the number of licensed premises had been accompanied by a rise in the number of assaults; between 1998 and 2001 the capacity of licensed premises increased by 240 per cent, while the number of assaults reported to the police increased by 225 per cent. The majority of hot spots for violence in city centre areas are those with high concentrations of licensed premises, with the number of incidents peaking between 9.00pm and 3.00am on Friday night into Saturday morning (Nelson et al. 2001; Bromley and Nelson 2002). Indeed, the major problem of the contemporary night-time city is the domination of alcohol and drunkenness and the impacts of this on public space (see Jayne et al. 2006; 2008; Roberts 2006; Roberts et al. 2006; Roberts and Eldridge 2007b).

There are a number of speculative policy implications for the future revitalisation of the evening and night-time city centre in line with the 24-hour city concept. Despite an increase and diversification of activities since the late 1990s, there remains a need to develop a wider range of facilities and activities to straddle the temporal divides, such as late-night shopping, café culture and a variability of types of pubs and restaurants. The range of visitor groups must be broadened to include the whole age spectrum and all gender, social and ethnic groups (Thomas and Bromley 2000; Bromley et al. 2003b; Roberts and Eldridge 2007a). Access improvements with a view to safety at the city centre core are essential, with safer public transport and car parks achieved through target hardening measures and environmental softening. There must also be a concerted management of risk perception and a re-imaging of the night-time city centre with both explicit policing through police presence, CCTV, pub- and club-watch schemes, and the tackling of street and circuit drinking, and implicit policing encompassing design, management, and staggering of closing hours, associated with the overhaul of the licensing laws in 2003 (Thomas and Bromley 2000; Roberts 2006). There are also strong arguments for breaking down concentrations and enclaves of entertainment, which tend to be expressed as problematic 'hot spots', although it is recognised that policing can be concentrated in these zones. The segregation versus dispersal of night-time activities debate continues (Bromley et al. 2000). Promoting residential and employment functions within the city centre appears fundamental to a thriving city centre during the night, with increased internal markets, improved perceptions of safety, and contributions to economic regeneration and social revitalisation (see Chapter 11). The process of change is slow and incremental as the problems of the night-time economy are as much social and cultural in nature as they are planning and regeneration policy-based. Subtle and innovative approaches are required and the same model cannot be replicated across cities.

The image of a vibrant and urbane 24-hour city is not widely apparent in reality and remains a romantic vision rather than current reality in the late 2000s. Despite high-profile reforms to licensing laws in 2003, it appears that the 'relentless expansion of nightlife' has transformed UK city centres from 'creative city' to 'no-go areas' (Roberts 2006). The state of the 24-hour city is perhaps fittingly summarised by Hadfield et al. (2001, p. 300):

> The concept of the '24-hour city' is not in good health. It's 2.15 on Saturday morning in an English city centre and the 'Mass Volume Vertical Drinker' has assaulted the concept and all its good intentions, leaving it for dead in streets splattered with blood, vomit, urine, and the sodden remains of take-aways.

URBAN TOURISM AND URBAN HERITAGE

Linked to the promotion of the leisure and cultural economy and the 24-hour city, urban tourism and heritage have emerged as a new form of activity since the late twentieth century. Tourism-led regeneration has emerged as a key facet of the post-industrial and postmodern city (Judd 1994; Holcomb 1999; Judd and Fainstein 1999). In the post-industrial economy, cities have become major tourist attractions rather than places to escape from (Williams 1998). Active and clever promotion of urban business tourism, sport and event-related tourism and development of new attractions focusing on leisure shopping or industrial heritage have permitted places with no tradition of tourism to develop a new industry that has revitalised local or regional economies. Urban areas have assumed a greater role as visitor destinations, encompassing not only those places with an obvious resource base such as global cities and historic towns, but also former industrial conurbations with, on the face of it, relatively modest appeal (Table 12.6), often linked to urban place marketing (Selby 2004; see Chapter 6). Large cities such as London, Edinburgh and York have enjoyed flourishing tourism typically centred on sightseeing at places of interest, and on visiting galleries and museums, theatres and concerts, and restaurants and clubs, attracting both foreign and domestic tourists. Top UK urban attractions in the late 2000s included the London Eye, Tower of London, Canterbury Cathedral, Tate Modern, Albert Dock in Liverpool, Edinburgh Castle and York Minster. Visitors are attracted to urban areas for a variety of motivations (Table 12.7). Law (2000) illustrated the complexity and diversity of ways in which tourism is linked to urban regeneration, particularly in terms of physical, economic and social regeneration. Regeneration can be achieved through increased visitor numbers, creation of a new image, increased income, expansion of other economic activities, population growth, enhanced civic pride, job creation, and further investment in attractions and environmental improvements (Law 2000, p. 120). The benefits of urban tourism for urban regeneration are not purely that more jobs will be created by the industry. The basis of the strategy is that tourists will be drawn to the area, generating income, which directly or indirectly creates jobs. Hopefully, profits from the industry will encourage further investment leading to a virtuous circle of growth. Increasingly, the arts, culture and entertainment (ACE) industries are employed by municipal authorities, often acting in partnership with private enterprise, as a catalyst for regeneration (see Law 1991; 1992; 1993; 2000; Page 1995; MacDonald 2000; Bradley et al. 2002).

Heritage tourism and development have been a component of wider urban regeneration

TABLE 12.6 A TYPOLOGY OF URBAN TOURIST DESTINATIONS

Capital cities (e.g. London, Paris, New York)
Metropolitan centres, walled historic cities, small fortress cities (e.g. York, Canterbury)
Large historic cities (e.g. Oxford, Cambridge, Venice)
Inner-city areas (e.g. Manchester)
Revitalised waterfront areas (e.g. London Docklands)
Industrial cities (e.g. nineteenth-century Bradford)
Seaside resorts and winter sports resorts (e.g. Lillehammer)
Purpose-built integrated tourist resorts
Tourist-entertainment complexes (e.g. Las Vegas, Disneyland)
Specialised tourist service centres (e.g. spas, pilgrimage destinations)
Cultural/art cities (e.g. Florence)

Source: **Page (1995, pp. 16–17).**

TABLE 12.7 A TYPOLOGY OF URBAN TOURISTS: MOTIVATIONS FOR VISITING URBAN AREAS

Visiting friends and relatives
Business travel
Conference and exhibition attendance
Educational reasons
Cultural and heritage tourism
Religious travel (e.g. pilgrimages)
Hallmark events attendance
Leisure shopping
Day trips

Source: **Page (1995, p. 48).**

initiatives linked to leisure and culture. Such development has aimed to regenerate the historic parts of cities that typically surround waterfronts, docks and cathedrals (Tiesdell et al. 1996). Synonymous with other approaches focused on leisure and culture, the transformation of urban image and job creation in the service sector are the central endeavour. Attraction of United Nations World Heritage Site designation is also associated with urban economic regeneration (see, for example, Jones and Munday 2001).

SPORT AND THE ENTREPRENEURIAL CITY

Along with festivals, spectacle and urban tourism, sports are an element of popular culture which seek to contribute to representing cities in a global world (Short and Kim 1999; see Chapter 6). Sports franchises, signature sports events and infrastructure for sport are increasingly important for facilitating leisure and cultural urban regeneration in the entrepreneurial city (Shropshire 1995; Short and Kim 1999). Many cities have made considerable efforts to build new sports stadia with the intent of attracting a professional sports team that would provide them with big city status, a vital and youthful image, and a powerful vehicle for economic development. In US cities, there has been 'stadium mania' to capitalise on popular culture through sports. In the UK, stadia have been employed to regenerate declining inner city and city centres areas (see Thornley 2002), recent examples being Cardiff's Millennium Stadium (Jones 2002; Davies 2008) and the City of Manchester Stadium (Davies 2008). There is also the role of sports in transforming city image (Smith 2005), attracting tourists to cities (Euchner 1999; Faulkner et al. 2001), and contributing to sustainable urban regeneration (Smith 2007).

Hosting large-scale sporting and other events is often justified by the envisaged regeneration benefits and have epitomised top-down approaches to regeneration for fifty years (Smith 2007). Large-scale events are sometimes referred to as 'peripatetic events', as they are staged regularly but by different host cities. This makes them different from one-off events or events that are staged at regular intervals at the same place (Smith 2007). Large-scale sporting events contribute to urban regeneration; act as vehicles for levering funds from regional, national and supra-national agencies; induce opportunities for funding wider regeneration programmes; generate publicity for host cities; and showcase opportunities in the city linking events and regeneration (Smith 2007). The ultimate sports event in terms of place competition, profile and potential for urban regeneration is the global spectacle of the Olympic Games (Short and Kim 1999, pp. 92–94). So-called 'wannabe world cities' compete to stage the games to confirm their position in the league of world cities, as they can give a city international recognition, a psychological boost, and a great opportunity for business, real estate development and increased investment (Short and Kim 1999, p. 100). Since the early 1990s, the competition to stage the games has become increasingly fierce, with the cities of Barcelona, Atlanta, Sydney, Athens, Beijing and London being beneficiaries (see Hill 1996; Senn 1999; Rutheiser 1996; Waitt 1999; Whitelegg 2000; Gold and Gold 2007). Many European cities such as Amsterdam, Birmingham, Manchester, Rome and Stockholm have sought unsuccessfully to stake their claim as a world city to attract foreign investment through staging the Olympics. The Winter Olympics perform a similar role in urban and regional policy (Essex and Chalkley 2004).

However, there are problematic issues surrounding the staging of the Olympics, as they often result in city debt and obsolete stadia and other facilities once the competition ends. High-profile failures such as the 1976 Montreal Olympics (Whitson 2004) and other large-scale events such as the 1992 Seville Expo (Gold and Gold 2005) were not successfully integrated with the rest of the city and this limited regeneration outcomes (Smith 2007). Sheffield's hosting of the World Student Games in 1991 was subsidised by Sheffield City Council and witnessed the regeneration of the Don Valley, but left the city in severe financial crisis afterwards (Foley 1991; Loftman and Nevin 1996; Henry and Paramio-Salcines 1999; Hall 2001, p. 82). There are other problematic issues relating to using large events as a tool of regeneration. 'Successful' events in themselves have often had limited wider effects on local communities. Uncertainty about whether mega-events such as the Olympics would actually occur in a city has arguably produced planning blight and obstructed private investment, and failed bids can prove costly (Cochrane et al. 1996; Essex and Chalkley 2004). Increased tourism subsequent to staging major events is not guaranteed in the long or short term (Balas 2004; Smith 2007). Smith (2007) suggested that urban regeneration directly attributable to an event remains rare.

Smith (2007) proposed ten key generic principles for host cities to maximise the regeneration legacy of large-scale sporting events, especially relating to community benefits in the context of the 2012 Olympics in London. These were derived from research regarding the impacts and legacies of events in various international contexts. These ten principles were:

- to embed event strategies within wider urban regeneration programmes;
- to use the event as a coherent theme and effective stimulus for parallel initiatives and more diverse regeneration projects;
- to ensure that regeneration planning is fully incorporated into the initial stages of planning for an event;
- to promote shared ownership and responsibility among all partners of the legacy and event programmes;
- to design effective organisational and structural arrangements between event regeneration agencies and event management representatives to ensure joint working towards clearly defined and shared goals;
- to allocate sufficient human and capital resources throughout the lifetime of event regeneration projects to achieve sustained effects;
- to design event regeneration projects to prioritise the needs and engagement of the most disadvantaged members of the target community;
- to try to ensure an even geographical dissemination of positive impacts among targeted areas;
- to ensure that event-themed social and economic regeneration initiatives built upon, and connect with, any physical and infrastructural legacy;
- to ensure community representation from the planning stage onwards to promote community ownership and engagement.

Cities remain desperate to attract large-scale sporting events, and other large-scale events in the age of urban entrepreneurialism and place promotion (Short and Kim 1999). Therefore, Smith (2007) argued that it is important to consider how cities can best use these to assist urban regeneration, particularly as related outcomes are usually heavily emphasised in the rationale for event strategies. Unless parallel initiatives are pursued, large-scale events should be regarded as branding, promotional tools or ways of leveraging, rather than as a genuine instrument for urban regeneration.

THE CREATIVE CITY

Developing in parallel with the cultural city has been the notion of the 'creative city' (Landry and Bianchini 1995; Landry 2000; Florida 2002; 2005). The Fordist production-based economy consisted of mass production, standardisation, large and inflexible production, a strict division of labour, and was rigid and uncreative (see Chapter 7). By contrast, post-Fordist cities are characterised by flexibility and variation, decentralised output, rapid change, and are knowledge-based and creative. The changing urban economy has progressively favoured knowledge over raw materials, short-term contracts, frequent changing of jobs, outsourcing, and more private-sector and less public-sector investment. Florida (2002, p. 6) argues that 'access to talented and creative people is to modern business what access to coal and iron once was'.

Many cities are now post-industrial (see Chapters 1 and 6) and are based on a service sector of mainly tertiary employment. Homogenised cities have evolved which have witnessed an assortment of similar strategies including tax breaks, grants, improved infrastructure, regeneration of former industrial spaces, and presentation of a new image through festivals and promotions. The same ideas often equal the same results, and a generic economic base composed of office work and call centres emerges. However, a small number of cities, initially in the US, are separating from the pack – the so-called 'creative cities'. According to Florida (2002, p. 237), cities in the US that display a high proportion of 'creatives' in the workforce include Washington (38 per cent), Boston (38 per cent), Austin (36 per cent), San Francisco (35 per cent) and Seattle (33 per cent). These cities exhibit a high rate of economic growth based on the quaternary sector and are becoming foci for innovation.

Creative classes who are driving the growth of creative cities embrace scientists, designers, engineers, university researchers, professionals, artists and engineers. Creative professionals consist of senior managers and business innovators. Florida (2002) differentiated this creative class from the service sector comprising tertiary workers such as junior managers, administrative workers, and those employed in call centres or data entry jobs. The creative class are attracted to interesting jobs and flexible work in vibrant cities where people with unconventional lifestyles fit into a diverse and tolerant community. Florida (2002) argued that eccentrics and non-conformists are at the very heart of the process of innovation in creative cities. Attractive cities display an old and authentic centre with a sense of 'place', local and independent bars, cafés and shops, and a university.

Creative cities have a different economic base focused on the quaternary sector; a different demographic base representing a younger population who are more likely to be single; and a different culture with a thriving alternative lifestyle and music scene, and fewer 'corporate' arts events. Creatives are fashion setters and innovators who have greater spending power, and are usually university graduates who are more likely to own a business. Places with a 'creative city' image are an attractive location for 'ordinary' businesses to locate. The reputation of creative cities attracts more creatives through agglomeration economies. Through a multiplier effect, even a small creative class can act as a magnet for more jobs.

Florida (2002, pp. 274–275) placed creative cities by identifying four categories of contemporary US cities, which can to an extent be applied to UK urban areas. The first was 'classic social capital communities', characterised by small towns, strong community, and traditional and family values. The second type termed 'organisational age communities' related to the former Fordist mass production centres of Detroit and Sheffield in the UK, which have experienced severe de-industrialisation. A third category of cities described as 'nerdistans' were those cities with a high proportion of jobs in information technology, but many jobs in basic data entry, and an abundance of standardised housing estates, shopping malls, and multi-national retail and entertainment environments. A UK example matching this description might be Swindon and other M4 corridor urban areas. Fourthly, 'creative cities' are distinguished by Florida (2002) from the other three categories by the focus on new creative post-Fordist production. 'Creativity and cities revolves around a simple formula, the 3T's of economic growth: technology, talent, and tolerance' (Florida 2005, p. 6; see Florida 2002; 2005; Chapter 6).

MEASURES TO PROMOTE CREATIVE INDUSTRIES

Linked to the idea of the creative city, governments at national and local levels are increasingly viewing the creative industries as the basis for a more secure, competitive and sustainable future (see Aitchison and Evans 2003; Dziembowska-Kowalska and Funk 1999; Hall 2000). Many cities are seeking to foster the creative industries as a foundation for economic and social success in the post-industrial era. A key reason for this interest is the growing importance of the 'knowledge economy', which sees innovation and ideas as the main source of added value (Leadbeater 1999). A further reason for the emphasis on the creative industries has been the 'culturalisation' of the economy (Scott 1999; 2000). Creativity has, especially in working with cultural materials, come to assume a critical economic and social role (Griffiths 2005). Although the main arguments supporting the creative industries have been economic in nature, non-economic benefits have also been put forward. For example, a report for the Social Exclusion Unit (DCMS 1999) drew attention to the ways in which the arts and creative activity can contribute to social development in deprived communities by developing people's potential and self-confidence. It is also argued that they can create a sense of common identity and interest, can foster networks of social connection, and help build positive links with the community (Griffiths 2005). The government has argued that the arts and creative activity can offer routes into employment as they are associated with rapidly growing industries (DCMS 1999). These arguments have given impetus to community-based arts provision (see Bassett et al. 2003).

Creative industries are difficult to define. The Creative Industries Task Force (1998, p. 10) stated that they are activities 'which have their origin in individual creativity, skill and talent and which have a potential for wealth and job creation through the generation and exploitation of intellectual property'. The main sub-sectors of the creative industries encapsulate advertising, architecture, the arts and antiques market, crafts, product and fashion design, film, leisure software, the music industry, the performing arts, publishing, and radio and television (on media industries and regeneration, see Montgomery 1996; Griffiths et al. 1999; Bassett et al. 2002a). The UK appears to have a major competitive advantage in the field of the creative industries, and creative enterprises seem to cluster in urban settings thus representing a potential contributor to urban regeneration.

Knowledge of the creative industries sector is still underdeveloped as the sector has unclear boundaries, is not effectively captured by official statistics, and is diverse in its nature. The sheer diversity of the sector makes it difficult for public policy makers who seek to foster and promote the creative industries (Griffiths 2005). Caves (2000) identified seven fundamental economic properties of creative activities. First, it is a field in which nobody knows which products will be successful. Second, there is an art-for-art's-sake culture which leads practitioners to place great importance on the qualities of their products that consumers might not even notice. Third, is the motley crew phenomenon which alludes to the immense diversity of the skills and aesthetic judgements required in the realisation of creative products which generates problems of co-ordination. Fourth, there is an infinite variety and high level of differentiation of products. Fifth, there is an A list/B list characteristic that enables top-ranked individuals to earn rewards that are greatly in excess of the minimum needed to secure their services. Sixth, the making of creative products is highly time-bounded in that the co-ordination of activities in time is critical. The seventh and final aspect is that products and rents in the field of the creative activities are highly durable and can generate incomes many years after they were first created (Caves 2000; Griffiths 2005).

Creative lifestyles often blur the boundaries between work and non-work for urban cultural entrepreneurs (Mommaas 2003), particularly within the fashion design industry in cities such as Manchester (Banks et al. 2000). Linked with this, there is a strong dependence on informal, highly localised social networks within the sector, a consequence of which is collaborative competition between businesses (Griffiths 2005).

Griffiths (2005) argued that creative enterprises can exist anywhere because of their typically small scale, their dependence on individual creativity and their capacity to exploit new media technologies. However, it is cities that have provided the most fertile conditions for the growth of such industries and they are embedded in a rich and varied cultural milieu. This is partly because the de-industrialisation of former manufacturing centres left behind a stock of redundant industrial spaces, which were ideal for the needs of creative industries until displaced by the effects of gentrification (Tallon et al. 2006; see Chapter 11). However, a principal reason put forward by Griffiths (2005) is the highly localised informal networks and flows of cultural value which are important to the working of these industries and which are abundant in cities. Creative workers are attracted to the mix of street level cultural and leisure opportunities offered by cities, and the acceptance of social diversity and experimentation is most readily provided by successful cities (Florida 2002; 2005). Examples are the cultural industries quarters in Sheffield (Brown et al. 2000); Dublin's Temple Bar (McCarthy 1998b) and Bristol's cultural industries cluster which shows linkages between art, media and design in the context of wider waterfront regeneration (Bassett 1993; Bassett et al. 2002b; 2003; Griffiths 1993; 1995a; Griffiths et al. 1999; 2003).

Florida (2002, p. 232) accounted for the urban conditions that attract the 'creative class' in terms of the 'quality of place':

> The quality of place a city offers can be summed up as an interrelated set of experiences. Many of them, like the street-level scene, are dynamic and participatory, you can do more than be a spectator; you can be part of the scene. And the city allows you to modulate the experience: to choose the mix, to turn the intensity level up or down as desired, and to have a hand in creating the experience rather than merely consuming it. The street buzz is right nearby if you want it, but you can also retreat to your home or other quiet place, or go into an urban park, or even set out for the country. This is one reason canned experiences are not so popular. A chain theme restaurant, a multimedia-circus sports stadium or a pre-packaged entertainment-and-tourism district is like a packaged tour: you do not get to help create your experience or modulate the intensity; it is thrust upon you.

Public authorities can engage with creative enterprises to assist urban regeneration, and there has been a shift in ways of using culture for this purpose. An early phase during the 1980s and 1990s witnessed a focus on consumption-based approaches to culture-led regeneration geared to attracting mobile capital and high spending visitors through place promotion. This resulted in cities directing massive investment towards flagship cultural buildings including museums, art galleries, concert halls and high profile public art installations (see Hall and Robertson 2001; Cameron and Coaffee 2005; Sharp et al. 2005). However, since the mid-1990s cities have adopted a broader perspective which has increasingly come to see the benefits of culture not just as a tool of place promotion but also as a field of production, and the benefits in terms of its capacity to forge shared identities and foster social cohesion (Griffiths 2006). Griffiths (2005) argued that this shift has led to difficulties for urban public authorities in intervening and engaging with the creative industries. Creative industries are driven by ideas, fantasies and individual passions, which are not attributes that square comfortably with conventional policy-making cultures. The diversity and fragmented nature of the creative industries also make it difficult to identify common sector-wide interests. In addition, micro firms and sole traders have little capacity in terms of time or inclination to invest in extended dialogue with public policy makers when the benefits are often diffuse and indirect.

As explored earlier, the period since the mid-1990s has seen the rising popularity of the 'cultural quarters' or 'cultural districts' approach among city authorities. This response has been an effort to engage with the creative industries and encourage the clustering of creative industries (Montgomery 2003; 2004; Tallon et al. 2006; McCarthy 2007b). This approach 'fits the long-standing technocratic mentality of order and control. Creative enterprises tend to be seen as a handy, relatively low cost, way of breathing life into marginal districts for which more "high-value" conversion is a long way off. There are also place-marketing and image-building gains to be achieved by adding "creative zones" to the mosaic of themed spaces' (Griffiths 2005, p. 5).

There are doubts about the extent to which viable spaces for creative production can be brought about by public policy initiatives where they have not already emerged through the 'organic' workings of the local cultural ecology. There are a number of ways in which cultural quarters have developed in different cities (McCarthy 2007b). Shorthose (2004) distinguished between 'engineered' and 'vernacular' approaches. An engineered approach takes its lead from professional and institutional perspectives and priorities. This approach has been a common feature of urban cultural policy exemplified by cities such as Amsterdam (Hitters and Richards 2002), Birmingham (Hubbard 1996a), Bristol (Bassett et al. 2002b) and Leicester (Shorthose 2004). By contrast, 'vernacular' approaches are characterised by bottom-up informal interactions, everyday social and cultural networks and mutual interdependence. Examples of this kind of smaller-scale and organic approach can be seen in Amsterdam (Buchel and Hogervorst 1997), Rotterdam (Hitters and Richards 2002), Manchester (McCarthy 2007b), Bristol (Griffiths 1999), Nottingham (Shorthose 2004) and Swansea (Tallon et al. 2006). Individual cities can display both engineered top-down approaches and bottom-up vernacular approaches. Cities draw upon a full repertoire of approaches rather than exclusively on engineered or vernacular approaches. The advantage of the former policy-led approach is that it is capable of developing neat, long-term plans for managing cultural resources and delivering support to the creative sector. It provides a framework within which decisions about cultural development can be made that take explicit account of diverse needs and competing demands. The latter approach is better able to cater for the interactivity and fluidity of creative practice, and is more consistent with a view of culture and creativity as 'ends in themselves' rather than as instruments for economic ends. The outcomes of vernacular approaches are also less likely to be susceptible to the vagaries of funding regimes, property markets and other macro-economic variables. Griffiths (2005, p. 7) concluded that policy-makers need to be able to 'read' the creative ecology of a particular place both in terms of its stage of development and the blend of its sub-sectors of the creative industries that make it up.

SELLING THE CITY OF SPECTACLE

Flagship venues have been common features of the landscapes of urban regeneration in post-modern cities, and usually have a central leisure or cultural element (see Smyth 1994).

These are constructed as attractions in their own right; as catalyst or marshalling points for further investment; as marketing tools for a city or destination; and as central features of large-scale urban regeneration projects. Key recent examples include the Guggenheim Museum in Bilbao, Spain, the Lowry Centre in Salford Quays, Manchester, the Baltic Centre and Winking Eye Bridge on the River Tyne at NewcastleGateshead, and the City of Manchester Stadium built for the Commonwealth Games. These are large-scale projects that aim to initiate a multiplier effect through jobs in the construction, and retail, leisure and entertainment sectors. The flagship venue sites can become magnets for further investment in housing, retail and office developments. Investment in flagship venues is often in central brownfield sites following remediation. Flagship cultural venues usually take on the characteristics of stylish post-modern architecture (Table 12.8; see Chapter 6). Essentially though, '[post-modern] urban design simply aims to be sensitive to vernacular traditions, local histories, particular wants, needs and fancies, thus generating specialised, even highly

TABLE 12.8 CHARACTERISTICS OF MODERN AND POST-MODERN ARCHITECTURE

Modern architecture (early twentieth century)

- Repetition of a simple shape – squares, rectangles, boxes
- Uniformity and mass production of design styles
- Complete lack of ornamentation
- Harsh, industrial-looking materials, especially reinforced concrete
- A flat roof
- Buildings designed to be constructed using latest techniques
- Dominance over the surrounding environment
- Functional
- Rational
- Logical
- Cheap
- Efficient
- Left-wing agenda (social housing)

Post-modern architecture (post-1970s)

- An eclectic 'collage' of architectural styles from past times
- A mixture of styles from different places, for example, American-style diner, Italian-style pizzeria
- Lots of play between different surfaces, materials and colours
- A high degree of 'fake', for example, floors and columns that look like marble may well be artificial
- A high degree of 'reinforcing' of local designs and materials as well as the past
- Spectacular
- Playful
- Ironic
- Produced for specialist markets
- Little social agenda

Source: **Hall (2006, p. 100); Mills (1990).**

customised architectural forms that may range from intimate, personalised spaces, through traditional monumentality, to the gaiety of spectacle. All of this can flourish by appeal to a remarkable eclecticism of architectural styles' (Harvey 1989, p. 66).

Post-modern, iconic, showpiece buildings associated with the 'design-led' urban renaissance agenda since the late 1990s (Urban Task Force 1999; Bell and Jayne 2003; Punter 2007) have emerged in the central areas of US, UK and other Western European cities. In world cities globally and in UK cities such as London (McNeill 2002), Manchester and Liverpool, tall buildings have also been employed recently to contribute to wider urban regeneration and place promotion aims (see Abel 2003; Costello 2005; Charney 2007). This sometimes involves the conversion of former council house tower blocks (Jacobs and Manzi 1998; Hoskins and Tallon 2004). These developments have been the subject of fierce debate linked to issues such as conservation, exclusion, privatisation and gentrification.

LEISURE AND CULTURAL REGENERATION: CRITICAL ISSUES AND POLICY IMPLICATIONS

Key issues and policy implications can be synthesised from the variety of approaches to urban regeneration which come under the banner of leisure and cultural regeneration. Specifically, this chapter highlighted the nature of the cultural city and cultural quarters in cities, new urban festivals and carnivals, the development of the evening and night-time economy and '24-hour city', the promotion of urban tourism and urban heritage, the use of sport in urban regeneration, the notion of the creative city and measures to promote creative industries, and the selling of the city of spectacle through flagship urban regeneration projects.

In general, positive experiences of regeneration through leisure and culture have been the diversification of economies; the conservation and promotion of heritage and culture; the acknowledgement of diverse cultures within cities; and the social and community as well as economic benefits of community-based cultural regeneration projects. Negative consequences have been that flagship cultural projects have replicated some earlier mistakes of property-led schemes. They appeal to similar markets, and are often in direct competition; there are examples of high profile 'white elephants' such as the Millennium Dome; there are concerns revolving around exclusion and excluded cultures; high culture rather than community culture is favoured; the selective nature of leisure and cultural schemes masks wider social problems; and the exploitation of local cultures and communities can occur. Seven connected issues that have implications for urban regeneration can be distilled.

A first issue relates to the commercialisation and commodification of local culture, for example, through the Glasgow Capital of Culture 1990, and in events such as the Handsworth carnival in Birmingham (Hall 2001, p. 164).

Second, there is often a diversion of resources from social and community facilities and services towards investment in flagship cultural projects and events. For example, Birmingham City Council invested over £100 million of public money in the International Convention Centre, which opened in 1991, and continued to underwrite its losses for several years subsequently (Loftman and Nevin 1996; Hall 2001, p. 82). Although cultural policies commonly aim to link cultural activities to economic and social policy objectives (Mommaas 2004), Long and Bramham (2006) argued that there has been a dubious contribution of cultural projects to social inclusion.

Third, selective narratives are often employed to sell and stage local events that can marginalise local and oppositional voices and favour elite groups (Boyle and Hughes 1994; Waterman 1998).

Fourth, gentrification can be encouraged through leisure and cultural regeneration, the staging of 'middle-class' events, and the development of flagship property-led buildings. Their relevance to local populations is often questionable.

Fifth, displacement of cultural producers and fracturing cultural networks are issues linked to the gentrification of the cultural landscape in city centres. It can be difficult to retain the element of bottom-up 'bohemianism' as an integral part of 'mainstream' cultural quarters, to prevent displacement to 'fringe' locations and 'substandard' premises (Tallon et al. 2006).

Sixth, serial replication and homogenisation of the urban landscape have been encouraged through the rolling out of similar types of leisure and cultural strategies, events and flagship buildings across cities. Evans (2003, p. 417) argued that the branding of commercial entertainment products and leisure-shopping presents a synthesis of physical and symbolic economies of urban consumption spaces. Hannigan (1998) further contended that there has been the 'fantasization' of the everyday through the exploitation of cultural assets in cities. These processes have led to predictable competition between cities, copycat corporate routines, and the extension of brand life geographically (Griffiths 1998a; Evans 2003). The new urban leisure and entertainment economy is characterised by the concentration of corporate ownership, and the increased use of branding and theming (Hannigan 2003; Hollands and Chatterton 2003). Within the night-time economy, a shift has been witnessed from locally based artists and entrepreneurs to a small number of dominant corporations (Hollands and Chatterton 2003). The UK has also witnessed the 'Europeanisation' and 'Americanisation' of cultural regeneration programmes (Bianchini and Schwengel 1991), and the emergence of a 'culture of nowhere' (Kunstler 1993).

Seventh and finally, volatility of the market for leisure and culture, brand decay over time and long-term sustainability are issues to consider for the future of urban regeneration through leisure and culture. Overall, significant doubts remain as to whether government-sponsored cultural policies can live up to their bold claims (Jones and Wilks-Heeg 2004; Miles and Paddison 2005).

In terms of policy for leisure and cultural regeneration in cities, Tallon et al. (2006) found support for a pub and club culture among urban authorities despite the well-documented and escalating problems of the night-time city (Hobbs et al. 2003); the need for balance and diversity in terms of leisure and cultural activities; and that the voluntary sector was significant to smaller-scale cultural and arts activities in cities. Policy lessons for the further stimulation of the leisure and cultural economy in city centres throughout the UK embraced the need to develop a wider range of leisure and cultural attractions; to attract a wider range of social groups to urban leisure and cultural spaces; to adopt an appropriate spatial policy; and to tackle conflict within the leisure and cultural economy (Tallon et al. 2006).

SUMMARY

Culture has been used as a tool for attracting inward investment to regenerate urban areas, and culture as a product in the market can be used to sell places globally as 'cultural' or 'creative' cities.

Cultural quarters offer the opportunity of contributing effectively to the wider urban regeneration of an area, however, there are dangers of gentrification and questions over their long-term sustainability.

The night-time economy has economic benefits but has developed as socially and culturally problematic.

Sports, urban tourism, and spectacular, flagship post-modern cultural developments can make a contribution to urban regeneration, but none offers an immediate panacea for the social, economic and environmental regeneration of cities.

In the post-industrial city, leisure and culture-driven urban regeneration is embraced as a central plank of the new urban entrepreneurialism, closely tied with cultural globalisation.
Increasing globalisation and diffusion of leisure and cultural regeneration strategies are encouraging the homogenisation of culture in city spaces.
Leisure and cultural regeneration has positive and negative impacts upon the ways in which cities are evolving.

STUDY QUESTIONS

1. Critically assess the contribution that the leisure economy and the creative industries can make to urban regeneration.
2. 'The emerging geography of the creative class is dramatically affecting the competitive advantage of regions' (Florida 2002, p. 243). Critically discuss this assertion.
3. Describe and account for the conversion of centrally-located urban real-estate into what Hannigan (1998) called 'fantasy cities'.
4. To what extent has the city centre night-time economy transformed from 'creative city' to 'no-go area' (Roberts 2006)?
5. 'The city has to appear as an innovative, exciting, creative, and safe place to live or visit, to play and consume in' (Harvey 1989, p. 8). To what extent does this argument continue to have relevance for UK cities?
6. Explain how tourists 'consume' cityscapes, and critique the use of spectacular architecture in the creation of new urban tourist spaces.

KEY READINGS ON LEISURE AND CULTURAL REGENERATION

Aitchison, Richards and Tallon (eds.) (2007) *Urban Transformations: Regeneration and Renewal through Leisure and Tourism* [contemporary collection of writings covering approaches to urban regeneration through leisure and tourism across Europe]
Bell and Jayne (eds.) (2004) *City of Quarters: Urban Villages in the Contemporary City* [examines the role of urban (cultural) quarters in terms of structural, political, economic, social, cultural and spatial change]
Creativity, Cities and the New Economy Special Issue: *European Planning Studies* (2001, vol. 9, no. 7) [special issue exploring the links between creativity, the new economy and city regeneration]
Cultural Policy and Urban Regeneration Special Issue: *Local Economy* (2004, vol. 19, no. 4) [theme issue revolving around the European Capital of Culture]
Florida (2005) *Cities and the Creative Class* [influential research on how the growth of the creative economy shapes the development of cities and regions; for cities to grow they must base their strategies on technology, talent and tolerance]
Gold and Gold (eds.) (2007) *Olympic Cities: City Agendas, Planning, and the World's Games 1896–2012* [18 essays covering key aspects of activity involved in staging the Olympics – finance, place promotion, managing spectacle and urban regeneration]
Law (1993) *Urban Tourism: Attracting Visitors to Large Cities* [covers a variety of city attractions including conferences and exhibitions; culture, sport and special events; hotels, shopping and evening activities; and links to urban regeneration]
Short and Kim (1999) *Globalization and the City* [Chapters 6, 7 and 10 look at cultural globalisation and the city, representing cities in a global world, and Sydney's Olympics]
The Rise and Rise of Culture-led Urban Regeneration Special Issue: *Urban Studies* (2008, vol. 42, no. 5–6) [critical review of culture-led regeneration drawing on UK and global case studies]
Urban Regeneration and Sports Stadia Special Issue: *European Planning Studies* (2002, vol. 10, no. 7) [case studies of the impacts of stadia development on urban regeneration]
Urban Space and the Uses of Culture Special Issue: *International Journal of Cultural Policy* (2004, vol. 10, no. 1) [evaluations of uses of culture to transform urban space]
Urban Vitality and the Culture of Cities Special Issue: *Planning Practice and Research* (1995, vol. 10, no. 2) [discussion of culture and the night-time economy]

KEY WEBSITE ON LEISURE AND CULTURAL REGENERATION

Department for Culture, Media and Sport: www.culture.gov.uk [government department with responsibility for culture]

13 Regenerating suburban and exurban areas of cities

INTRODUCTION

The academic and policy focus of urban regeneration, and indeed the content of this textbook, have concentrated principally on the inner city and city centre. This is because spatially these are the areas that have experienced the most severe economic, social and environmental decline. Additionally, most urban regeneration research has focused on transformations in the central geographical area. This chapter redresses the balance to a degree by exploring and tracing the extension of the urban regeneration agenda beyond central cities, to suburban and exurban developments. The first section provides a context by examining the changing geographical nature of suburban areas and exurban developments, with reference to wider urban spatial models applicable to the US and the UK. It then focuses on regeneration initiatives aimed at the suburbs and at edge cities in the UK.

URBAN SPATIAL MODELS AND CHANGING SUBURBS

Chapters 10 to 12 focused predominantly on approaches to regeneration in the central business districts of UK cities covering retail, housing, and leisure and cultural regeneration. This chapter focuses on urban regeneration in the zone of transition between the city centre and suburbs, the suburban fringe, the suburbs, the rural–urban fringe and exurban or edge-city developments, representing the outer rings of the conventional model of the industrial city (see Figure 1.1). Traditional geographical models of urban space include the Burgess model and its variants based on Chicago, and the Harris Ullman model and its variants based on Los Angeles (see Hall 2006, Chapter 2; Pacione 2005, Chapter 7). These models are constantly evolving resulting in new spatial patterns of UK cities (see Chapter 1). This chapter looks at the characteristics, problems and policies related to urban areas beyond the centre and reaching exurban developments.

Zones of transition in UK cities were historically dominated by housing and warehousing. Small terraced houses constructed for industrial workers were situated near to industrial, warehousing and dock activities that still dominate the eastern sides of many former industrial cities. Grand Georgian houses designed for the merchants were also situated close to working class-housing and industry, but spatially segregated and often to the west of the city centre. Through the processes of de-industrialisation and decentralisation, zones of transition were abandoned by people and capital. Subsequently a greater mix of activities was introduced comprising residential, office, retail, leisure and infrastructure uses (see Coupland 1997). Some buildings were put to multiple use and often supported the economy of the city centre. The edge of the city centre was characterised by lower rents, reflected in a landscape of, for example, small-scale financial services firms, stationery shops, staff recruitment agencies, sandwich and coffee bars, and car parks. Over time, the Victorian terraced housing of the zone of transition has witnessed a gentrification process (see Lees et al. 2008), and warehouses have been converted to gentrified housing, arts and leisure spaces, and open-plan office uses (see Chapter 11). The area of the city between the central business district and the suburban fringe has therefore been subject to a transition of uses connected with the shift from an industrial to post-industrial urban economy since the mid-twentieth century.

The suburban fringes of UK cities were often initially based on the Victorian railway network. By the middle of the twentieth century, the outer suburbs of cities were based on the private car and the built landscape became dominated by the archetypal three-bedroom semi-detached house. These areas also contained public functions such as schools, shops, churches, libraries and parks. A brief history of the geographical evolution of suburbs places the contemporary issues into context (English Heritage 2007b).

Suburbs can be defined as 'outgrowths or dependencies of larger settlements – somewhere with a clear relationship with a city or town but with its own distinct character' (English Heritage 2007a, p. 2). Most urban places were at one time suburbs until they were overtaken by the outward expansion of the original settlement, so that they are now suburbs only in a historical sense (English Heritage 2007b). Many suburbs in the UK have become historic and valued through proving their long-term sustainability reflecting popularity, adaptability, stability, with maturity and distinctiveness now added qualities.

In the UK, the suburbs took shape from the early nineteenth century; before 1800 suburbs possessed no uniform character or special building types. Some were dense, unsafe and poorly managed, and others were looser entities with better amenities. One common physical pattern on the edge of cities was ribbon development along major arteries (see Chapter 2). The definition of suburban character took place in the nineteenth century when wealthy urban residents were able to escape the peripheries of polluted and dangerous city centres due to rapid changes in transport. Popularity and momentum were given to the suburbs, which were seen by urban planners and reformers as salubrious places to live. Suburbs were perceived as places of order, safety and health between 1804 and 1939. The suburb varied depending on whom it was built to house, its relationship and transport links with the original settlement, and its topography (English Heritage 2007b). The ideal vision was one of detached villas in natural parkland setting with tree-lined avenues and large gardens. More common in Victorian times was development set out in hierarchical fashion off straight streets on plots with space between houses or groups with ample gardens, and the semi-detached type became commonplace. Houses in rows or terraces were standard for lower middle-class or working-class communities, located at closer proximity to urban jobs, not differing to a great degree from their inner city equivalents.

From 1875 suburbs evolved with architectural style being reformed to a consciously vernacular character based on the traditional English cottage, exemplified by the garden city or suburban development. Suburbs became larger and more numerous as a result of the growing railway network, which characterised the first half of the twentieth century, though they were later adapted to cars. Few suburbs of this time were built to cater for the car and were developed without adequate public facilities to meet the demands of the residents; however, their layout was often sophisticated. Part of the ideal of the suburb between 1840 and 1939 was that it was for family and not economic life; however, factory villages were developed such as Saltaire in West Yorkshire, Bournville in Birmingham and Port Sunlight in Wirral, and these operated as self-sufficient communities. Suburbs also grew up surrounding mills and factories that had been outside city centres.

Since 1945, suburbs have started to become harder to define; increased mobility following the Second World War and the blurring of distinctions between social class, jobs and lifestyle made it increasingly difficult to generalise about suburbs and suburban lifestyle, and terms such as subtopia and exurbia emerged. The widespread development of suburbs and the number of people living in suburbs demonstrated their importance in terms of planning and regeneration in the future. English Heritage (2007b) argued that many historic suburbs have proved themselves to be sustainable, and after a century of existence continue to exhibit a thriving mix of residential, retail and commercial elements contributing to a strong sense of community. However, this success has attracted development pressures and has focused attention on the suburban capacity to accommodate change. Further, Gwilliam et al. (1999) argued that some suburbs are becoming less sustainable than when they were built, particularly in terms of land and energy use, which make them collectively unsustainable.

Suburbs are coming under ever-increased development pressures in the early twenty-first century. A variety of demographic trends, changes to planning policy and housing market conditions have combined to threaten relatively spacious, low-density suburbs. Successive waves of new development and incremental change are putting local character and distinctiveness at risk (English Heritage 2007a; 2007b), and local authorities must address specifically suburban issues as well as inner city and city centre problems, to facilitate a suburban renaissance (see Gwilliam et al. 1999; Kochan 2007a; 2007b; Potts et al. 2008).

The rural–urban fringe beyond the suburbs is constantly moving outwards as the evolving and expanding city reaches out towards the countryside (see Chapter 9). This zone of the city often represents the limits of commuters, is often connected to high-speed rail links, and exhibits expensive housing and dormitory villages. To this pattern, more recent additions to the space of the rural–urban fringe have been out-of-town shopping malls, business parks and university campuses. These areas are characterised by plentiful cheaper space and easy access by car.

Since the 1970s, the emergence of 'edge cities' has characterised exurban development (see Garreau 1991; Hall 1997a; Kloosterman and Musterd 2001; Lambert and Smith 2003). These developments have been consistent with the emergence of multi-nucleated or polycentric city-regions, and creeping urbanisation. These have developed along the lines of the US experience and are essentially city centres in the suburbs or 'suburban downtowns' (Hartshorn and Muller 1989). Edge cities are characterised by major transport infrastructure, housing, retailing, offices and leisure facilities (Garreau 1991). They essentially contain all of the functions of a city but are located several miles outside of the old city or 'host city'. Edge cities are based on mass car ownership and service sector economic activities. Garreau (1991) defined an edge city as containing over 5 million square feet of office space, over 600,000 square feet of retail space, more jobs than homes, located on sites that had been mainly farmland in the 1960s, equating to around 150 edge cities in the US. Indeed, over 50 per cent of the US population are defined as living in wider suburban areas compared with 15 percent in the 1920s (see Fishman 1987; Nicolaides and Wiese 2006). Edge-city developments have emerged in the UK along the M4 corridor (Lambert and Smith 2003), in the north-east (Byrne 2001), in the West Midlands (Lowe 2000b) and in central Scotland (Bramley 1999). Lambert and Smith (2003) argued that while parallels can be drawn with the US model of edge cities, UK examples are generally much more modest in scale. They might also be an expression of general decentralisation rather than independent alternatives to the city as the US experience indicates (Hartshorn and Muller 1989; Garreau 1991). In the UK context, it can be argued that 'post-suburban' development (Hall 1997a) might be a more appropriate term for these new urban spaces (Lambert and Smith 2003). Edge cities have more recently been termed variously 'non-centred metropolis' and 'boomburbs' (Lang and LeFurgy 2007), which are characteristically large, rapidly growing cities that remain essentially suburban in character even when reaching populations more typical of urban core cities.

Edge cities are unplanned, unco-ordinated, develop incrementally and are wholly private sector initiatives (Lambert and Smith 2003). Their stages of development consist of office space being developed near to existing suburbs; followed by residential space developed for office workers; shops, cinemas and bars established for leisure and entertainment use; and finally some public sector service provision follows. In US cities, many 'public sector' services are provided by developers who then 'tax' the residents consistent with the gated community ethos (Blakely and Snyder 1997). Edge cities in the US and the UK have grown near successful cities in response to problems such as shortage of city centre space, pressures on house and office prices, congestion and pollution, crime, pressure on public services and infrastructure, and shortage of green space. Affluent workers in the tertiary and quaternary sectors leave the main city for edge-city environments. Edge cities offer cheap office space and housing, scope for expansion, reduced travel time, less crime and pollution, and access to green space and the countryside.

However, edge cities have developed an array of problems that in many ways reflect their 'success'. Successful edge cities simply recreate the same problems of the existing old city, such as high prices and congestion. Edge cities can be seen as a variant of the Harris Ullman spatial model, based on mass transport and some zonings of spatial use. However, they tend to be more fragmented due to less strategic planning and zoning, and less regulation due to their private sector nature (Lambert and Smith 2003). Edge cities display less variance in use and are devoid of industrial space and social housing.

Edge cities raise a number of political issues, some of which reflect criticisms of gentrification and gated communities (Lees et al. 2008; Blakely and Snyder 1997; see Chapter 11). In the US, developers often form a company to manage the area and contract the private sector to carry out activities such as rubbish collection, street cleaning, security, and management of services such as libraries. Taxes are 'ring-fenced' or hypothecated; they have to be spent on the area in which they are raised. Issues of privatised cities and 'shadow government' are introduced by edge city development resembling the characteristics of gated communities. This is because private sector corporations own the land, and the purchase of any site or property includes a contract including obligations to and rights of the corporation, which can have the power to tax, regulate and enforce. In essence, the rights of government are transferred to the private sector. The consequences of this in the US have been that corporations can manage schools, hospitals, fire stations, museums, parks, private police forces, and the planning system. These corporations collect a management fee, which in effect represents a tax. Potential problems with private management include profit orientation with no public sector welfare ethos; corporate interests being favoured in the planning system; the development of a poor-quality built environment of malls and roads, and urban sprawl; a lack of green space; and less accountability and transparency than elected governments.

Edge cities also result in problems for their host city. Edge cities avoid incorporation to avoid taxation which leads to a loss of revenue; they lose tertiary and quaternary sector employees who are often the best educated and highest earners, so host cities lose out in the 'new economy'; and city centres lose prestige retail and office business. City centre authorities have responded with a range of policies in the US such as BIDs to address this competition from edge cities (see Chapter 10).

An example of edge city development in the UK is the rapid expansion of the northern fringe of the Bristol city-region in South Gloucestershire. Since the late 1970s, this has been a key component of the growth and success of the Bristol city-region as a whole (Boddy 2003b; Boddy et al. 2004) and has been characterised as an emerging 'edge city' (Garreau 1991; Lambert and Smith 2003; Tallon 2007a). The northern fringe edge city is the focus of much economic activity, as shown by population and employment statistics, and contains many key sites in the city. These include Filton Airfield; the Aztec West Science Park, which houses 86 businesses in computers and software employing over 5000 people (Bassett 2001); Abbey Wood Ministry of Defence office complex, opened in the mid-1990s and employing 6000 people; the main of four campuses of the University of the West of England, one of the city's two universities, which has over 30,000 students and 3000 staff in total; the Bradley Stoke housing development of 8500 homes constructed by volume house builders; and major transport nodes such as junctions on both the M4 and M5 motorways, and Parkway and Filton Abbey Wood railway stations, which significantly contribute to the attractiveness of the area for both commercial activity and residents (Lambert and Smith 2003; Tallon 2007a).

The north Bristol edge city has generated a range of problems including loss of greenfield sites, orientation towards car users, and difficulties of access by public transport. Additionally the focus of employment growth in a peripheral area does little to combat unemployment problems elsewhere in the city-region such as in the inner city and the southern local authority estates. In addition, The Mall regional shopping centre at Cribbs Causeway signifies a 'new urban form' (Lowe 2000b), and has been in direct competition with other shopping and entertainment facilities in the city-region, including the city centre (see Chapter 10). Local government reorganisation in 1996 resulted in the northern fringe area being included as part of an enlarged South Gloucestershire rather than an extended Bristol city. This has resulted in some conflict concerning aspects of the strategic development of the city-region (see Lambert and Oatley 2002; Lambert and Smith 2003; Stewart 2003; Boddy et al. 2004).

A next stage in the evolution of urban form in the UK beyond the suburbs and edge cities might witness the emergence of 'edgeless cities' (Lang 2003; Lang and LeFurgy 2003). These are symbolised by highly dispersed office clusters on main interstate highways in the US. They contrast with edge cities in that they offer no residential space, no retail space, no facilities, and no public transport connections. They might represent the next generation of edge cities or a wholly new form of exurban space. Similarly, technopoles exhibit a clustering of 'new technology' activities in the quaternary sector (see Castells and Hall 1994). They are new growth poles in the 'knowledge economy', representing the extreme high end of the sector and have links to edgeless cities. Exurban technopoles are inclined to be located at motorway junctions, near to airports or university campuses, and close to major cities. These all offer proximity to pools of skilled labour. Technopoles have links to residential environments, which have been described as 'elite corridors' of quaternary sector, or elite housing. These corridors tend to be located along the motorway network, in small estates of large houses that are often gated with private security, often close to airports and poorly served by public transport. In essence, elite residential environments, which are prevalent in the US and increasingly so in the UK, represent 'prestige gated cities' (Blakely and Snyder 1997; see Chapter 11).

REGENERATING SUBURBAN AND EXURBAN AREAS: PROBLEMS AND RESPONSES

City centres have been the focus of urban regeneration initiatives since the 1970s. However, in recent years it has been recognised by politicians, practitioners and academics that the suburbs are also in need of social, economic and environmental regeneration (Potts 2007b). Urban Task Force (1999) and Urban White Paper (DETR 2000a) illustrated a wider policy focus than just the city centre, embracing the city, town and suburb as a form of post-innercity urban policy (see Johnstone and Whitehead 2004b). There are challenges and opportunities facing suburban areas in providing communities suitable for the future (Potts et al.

2008). Suburbs in the context of the UK are areas of generally low-rise, low-density, semi-detached housing with front and back gardens. Housing is the principal land use and around 86 per cent of the population lives in suburbia (Potts 2007b; English Heritage 2007b). Suburbs are also peppered with light industrial and retail centres, but typically residents work elsewhere in city centres or edge city business parks. However, despite their importance, Gwilliam et al. (1999) stated that few recent studies of suburbs have been undertaken in the UK compared with international analysis (see Potts et al. 2008).

Despite the idyllic image of the suburban dream, many suburbs throughout the UK are reaching the stage of their life when the built and natural environment begins to show signs of decay and lack of upgrading. Gwilliam et al. (1999) pointed out that some suburbs are experiencing significant 'stress', with deteriorating community facilities, declining local centres, dominance of the car, and monotone housing that fails to reflect population and social change. Suburbs face challenges in attracting investment and residents, and connecting dormitory suburban areas to city centres through new public transport links, which can have regenerative effects (Potts et al. 2008). International examples are highlighted by Potts et al. (2008) such as buses connecting suburbs in Rouen in France, Metropolitan Rapid Transit in Toronto in Canada, and the coastal train in Buenos Aires in Argentina.

English Heritage (2007b) argued that the higher density and sustainability agendas have contributed to an increasing number of problems specific to the suburbs. Pockets of suburbs suffer significant social-economic deprivation (Potts 2007b), consistent with the experience of many inner city areas. Suburban residential regeneration in Paris and Rome, the protection of company housing in Germany, gentrification and social polarisation in Paris, the reuse of warehouses in Paris, and cultural heritage protection in Stockholm were European case studies highlighted by Potts et al. (2008), and these hold lessons for the regeneration of the UK's suburbs. The growth in car ownership has led to increasing traffic levels in suburbs and the implementation of intrusive traffic management and congestion measures. An aim has therefore been to minimise the commute from home to work (see Chapter 9). There is also potential for more economic activity in suburbs without negatively affecting their character. Economic activity impacting on suburbs internationally was illustrated by Potts et al. (2008) based around an arts cluster in Montreuil, Paris; business and transport links in Sydney; retailing and commerce in Silver Springs, Maryland, USA; university-led regeneration in Milan; and economic intensification in Toronto.

The growth in car ownership has had economic implications for historic suburbs, leading to the decline of local services, vacant retail units, environmental deterioration, and a gradual downward spiral of neighbourhood and district centres (see Filion 2001; Chapter 10). In this context, more suburban job opportunities would contribute to both the regeneration of suburbs and sustainability aims, particularly if high-quality residential developments can encourage people to live and work locally, as has been pursued in city centre environments (Bromley et al. 2005). Suburban advantages are proximity to the motorway network and cheaper office rents than in the city centre. 'Suburban growth poles', such as in Bury near Manchester (Potts 2007b) could capitalise on creating strong knowledge economies, warehouse-style nightclubs, cutting-edge fringe theatre, business parks for green industry, and cheap space for arts and design graduates, similar to city centre arts and creative quarters (see Chapter 12; Potts et al. 2008). Potts (2007b) also advocated suburbs trying out green wedges to introduce green space and to link them to green belts.

Because suburbs are low density and spacious, they are also potentially the location of some of the new housing required to meet demand through infill development, especially in the south-east of England (CLG 2006a; 2007a; English Heritage 2007a). The agenda to protect greenfield areas, redevelop brownfield sites and increase housing densities aims to increase supply and deliver compact cities (see Chapters 9 and 11). Infill and intensification of development have been achieved in cities such as Ontario, Canada; Lille, France; Milan, Italy; Richmond, British Columbia; and Melbourne, Australia (Potts et al. 2008).

Pressures of high housing demand in many of the more attractive suburbs threatens their character and distinctiveness, as new development increases density and reduces green space (CLG 2006a; English Heritage 2007a). Gwilliam et al. (1999) suggested that suburbs are not generally suitable for large-scale restructuring, and do not provide an easy solution to meeting housing needs. The designation of gardens as brownfield sites has encouraged the trend towards 'garden grabbing' for new small-scale new-build housing development in suburban gardens to realise revenue for their owners. English Heritage (2007a) has argued that the loss of private gardens, mature trees and hedges in suburbia are detrimental effects of housing regeneration. Conversion of suburban property in response to the trend towards smaller households also threatens historic suburbs, although increasing the supply of housing.

Potts (2007b) has argued that any economic or environmental developments in suburbs should not alienate existing residents, particularly through affecting their high housing value and environmental amenity. In terms of regeneration, conversions and infill should take precedence over demolition and comprehensive redevelopment. Potts (2007b) has also argued that high-density housing in suburbs can be reconciled with popularity and family living, and that higher densities around transport hubs have the potential to reinvigorate suburban centres and reduce the commuting of suburbanites, particularly by car.

Conversely, in areas of low housing demand in the UK, the character of historically distinctive suburban housing areas, often in inner suburbs, is at risk as a result of abandonment and calls for demolition and selective redevelopment (English Heritage 2005). Issues of demolition of property in suburbs are highlighted by the Housing Market Renewal Pathfinders (see Cameron 2006; Townshend 2006; Chapters 5 and 11). Preserving housing of historic value, and reducing the displacement of communities and gentrification process must be considerations (Atkinson 2004b).

REGENERATING SUBURBIA: POLICY IMPLICATIONS

Drawing upon a review of the character of suburban communities, Gwilliam et al. (1999) found resistance to change, as it was usually perceived as a threat. Few parts of the suburbs receive direct government assistance from urban policy funds and there is no suburban regeneration fund as such. Therefore, Gwilliam et al. (1999) identified four general topics that were regularly found to need action:

- transport integration, particularly through public transport enhancements;
- reshaping local centres through the provision of more facilities and services;
- improving community facilities;
- housing renewal and adaptation.

English Heritage (2007a; 2007b) called on local authorities to assess their suburbs to identify design of architectural and historical significance and local community value to inform future regeneration and planning strategies. Potts et al. (2008) presented examples of good design in suburban regeneration schemes in Freiburg, Germany; Perth, Australia; Rennes, France; and Portland, Oregon, USA.

Potts et al. (2008) synthesised the main challenges for policy makers as relating to:

- city layout;
- transport and connection;
- economic activity;
- age and heritage;

- housing;
- place-making;
- cohesion, inclusion and diversity.

In terms of city layout, international case studies showed that suburban areas can move away from the traditional 'dormitory' model, in which residents travel to work, retail, and leisure opportunities in city centres. Transport and connection strategies aim to develop 'collective' transport such as trains and buses that connect the suburbs to the city centre and to each other; transport can have an effect on suburban regeneration. Economic activity in suburban areas can be achieved through attracting certain sectors and functions from city centres to clusters, or by newly generating them. Age and heritage are a key issue for older suburbs that have to deal with decay and development pressure. Housing can be developed in existing suburbs through densification and development of brownfield sites. Place-making can be achieved through good design in suburban regeneration schemes. Finally, cohesion, inclusion and diversity can be embraced, counteracting trends towards gated suburban communities, segregation, polarisation and gentrification.

English Heritage (2007a) put forward six points of policy advice for local authorities wishing to regenerate historic suburbs, in the context of incremental infill and the density and sustainability agendas of the late 2000s:

- Change is inevitable and can be economically beneficial, however, only limited new development can be accommodated in historic garden cities and suburbs without destroying their distinctive nature.
- Analysis of the suburbs should be encouraged for future sustainable planning and regeneration strategies, as they have the potential to adapt to more sustainable patterns (Gwilliam et al. 1999).
- A long-term strategy for suburbs is required, which ensures that significance, integrity, character and value to local community are retained.
- The diversity of the suburbs should be recognised, and higher density and infill housing development are more appropriate in some suburbs than others.
- The relationship between elements of the suburban landscape should be valued; public parks, open spaces, street trees, private gardens, and views all combine to create a local suburban identity.
- Proposals for significant change should conserve and enhance existing character and identity, and recognise and address the historic nature of the suburban built environment in cities; regeneration policy and planning should take this into account, and combine housing regeneration with encouragement of new retail development and green spaces.

English Heritage (2007a) further proposed a checklist of eight principles for the future regeneration of historic suburbs in the UK:

- A comprehensive vision for local suburbs that includes the historic environment should be developed, with long-term and strategic regeneration plans that involve the local community (see also Gwilliam et al. 1999). Shops and other services and green space should be designed in.
- The historic and architectural significance of suburbs should be recognised when local authorities consider proposals for future development.
- Decisions taken on historic suburbs should be made on the basis of an understanding of their significance and character, and the value placed on them by the local community.
- When a suburb is identified as of significant or historic interest, the local authority

should give consideration to designation as a conservation area so that new development is sympathetic to the surroundings.
- An assessment of the historic suburb can inform the development of Supplementary Planning Documents and give guidance on which areas should be covered and what development can and cannot be undertaken.
- The local community should be engaged in developing a strategy for the future of their historic suburbs, particularly residents' associations, tenants' groups, and amenity and local history societies.
- The public realm in suburbs should be preserved and enhanced as a key element of the character and distinctiveness of suburbs.
- The impact of traffic and parking on historic character can be significant and should be considered and managed.

Policy must also address problems stemming from economic and population growth in edge cities in the UK. Social and economic problems result from growing edge cities, as local communities have been faced with the negative externalities that strong economic growth brings, particularly linked to sustainability and affordability (see Raco 2004). Urban regeneration policies should focus on addressing growth problems including excess demand for housing, skills shortages and problems in recruiting 'key workers', pressure from housebuilders to develop on greenfield sites, and pressures on the public infrastructure. Strong and pro-active management of growth is required, especially concerning the jobs–housing imbalance in cities and transport infrastructure, both of which possibly pose longer-term environmental threats if current trends continue. These pressures are combined with local political resistance to sustained growth and development in edge cities in many cities, especially regarding new housing development in the south of the UK (Bramley and Lambert 2002; Lambert and Oatley 2002). Edge cities represent recent economic success stories, but are environmentally problematic. The restructuring of edge cities requires attention, particularly in terms of the infrastructure deficit. There is scope for intensifying use, more mixed and compact forms of development, reclaiming space from the car, a better quality environment in which to live and work, and creating more of a sense of 'place' in exurban environments.

Gwilliam et al. (1999) recommended that local authorities and other agencies should encourage and assist suburban adaptation more positively. Regeneration strategies for suburban and exurban areas of cities will inevitably have substantial financial and logistical implications, and Potts (2007b) suggested the strategies be pursued as long-term goals rather than as short-term urban transformations. A city-region-wide urban renaissance means that suburban and exurban regeneration should be more fully explored and understood. Much attention has focused on city centre and inner city areas or on the urban edge to the extent that research on the suburbs in between has been neglected. The scope for the suburbs to play a positive role in the government's urban renaissance and sustainable communities agendas should not be overlooked (Gwilliam et al. 1999; Unsworth and Nathan 2006).

SUMMARY

In terms of models of urban form, the zone of transition between the city centre and suburbs, the suburban fringe, the suburbs, the rural-urban fringe and exurban or edge city developments represent the focus of urban regeneration beyond the central city.

Edge cities have developed along with multi-nucleated or polycentric city-regions, associated with advancing urbanisation, and are essentially city centres in the suburbs or 'suburban downtowns', which experience the problems of booming places.

In the early twenty-first century, many suburbs experience significant problems of deteriorating community facilities, declining local centres, dominance of the car, and monotone housing which fails to reflect population and social change.

Policy responses put forward for regenerating suburbia include transport integration, attracting economic activity, reshaping local centres, enhancing the public realm, improving community facilities, and housing infill, renewal and adaptation.

Compared with the focus on the central city's urban renaissance, there is a dearth of research on suburban and exurban regeneration in the UK.

STUDY QUESTIONS

1. To what extent do the problems of edge cities result from their success?
2. What are the challenges facing suburban areas and how can urban regeneration policy ameliorate these?
3. Assess whether the focus of urban regeneration on central cities has been to the detriment of a wider urban renaissance.

KEY READINGS ON REGENERATING SUBURBAN AND EXURBAN AREAS OF CITIES

Gwilliam, Bourne, Swain and Prat (1999) *Sustainable Renewal of Suburban Areas* [examines the conditions and future for suburbs in the context of the sustainable cities debate]

Potts with Besussi, Gaus, Hassler, Lesteven, Markovich, Munoz-Rojas Oscarsson, Parham and Porthe (2008) *Suburban Regeneration: The Real Challenges* [contemporary assessment of challenges facing the suburbs; covers city layout, transport, economic activity, heritage, housing, place-making, and social cohesion; illustrates international good practice]

KEY WEBSITE ON REGENERATING SUBURBAN AND EXURBAN AREAS OF CITIES

Joseph Rowntree Foundation: www.jrf.org.uk [contains research reports on aspects of suburban regeneration and change]

SECTION IV

Conclusion

14 Urban regeneration into the future

INTRODUCTION

This concluding chapter summarises and integrates the overarching key themes from the two substantive sections of the textbook. First, it highlights how towns and cities have changed over the past 60 years as explored in Section I. Second, based on Section II, it gives an overall critique of successive periods of urban regeneration policy looking at problems, lessons and potential barriers to future success. Third, an assessment of thematic approaches to urban regeneration in the UK derived from Section III is presented. Finally, the chapter seeks to identify forthcoming developments in urban regeneration within policy and academic spheres.

URBAN REGENERATION AND TRANSFORMING CITIES

There is striking evidence to show that towns and cities across the UK have undergone dramatic transformations over the last six decades, arguably for the better. Six major processes have been at work that have driven and shaped the changes, which have benefited some places and groups, but not others. These processes are post-industrialism, globalisation, changing inequalities and exclusion, growing social and cultural diversity, the emergence of new forms of urban governance, and the creation of new types of urban space.

The first major process has been the shift to post-industrialism, which has seen a decline in production and an increasing emphasis on consumption and knowledge services. There has been a massive shift in the balance between the major sectors, and not just a decline in economic activity. De-industrialisation has seen the loss of manufacturing jobs. At the same time, there has been a 'tertiarisation' of urban economies which combines producer services including advertising, banking and insurance, also termed the FIRE industries of finance, insurance and real estate; cultural and creative industries including film, television, music, fashion and design; and consumption and entertainment services such as festival shopping, the night-time economy, museums and galleries. Post-industrialism has also witnessed a shift towards the informational economy which in essence symbolises a shift in the factors of production from physical resources to informational resources, including the networks, channels and technologies through which information flows.

Major cities have emerged as nodes in networks of information flow (see Graham and Marvin 1996; Short and Kim 1999). Conditions for maintaining competitive success in post-industrialising cities include attracting and retaining members of the 'creative class' (Florida 2002; 2005).

A second major process that has improved the economy, society and built environment of cities, but has also been uneven and inequitable, is globalisation. In general terms this means 'the intensification of worldwide social relationships which link distant localities in such a way that local happenings are shaped by events occurring many miles away, and vice versa' (Giddens 1990, p. 64). Short and Kim (1999) looked at how economic, cultural and political globalisation have led to changes in cities. Economic globalisation relates to how units of economic decision-making have become increasingly global in the scope of their operations leading to instability and insecurity at the local level, and to inter-place competition and place marketing (see Griffiths 1998a; Dicken 2007). Cultural globalisation has seen symbols, genres, values and so on proliferate around the world, leading to concerns about the erosion of local and regional identities and the homogenisation of urban landscapes. However, reactions against this have seen the more recent development of diasporic groups, trans-national social networks, multiple identities and regional resurgence. Political globalisation has observed the global spread of political movements and the growing significance of supra-national institutions.

A third major process of change in cities has been the changing landscape of social inequality and exclusion, which is closely linked to post-industrialisation and globalisation. Tertiarisation and the emergence of a dual labour market have resulted in the loss of middle-income jobs. Consequently significant sections of the population have been cut off from mainstream labour markets and have become dependent on welfare. 'Exclusion' is a process of essentially not being able to be part of the mainstream, and is multi-dimensional incorporating being excluded from employment, finance and social networks (see Levitas 2005), and can be spatial, political and cultural.

Fourth, a growing social and cultural diversity has shaped urban change and can be related to urban regeneration initiatives. Social structure relates to characteristics of class, race, gender and religion; culture relates to identity and expression. Aspects of diversity relate to household forms, immigration and diasporic communities, more complex patterns of ethnic belonging, and diverse sexualities. The relationship between social structure and culture has become more fluid and blurred. In terms of implications for urban change and regeneration, a range of needs and demands must be catered for, urban spaces and neighbourhoods must take account of diversity, and city authorities have to engage with an increasingly varied population.

A fifth process of change has been the emergence of new forms of urban governance, which have been connected to wider processes of restructuring and reorientation of state power. State functions have been lost to the market, to supra-national institutions and to local communities since the 1980s. In an era of inter-city competition at the global level, city governance has been transformed from being 'managerial' to 'entrepreneurial' to mobilise assets and regenerate cities.

A sixth and final way in which urban places have changed, linked to regeneration, is in terms of the creation of new urban spaces. Central areas and the peripheries of cities have experienced the most striking and dramatic changes. In city centres, high profile 'flagship' developments have emerged, often based on culture and consumption; themed enclaves have been developed; and repopulation of central areas has led to zones of gentrification. In peripheral urban fringe 'edge city' areas, campus-style office parks, multi-retail and leisure developments, and warehouse and distribution complexes have appeared. These spaces have a distinct morphology of self-contained cells, a poorly developed public realm, disconnected elements, no obvious centre, and no evident spatial logic (see Sorkin 1992). It is reasonable to talk of a 'post-modern urban landscape' reflected in the new urban spaces comprising spectacular central city regeneration developments and out-of-town consumption and service industry spaces. However, they lead to the fragmentation of urban space, are exclusionary, and erode the public realm.

Each of these six major processes of change at work since the mid-twentieth century is inextricably linked with urban regeneration and illustrates how urban landscapes have changed, who the beneficiaries are, and what the implications are for the future of urban space. It is clear that over the last 30 years, urban regeneration has been underpinned by a focus on economic and property development, with intermittent attention paid to social and cultural issues. Particularly since the early 1990s, urban regeneration has been based on a series of assumptions linked to the macro-economy including continued economic growth in the context of globalisation, low unemployment, low inflation, low interest rates, rising property values, and increasing owner-occupation of housing. During the late 2000s it is emerging that this period, upon which much successful urban regeneration has been based, is coming to an end. Economic recession and its impacts on the private sector and consumers, and a reduction in public spending are likely to have severe impacts upon economic and social regeneration for many years to come. However, lessons can be drawn from past central government policies and from the strategies initiated by city authorities.

CRITIQUE OF CENTRAL GOVERNMENT URBAN REGENERATION POLICY

As is evident from the review of successive periods of urban policy since the 1940s (Table 2.1), each had its distinctive approaches, achievements and weaknesses, which governments of subsequent policy periods sought to address. Clearly, there is a vast array of problematic issues revealed by Sections I and II of the textbook. Key points can be made which highlight the general errors made by central government urban policy across the 60-year period. Some of the weaknesses of urban policy were operational in nature related to particular policies or programmes, and other criticisms were wider and strategic in nature (see Turok 2005, pp. 61–62). Six spheres of failure can be synthesised from an evaluation of the periods of urban regeneration in the UK since the mid-1940s, and particularly from the late 1960s when a distinct urban policy emerged. Significant reviews from both academic and public policy circles largely reveal a similar set of concerns (see, for example, Stewart 1987; Audit Commission 1989; Parkinson 1989; Robson 1994; Robson et al. 1994; Lawless 1996; Shaw and Robinson 1998; Carley 2000; Gripaios 2002; Atkinson 2003; Johnstone and Whitehead 2004b; Turok 2005; ODPM 2006a; Centre for Cities 2008; Leunig and Swaffield 2007; 2008).

1. Lack of clarity and purpose of urban policy

Objectives of urban policy initiatives driven by central government have often been too broad and ill-defined. Area-based initiatives are unable to realistically tackle the multiple problems of society (Turok 2005). Imprecision of purpose has caused unrealistically high expectations, especially among local communities. Multiple objectives have meant that it has been difficult to set priorities and focus on feasible targets. Related to this problem with its remit, urban regeneration policy has seen insufficient commitment and resources from central government over timescales that are too short-term for realistic change to be achieved. Some policies can be seen as token and a diversion from mainstream policies. 'Quick fix' rather than long-term sustainable commitment has been an accusation levelled at central government urban policy (Shaw and Robinson 1998; Johnstone and Whitehead 2004b; Turok 2005).

2. Excessive central government control of urban policy

Governments have often been accused of being dictatorial about urban regeneration, as it has been conducted in top-down fashion reducing the scope for local solutions to local problems (Turok 2005). Little analysis of local geographical context and the dominance of 'one-size-fits-all' policies have resulted in little potential for flexibility, creativity or innovation in local urban regeneration. Central government has set a plethora of targets, given little room for local discretion, and caused friction among local partnerships, especially when government priorities have differed significantly from those at the local level (Johnstone and Whitehead 2004b). Often partnerships have been forced together and have failed to cohere, and partnerships have not been given enough time to develop and flourish under many policy initiatives (Atkinson 2003).

3. Poor co-ordination and coherence of urban policy

Urban policy under successive governments has suffered a lack of horizontal and vertical co-ordination across departments of central and local government (Atkinson 2003; Johnstone and Whitehead 2004b). Many local initiatives have been largely isolated from mainstream programmes and services. Government has pursued separate short-term initiatives rather than taking a longer-term perspective driven by changes in mainstream services with greater local co-ordination and increased community involvement (Atkinson 2003; Turok 2005). Fragmented and piecemeal projects and limited resources have resulted in marginal achievements in many cases and policy has not been co-ordinated or strategic (Shaw and Robinson 1998). Turok (2005) noted that strategic leadership and local capacity to take difficult decisions have often been weak.

4. Implementing one-dimensional urban policies

The urban problem is multi-dimensional (see Chapter 1), but policy responses have often been too cautious and narrow. Successive policy periods have witnessed a property-based approach (Turok 1992; Healey et al. 1992), or limited social programmes, or a focus on business development, or a concentration on community regeneration (Taylor 2003). Successful urban regeneration requires the connection to be made between each of these physical, economic and social dimensions.

5. Dealing with neighbourhood as an isolated unit

Neighbourhoods have often been considered as units in isolation from their wider urban context (see Smith et al. 2007a). There has been insufficient understanding of the function played by the area in the wider housing and employment markets, and of the relationship with surrounding areas. Artificial and arbitrary boundaries have been drawn around areas designated for neighbourhood regeneration, and issues of displacement, gentrification and leakage have been overlooked (Turok 2005). It has also been argued that the explicit targeting of deprived areas has increased problems related to stigma and poor place image (see Hall 2006, Chapter 6).

6. Failure in realising community potential

Since the late 1990s in particular, communities have been placed at the heart of regeneration, which has placed a burden on them through this empowerment (Atkinson 2003). This reflected a communitarian view and a belief in self-help and social capital (Putnam 1993; 2000; Taylor 2003). This can be seen as a romantic notion considering the tensions and conflicts within communities, the tendency for burn-out in these communities because of the demands of successive initiatives, and the challenges that face such communities (Atkinson 2003; Turok 2005). Empowering communities remains a major challenge.

However, despite these six categories of central government urban policy problems, it is important to acknowledge that urban regeneration policies since the end of the Second World War have attempted to solve some of the most deeply entrenched and difficult social and economic issues facing the UK. Therefore, as Turok (2005, p. 61) stated, it is hardly surprising if they are not always entirely successful.

Shaw and Robinson (1998) proposed ten lessons that have emerged from the operation of urban policy since the late 1970s:

- The physical transformation of declining urban areas is only one part of the wider process of regeneration.
- Everything is inter-related and regeneration is economic, social and environmental in nature.
- The 'trickle-down' effect does not work.
- Regeneration is too important to be left to non-elected quangos.
- Partnerships are vital but need to be 'sustainable'.
- Resources are never sufficient.
- It is important to have clear aims and realistic objectives.
- Image matters in transforming cities.
- Regenerating people, rather than just places, is difficult to achieve.
- Sustainability is the key.

These ten lessons in addressing inner city and wider urban regeneration prove to be equally appropriate in the late 2000s. Finally, Shaw and Robinson (1998) argued that urban policy since the 1977 Urban White Paper on the inner cities had been characterised by a form of 'policy amnesia' – a failure to learn from past experiences. Within this context, 'research is rapidly forgotten ensuring that wheels have to be re-invented and long established truths have to be rediscovered' (Wilks-Heeg 1996, p. 1264).

Added to the six problems with past urban regeneration policy and ten lessons for future urban policy, five further barriers might preclude urban regeneration from ultimately meeting its ambitious aims (Johnstone and Whitehead 2004b, pp. 11–15).

1. The scale and intensity of urban problems

The nature and scale of urban problems create a major barrier. Despite regeneration efforts since the 1980s encompassing physical regeneration, city centre gentrification, and the mass disposal of the highest quality council housing, significant concentrations of deprivation remain in the inner suburbs and peripheral council housing estates in old industrial cities in particular. Unemployment, welfare dependency, poor educational attainment, high rates of crime and anti-social behaviour, poor health, and abandoned housing have proved difficult to tackle, and the stigmas attached to these areas remain deeply entrenched (see Gripaios 2002).

2. Community participation and integration

Encouraging community to participate and integrate within urban regeneration programmes remains problematic as communities cannot be forced to engage. There is the danger of 'regeneration fatigue' and disillusionment with past and often failed initiatives. Community involvement remains fraught with difficulties (Atkinson 2003; Taylor 2003).

3. Growing inequality between regions and cities

Inequalities especially reflect the north–south divide and divides within cities. For example, the Centre for Cities (2008) found that despite a decade of urban regeneration from the late 1990s under New Labour, big inequalities have persisted in cities as exemplified by dual landscapes such as London's Docklands. Leunig and Swaffield (2008) argued that regeneration towns and cities in the north could never converge with towns and cities in the south of the UK. Spatial and social polarisation and exclusion are enduring, intractable and insoluble in the early twenty-first century.

4. The monumental complexity of urban policy

The complexity issue continues to hamper the effectiveness of urban regeneration policy. Strategy after strategy and initiative after initiative conspire to result in replication and confusion. The policy arena has moved from 'patchwork quilt' to 'bowl of spaghetti' (Johnstone and Whitehead 2004b). Despite attempts by New Labour to streamline, co-ordinate and reduce the complexity of urban regeneration policy, it remains as complicated as ever (Weaver 2003).

5. The managerial and performance indicator culture

The approach of New Labour's administration since the late 1990s has seen a proliferation of strict controls, centralised management, league tables, evaluation and the like, which are largely unresponsive to local geographical variations and act against some of the aims of recent urban policies.

CITIES IN TRANSITION: THEMATIC APPROACHES TO REGENERATION

The thematic approaches to urban regeneration in the UK have emerged in the context of the focus on city competitiveness, social cohesion and inclusion, and the shift to new forms of urban governance associated with entrepreneurialism and the increased role of communities. These are strongly set in the context of urban sustainability. The principal thematic approaches to urban regeneration in the UK pursued by cities themselves are based on retail, housing, and leisure and culture. Three key issues stemming from the experience of UK cities mobilising thematic strategies can be elucidated.

1. Gentrification, exclusion and the privatisation of urban space

Many projects centred on retail, housing or leisure and cultural uses have promoted trends towards the commercial and residential gentrification of city centres and inner cities (Smith 2002; Lees et al. 2008). The lack of social or affordable housing in the landscape of housing-led renaissance has been the subject of much criticism (Slater 2006; Bromley et al. 2007b). Similarly, the appropriation of local festivals and carnivals by city authorities, the commodification of local cultures, and the exclusion of local groups have been issues of concern (see Boyle and Hughes 1994; Hall and Hubbard 1996; Waterman 1998).

2. Involvement of local communities

Consistent with the experience of lack of community involvement in central government-funded urban policy initiatives, involvement of local communities in projects and approaches undertaken by cities themselves has been equally patchy. Communities have been negatively affected by the gentrification process associated with housing-led renaissance, have often been excluded from giving a voice in new mega-project proposals and developments in central cities, and have often seen their cultures hijacked or commodified through city-authority-led cultural festivals and initiatives.

3. Globalisation and the focus on economic and physical regeneration

Prestige projects and mega-projects favour the city centre rather than inner city residential areas. Cloned, banal, branded landscapes have typically been a product of new central city malls and regenerated spaces. Loss of identity and the dominance of multi-national firms at the expense of local independent operators have been trends across the UK. Place competition for major retail or leisure projects and to attract signature events or city designations has been in the spotlight of activity, often at the expense of focusing on local economic development and social disadvantage. Anything innovative in terms of flagship or spectacular regeneration is quickly copied by other cities, meaning that all cities end up being the same through a process of 'serial replication' (see Griffiths 1998a). Cities are characteristically marketed to a niche, wealthy elite with the needs of less well-off groups being overlooked. Spectacular and post-modern urban regeneration projects have had a tendency to act as 'masks' hiding the reality of wider urban problems (Hall 2006). Many projects associated with promoting leisure and cultural regeneration are highly speculative ventures as only a select few cities can ever host the Olympic Games or be designated Capital of Culture; similarly there is a limit to the number of sports stadia, theatres or convention centres required. However, it appears that such approaches are considered essential by city authorities and remain *de rigueur* so that cities maintain a profile in the context of intensifying competition between them for public and private investment in the increasingly globalised economy.

FUTURE DIRECTIONS FOR URBAN REGENERATION POLICY

New Labour's urban policy in the late 2000s under Gordon Brown remains focused on sustainable communities, with a subtle shift away from the language of urban renaissance and from the emphasis on area-based initiatives associated with the late 1990s and early 2000s. Central government continues to work with English Partnerships and other agencies including the Regional Development Agencies to facilitate residential and commercial development, provide resources to stimulate regional economic growth and the creation of jobs, and provide strategic direction and investment to coalfield communities (www.communities.gov.uk 2008; see Chapter 5). A future Labour government from the late 2000s would conceivably continue this approach, in concert with wider social and economic policy linked to contemporary concerns of sustainability and affordability of housing.

In terms of urban policy approaches from any future Conservative administration, the most illuminating document in the late 2000s was the inner cities policy group report published in 2006. This was fronted by Michael Heseltine, the Conservative Secretary of State responsible for urban regeneration in the 1980s. During this period, Heseltine and the Conservatives believed that the private sector had to be persuaded to finance the regeneration of the inner city, combined with a relaxation of planning requirements, availability of urban development grants and exemption from industrial and commercial rates for properties (see Chapter 3). The Conservative task force looked at large-scale urban regeneration and in particular at capital projects that might be undertaken by a future Conservative government. This approach argued that urban regeneration must be more than just a makeover for city centres and must improve life in the neighbourhoods surrounding city centres.

The task force (Conservative Party 2007) recommended transfer of power from regional agencies downward to local government; directly-elected mayors for all top-tier authorities, with four-year terms; and 'pan-city' executive mayors for Birmingham, Newcastle, Manchester and Liverpool, who would sit above existing councils, with powers over regeneration, transport, skills, fire, waste and police services, and greater local control over finances, including retained business rates from new developments, more capital funding, and freedom to issue local bonds. These suggestions remain elements of possible Conservative urban policy, which marks a changed emphasis from their 1980s and 1990s approaches, and to an extent resonates with the philosophies of New Labour.

Whichever party assumes power in the late 2000s faces a plethora of challenges and issues to address in achieving the successful economic and physical regeneration across all of the UK's regions and urban areas, especially in the context of a global economic downturn, and in addressing social problems being experienced by those continuing to suffer multiple deprivation, in the context of the sustainable cities and communities agenda.

SUMMARY: KEY ISSUES AND DEBATES

At the centre of all regeneration policies and strategies is the pursuit of economic, social and environmental upgrading and change, to mitigate the negative consequences of urban decline. Transforming the image of the city, and enhancing economic competitiveness and social cohesion are key aspects of 'sustainable' urban regeneration strategies in the UK, as in other advanced capitalist societies, and increasingly in the global south. Urban regeneration has tended towards either the redistribution of resources to disadvantaged areas and populations or to promoting economic growth through property-led and market-oriented approaches, shifting towards the latter over time.

As with any process of urban change, the benefits and problems associated with urban regeneration are not evenly distributed. Urban regeneration is often highly selective, favouring particular spaces and social groups. Overall, urban regeneration has been 'partial' both spatially and socially (Carley 2000; Hall 2006).

The 'spectacularisation' of urban space has seen a transformation from industrial manufacturing landscapes to spaces of consumption dominated by housing, retail, and tourism, leisure, cultural and entertainment functions. Spectacular, post-modern, flagship mega-projects increasingly dominate the skylines of city centres in the UK, but these are juxtaposed with continued social and economic problems in many inner cities. Urban regeneration should move to focus on areas beyond the city centre, encompassing inner suburban and outer suburbs in particular. No single policy, strategy or approach should be seen as a panacea or 'magic solution'; they should be integrated and combined subtly to avoid a one-size-fits-all approach.

Urban regeneration remains a diverse, interdisciplinary and topical issue in the UK, and globally. Table 14.1 displays the huge variety of themes, features and policies related to the urban landscape, which are in some way intertwined with urban regeneration, many of which are not even mentioned in this textbook. This is a selective list that is presented in no particular order, and covers government policy and approaches taken within cities since the 1980s.

Since the 1970s, urban regeneration has contributed to some spectacular transitions in terms of the built environments, economies and social and cultural facilities available to urban residents. However, the results have often been very uneven within cities and between cities and regions, and this has not been helped by the nature of urban regeneration policy in terms of its experimental, short-term nature, with limited evaluation and constantly changing priorities. If urban regeneration is ultimately going to achieve its ambitious and laudable aims, problems of spatially extensive and deeply entrenched poverty and exclusion

TABLE 14.1 ILLUSTRATING THE VARIETY OF POLICIES AND APPROACHES TO URBAN REGENERATION IN THE UK

Urban Development Corporations
Enterprise Zones and Simplified Planning Zones
Task Forces and City Action Teams
English Partnerships
City Challenge
Single Regeneration Budget
Urban Regeneration Companies
Housing Market Renewal Pathfinders
City Pride
City marketing and promotion
National Lottery funding
Urban villages and mixed-use developments
Community involvement in regeneration
Partnership working in regeneration
Housing and regeneration
City centre living
Gentrification and gated communities
Urban regeneration, health and education
Local economic development
Training and Enterprise Councils and Learning and Skills Councils
Social exclusion and welfare to work (New Deal)
Social exclusion and disability
Employment programmes and regeneration
Town, district and city centre regeneration
Retail-led regeneration
Food superstores as town centre regeneration projects
Office development and employment
Business Improvement Districts
Town and city centre management
The evening and night-time economy and the 24-hour city
Live–work units and regeneration
Adaptation and reuse of buildings
Crime, safety and planning
Listed buildings, heritage and conservation-led regeneration
Heritage and regeneration
Urban tourism
Urban leisure planning, open space and recreation
Road transport, health and regeneration
Casinos and regeneration
Urban allotments and regeneration
Urban cemeteries
Creative cities
Culture-led regeneration
Sport-led regeneration and sports stadia
Music festivals and regeneration
Cinemas and multi-leisure parks
Carnivals and regeneration
Urban forestry
Cultural regeneration and public art

TABLE 14.1 CONTINUED.

Regeneration and the public realm
Black and minority ethnic groups, urban regeneration, housing and planning
Cities, telecommunications and information technology
Urban development finance
Waterfront, waterway and canal regeneration
Compulsory purchase and urban regeneration
Regeneration of market towns
Regenerating coastal and seaside resorts
European urban regeneration and planning
North American urban regeneration and planning
Third World urban regeneration and planning

need to be resolved; this is something that despite several decades and dozens of policies and approaches has remained beyond urban regeneration so far.

STUDY QUESTIONS

1. How could approaches to urban regeneration policy be improved to better tackle urban social, economic and environmental problems?
2. Is it a fair judgement that UK urban policy has been characterised by short-term, largely symbolic actions with little real continuity?
3. Assess the impact of urban policy since the 1980s drawing on academic critiques such as Stewart (1987); Parkinson (1989); Robson (1994); Robson et al. (1994); Lawless (1996); Shaw and Robinson (1998); Carley (2000); Gripaios (2002); Atkinson (2003); and Johnstone and Whitehead (2004b).
4. Critically evaluate an urban policy period taken from one of the following: 1945–1968, 1968–1977, 1977–1979, 1979–1991, 1991–1997, or 1997–the late 2000s, taking into account philosophies, policies and case study examples.
5. Critically evaluate a post-1979 urban regeneration policy initiative such as Enterprise Zones, Urban Development Corporations, Single Regeneration Budget, City Challenge, New Deal for Communities, Urban Regeneration Companies, New Urban Development Corporations, or Business Improvement Districts.
6. Conduct an appraisal of national government urban regeneration policies since the 1980s in a UK city of your choice; significant research evidence is available on London, Birmingham, Manchester, Liverpool, Leeds, NewcastleGateshead, Glasgow, Belfast, Bristol, Cardiff and Swansea.
7. Critically evaluate the contribution that gentrification, or the retail economy, or the leisure economy, or the creative industries can make to urban regeneration.
8. Discuss whether a post-modern form of urban landscape has been emerging and comment on whether such a landscape is desirable.
9. Conduct a critical literature review on a theme relating to urban regeneration in the UK, synthesising the issues, debates and controversies with case study examples, ending with research questions that require attention. A topic can be selected from Table 14.1.

KEY READINGS ON URBAN REGENERATION INTO THE FUTURE

Atkinson (2003) 'Urban policy' [brief overview of urban policy pre- and post-1997, including key challenges]

Gripaios (2002) 'The failure of regeneration policy in Britain' *Regional Studies* 36: 568–577 [looks at problem localities and problem towns and problems with regional, urban and planning policy]
Hall (2006) *Urban Geography* [Chapter 5 gives an accessible overview of urban policy and regeneration issues]
Johnstone and Whitehead (2004b) 'Horizons and barriers in British urban policy' [focus on the early years of New Labour's urban policy; puts forward several barriers to future success]
Journal of Urban Regeneration and Renewal [journal on contemporary urban regeneration practice and policy, encompassing physical regeneration, economic development and community renewal, with case studies]
Neighbourhood – The International Journal of Neighbourhood Renewal [international research and analysis in the field of neighbourhood renewal]

KEY WEBSITES ON URBAN REGENERATION INTO THE FUTURE

British Urban Regeneration Association: www.bura.org.uk [champions the exchange of ideas, experience and solutions within urban regeneration]
Centre for Cities: www.centreforcities.org [part of the Institute for Public Policy Research; carries out independent research on urban policy and the economies of cities]
Communities and Local Government: www.communities.gov.uk [central government department responsible for urban regeneration in the UK]
Online Planning Resources: http://planning.rudi.net [contains extensive reading lists on the vast array of central government polices and thematic approaches to regeneration in the UK, including many not covered in this textbook]
Planning Portal: www.planningportal.gov.uk [guide to planning in the UK; features regeneration policy]
Regen.net: www.regen.net [latest news in the regeneration sector]

References

Abel, C. (2003) *Sky High: Vertical Architecture*, London: Thames and Hudson

Abercrombie, N. and Warde, A. (2000) *Contemporary British Society*, Cambridge: Polity Press, 3rd Edition

Adcock, B. (1994) 'Regenerating Merseyside Docklands: the Merseyside Development Corporation 1981–84', *Town Planning Review* 55: 265–289

Aglietta, M. (1979) *A Theory of Capitalist Regulation*, London: New Left Books

Aitchison, C. and Evans, T. (2003) 'The cultural industries and a model of sustainable regeneration: manufacturing "pop" in the Rhondda Valleys of South Wales', *Managing Leisure* 8: 133–144

Aitchison, C. and Pritchard, A. (eds.) (2007) *Festivals and Events: Culture and Identity in Leisure, Sport and Tourism*, Eastbourne: Leisure Studies Association

Aitchison, C., Richards, G. and Tallon, A. (eds.) (2007) *Urban Transformations: Regeneration and Renewal through Leisure and Tourism*, Eastbourne: Leisure Studies Association

Alden, J. and Thomas, H. (1998) 'Social exclusion in Europe: context and policy', *International Planning Studies* 3: 7–13

Allen, C. (2007) 'Of urban entrepreneurs or 24-hour party people? City-centre living in Manchester, England', *Environment and Planning A* 39: 666–683

Allen, C. (2008) *Housing Market Renewal and Social Class*, London: Routledge

Allinson, J. (1999) 'The 4.4 million households: do we really need them anyway?', *Planning Practice and Research* 14: 107–113

Allinson, J. (2005) 'Students and the urban renaissance', *Town and Country Planning* 74: 274–275

Allinson, J. (2006) 'Over-educated, over-exuberant and over here? The impact of students on cities', *Planning Practice and Research* 21: 79–94

Ambrose, P. (1994) *Urban Process and Power*, London: Routledge

Amin, A. (ed.) (1994) *Post-Fordism: A Reader*, London: Blackwell

Amin, A., Cameron, A. and Hudson, R. (1999) 'Welfare as work? The potential of the UK social economy', *Environment and Planning A* 31: 2033–2051

Amin, A., Cameron, A. and Hudson, R. (2002) *Placing the Social Economy*, London: Routledge

Amin, A. and Thrift, N. (1995) 'Globalisation, institutional thickness and the local economy', in P. Healey, S. Cameron, S. Davoudi, S. Graham and A. Madanipour (eds.) *Managing Cities: The New Urban Context*, Chichester: Wiley, 91–108

Anastacio, J., Gidley, B., Hart, L., Keith, M., Mayo, M. and Kowarzik, U. (2000) *Reflecting Realities: Participants' Perspectives on Integrated Communities and Sustainable Development*, Bristol: Policy Press

Anderson, J. (1983) 'Geography as ideology and the politics of crisis: the Enterprise Zone experiment', in J. Anderson, S. Duncan and R. Hudson (eds.) *Redundant Spaces in Cities and Regions: Studies in Industrial Decline and Social Change*, London: Academic Press, 313–350

Anderson, J. (1990) 'The New Right, Enterprise Zones and Urban Development Corporations', *International Journal of Urban and Regional Research* 14: 468–489

Andrew, C. and Goldsmith, M. (1998) 'From local government to local governance and beyond', *International Political Science Review* 19: 101–117

Arlidge, J. (2006) 'Mall developers move back into town centres', *The Sunday Times*, 1 October

Arnstein, S.R. (1969) 'A ladder of citizen participation', *Journal of the American Institute of Planners* 35: 216–224

Arnstein, S.R. (1971) 'A ladder of participation in the USA', *Journal of the Royal Town Planning Institute*, April: 176–182

Ashworth, G.J. and Tunbridge, J.E. (1990) *The Tourist-Historic City*, London: Belhaven Press

Ashworth, G.J. and Tunbridge, J.E. (2000) *The Tourist-Historic City: Retrospect and Prospect of Managing the Historic City*, Oxford: Pergamon

Ashworth, S. (2003) 'Business improvement districts: the impact of their introduction on retailers and leisure operators', *Journal of Retail and Leisure Property* 3: 150–157

Atkinson, R. (1998) 'Contemporary English urban policy and its implications for the development of an "urban policy" in the European Union', Paper presented at a conference '3 European Capitals Facing the Future: Helsinki–Berlin–Stockholm', Hanasaari – Swedish–Finnish Cultural Centre, Espoo, Finland, 5–7 November

Atkinson, R. (1999a) 'Discourses of partnership and empowerment in contemporary British urban regeneration', *Urban Studies* 36: 59–72

Atkinson, R. (1999b) 'Urban crisis: new policies for the next century', in P. Allmendinger and M. Chapman (eds.) *Planning Beyond 2000*, Chichester: Wiley, 69–86

Atkinson, R. (2000a) 'Combating social exclusion in Europe: the new urban policy challenge', *Urban Studies* 37: 1037–1055

Atkinson, R. (2000b) 'Narratives of policy: the construction of urban problems and urban policy in the official discourse of British government 1968–1998', *Critical Social Policy* 20: 211–232

Atkinson, R. (2003) 'Urban policy', in N. Ellison and C. Pierson (eds.) *Developments in British Social Policy 2*, Basingstoke: Palgrave Macmillan, 160–176

Atkinson, R. (2004a) 'Creating the city region in the UK: the challenge of neighbourhoods', Paper presented at a Regional Studies Association Conference: City Regions: Creating New Urban Futures?, London, 18 October

Atkinson, R. (2004b) 'The evidence on the impact of gentrification: new lessons for the urban renaissance?', *European Journal of Housing Policy* 4: 107–131

Atkinson, R. (2007) 'Under construction – the city-region and the neighbourhood: new actors in a system of multi-level governance?', in I. Smith, E. Lepine and M. Taylor (eds.) *Disadvantaged By Where You Live? Neighbourhood Governance in Contemporary Urban Policy*, Bristol: Policy Press, 65–82

Atkinson, R. and Blandy, S. (eds.) (2006) *Gated Communities*, London: Routledge

Atkinson, R., Blandy, S., Flint, J. and Lister, D. (2005) 'Gated cities of today? Barricaded residential developments in England', *Town Planning Review* 76: 401–422

Atkinson, R. and Bridge, G. (eds.) (2005) *Gentrification in a Global Context: The New Urban Colonialism*, London: Routledge

Atkinson, R. and Cope, S. (1997) 'Community participation and urban regeneration in Britain', in P. Hoggett (ed.) *Contested Communities: Experiences, Struggles, Policies*, Bristol: Policy Press, 201–221

Atkinson, R. and Flint, J. (2004) 'Fortress UK? Gated communities, the spatial revolt of the elites and time–space trajectories of segregation', *Housing Studies* 19: 875–892

Atkinson, R. and Helms, G. (eds.) (2007) *Securing an Urban Renaissance: Crime, Community and British Urban Policy*, Bristol: Policy Press

Atkinson, R. and Moon, G. (1994a) *Urban Policy in Britain: The City, the State and the Market*, London: Macmillan Press

Atkinson, R. and Moon, G. (1994b) 'The City Challenge initiative: an overview and preliminary assessment', *Regional Studies* 28: 95–99

Audit Commission (1989) *Urban Regeneration and Economic Development: The Local Government Dimension*, London: HMSO

Badcock, B. (1991) 'Neighbourhood change in inner Adelaide: an update', *Urban Studies* 28: 553–558

Badcock, B. (1995) 'Building on the foundations of gentrification: inner city housing development in Australia in the 1990s', *Urban Geography* 16: 70–90

Badcock, B. (2001) 'Thirty years on: gentrification and class changeover in Adelaide's inner suburbs, 1966–96', *Urban Studies* 38: 1559–1572

Bailey, C., Miles, S. and Stark, P. (2004) 'Culture-led urban regeneration and the revitalisation of identities in Newcastle, Gateshead and the North East of England', *International Journal of Cultural Policy* 10: 47–65

Bailey, J.T. (1989) *Marketing Cities in the 1990s and Beyond: New Patterns, New Pressures, New Promises*, Cleveland, OH: American Economic Development Council

Bailey, N. (ed.) (1993) 'Picking partners for the 1990s – special feature on partnerships in regeneration', *Town and Country Planning* 62: 136–147

Bailey, N., Barker, A. and MacDonald, K. (1995) *Partnership Agencies in British Urban Policy*, London: UCL Press

Bailey, N., Haworth, A., Manzi, T., Paranagamage, P. and Roberts, M. (2006) *Creating and Sustaining Mixed Income Communities: A Good Practice Guide*, York: Joseph Rowntree Foundation

Balas, C. (2004) 'City centre regeneration in the context of the 2001 European Capital of Culture in Porto, Portugal', *Local Economy* 19: 396–410

Banham, R., Barker, P., Hall, P. and Price, C. (1969) 'Non-plan: an experiment in freedom', *New Society* 25: 435–443

Banks, M., Lovatt, A., O'Connor, J. and Raffo, C. (2000) 'Risk and trust in the cultural industries', *Geoforum* 31: 453–464

Bannister, J., Fyfe, N. and Kearns, A. (1998) 'Closed circuit television and the city', in C. Norris, J. Morgan and G. Armstrong (eds.) *Surveillance, Closed Circuit Television and Social Control*, Aldershot: Ashgate, 21–39

Barber, A. (2007) 'Planning for sustainable re-urbanisation: policy challenges and city centre housing in Birmingham', *Town Planning Review* 78: 179–202

Barker, K. (2004) *Review of Housing Supply – Delivering Stability: Securing our Future Housing Needs – Final Report: Recommendations*, London: HM Treasury

Barlow, J. and Gann, D. (1993) *Offices into Flats*, York: Joseph Rowntree Foundation

Barlow, J. and Gann, D. (1994) 'Living at the office', *Town and Country Planning* 63: 100–101

Barnekov, T., Boyle, R. and Rich, D. (1989) *Privatism and Urban Policy in Britain and the United States*, Oxford: Oxford University Press

Barnes, I. and Campbell, J. (1988) 'From planners to entrepreneurs: the privatisation of local economic assistance', *Public Policy and Administration* 3: 24–28

Baron, S., Harris, K., Leaver, D. and Oldfield, B.M. (2001) 'Beyond convenience: the future for independent food and grocery retailers in the UK', *International Review of Retail, Distribution and Consumer Research* 11: 395–414

Barton, H. (ed.) (2000) *Sustainable Communities: the Potential for Eco-Neighbourhoods*, London: Earthscan

Bassett, K. (1993) 'Urban cultural strategies and urban regeneration: a case study and critique', *Environment and Planning A* 25: 1773–1788

Bassett, K. (1996) 'Partnerships, business elites and urban politics: new forms of governance in an English city?', *Urban Studies* 33: 539–556

Bassett, K. (2001) *Discovering Cities: Bristol*, Sheffield: Geographical Association

Bassett, K., Griffiths, R. and Smith, I. (2002a) 'Cultural industries, cultural clusters and the city: the example of natural history film-making in Bristol', *Geoforum* 33: 165–177

Bassett, K., Griffiths, R. and Smith, I. (2002b) 'Testing governance: partnerships, planning and conflict in waterfront regeneration', *Urban Studies* 39: 1757–1775

Bassett, K., Griffiths, R. and Smith, I. (2003) 'City of culture?', in M. Boddy (ed.) *Urban Transformation and Urban Governance: Shaping the Competitive City of the Future*, Bristol: Policy Press, 52–65

BCC (Bristol City Council) (2003) *Indicators of the Quality of Life Report 2003*, Bristol: BCC

Beauregard, R.A. (1986) 'The chaos and complexity of gentrification', in N. Smith and P. Williams (eds.) *Gentrification of the City*, London: Allen and Unwin, 35–55

Beck, A. and Willis, A. (1995) *Crime and Security: Managing the Risk to Safe Shopping*, Leicester: Perpetuity Press

Begg, I. (1999) 'Cities and competitiveness', *Urban Studies* 36: 795–809

Begg, I. (ed.) (2002a) *Urban Competitiveness: Policies for Dynamic Cities*, Bristol: Policy Press

Begg, I. (2002b) 'Introduction', in I. Begg (ed.) *Urban Competitiveness: Policies for Dynamic Cities*, Bristol: Policy Press, 1–10

Bell, D. and Jayne, M. (2003) '"Design-led" urban regeneration: a critical perspective', *Local Economy* 18: 121–134

Bell, D. and Jayne, M. (eds.) (2004) *City of Quarters: Urban Villages in the Contemporary City*, Aldershot: Ashgate

Bell, J.L. (1993) *Key Trends in Communities and Community Development*, London: Community Development Foundation

Bennett, J. (2005) *From New Towns to Growth Areas: Learning from the Past*, London: Institute for Public Policy Research [www.ippr.org.uk/ecomm/files/housing.pdf]

Bennett, R.J. (1990) 'Training and enterprise councils (TECs) and vocational education and training', *Regional Studies* 24: 65–82

Bianchini, F. (1990) 'Urban renaissance? The arts and the urban regeneration process', in S. MacGregor and B. Pimlott (eds.) *Tackling the Inner Cities: The 1980s Reviewed, Prospects for the 1990s*, Oxford: Clarendon, 215–250

Bianchini, F. (1994) 'Night cultures, night economies', *Town and Country Planning* 63: 308–309

Bianchini, F. (1995) 'Night cultures, night economies', *Planning Practice and Research* 10: 121–126

Bianchini, F. (2004) 'The cultural impacts of globalisation and the future of urban cultural policies', in C. Johnstone and M. Whitehead (eds.) *New Horizons in British Urban Policy: Perspectives on New Labour's Urban Renaissance*, Aldershot: Ashgate, 215–228

Bianchini, F., Dawson, J. and Evans, R. (1992) 'Flagship projects in urban regeneration', in P. Healey, S. Davoudi, S. Tavsanoglu, M. O'Toole and D. Usher (eds.) *Rebuilding the City: Property-Led Urban Regeneration*, London: E & FN Spon, 245–255

Bianchini, F. and Parkinson, M. (eds.) (1993) *Cultural Policy and Urban Regeneration: The West European Experience*, Manchester: Manchester University Press

Bianchini, F. and Schwengel, H. (1991) 'Re-imagining the city', in J. Corner and S. Harvey (eds.) *Enterprise and Heritage: Crosscurrents of National Culture*, London: Routledge, 212–234

Blackman, T. (1995) *Urban Policy in Practice*, London: Paul Chapman

Blair, F. and Evans, R. (2004) *Seeing the Bigger Picture: Delivering Local Sustainable Development*, York: Joseph Rowntree Foundation

Blakely, E.J. and Snyder, M.G. (1997) *Fortress America: Gating Communities in the United States*, Washington DC: Brookings Institution and Lincoln Institute of Land Policy

Blowers, A. (1997) 'Society and sustainability: the context of change for planning', in A. Blowers and B. Evans (eds.) *Town Planning into the 21st Century*, London: Routledge, 153–167

Blowers, A. and Pain, K. (1999) 'The unsustainable city?', in S. Pile, C. Brook and G. Mooney (eds.) *Unruly Cities? Order/Disorder*, London: Routledge, 247–298

Boddy, M. (ed.) (2003a) *Urban Transformation and Urban Governance: Shaping the Competitive City of the Future*, Bristol: Policy Press

Boddy, M. (2003b) 'The changing city', in M. Boddy (ed.) *Urban Transformation and Urban Governance: Shaping the Competitive City of the Future*, Bristol: Policy Press, 4–19

Boddy, M. (2003c) 'Social exclusion and the polarised city', in M. Boddy (ed.) *Urban Transformation and Urban Governance: Shaping the Competitive City of the Future*, Bristol: Policy Press, 66–75

Boddy, M. (2003d) 'Conclusions: shaping the urban future', in M. Boddy (ed.) *Urban Transformation and Urban Governance: Shaping the Competitive City of the Future*, Bristol: Policy Press, 90–96

Boddy, M. (2007) 'Designer neighbourhoods: new-build residential development in nonmetropolitan UK cities – the case of Bristol', *Environment and Planning A* 39: 86–105

Boddy, M. and Parkinson, M. (eds.) (2004) *City Matters: Competitiveness, Cohesion and Urban Governance*, Bristol: Policy Press

Boddy, M., Bassett, K., French, S., Griffiths, R., Lambert, C., Leyshon, A., Smith, I., Stewart, M. and Thrift, N. (2004) 'Competitiveness and cohesion in a prosperous city-region: the case of Bristol', in M. Boddy and M. Parkinson (eds.) *City Matters: Competitiveness, Cohesion and Urban Governance*, Bristol: Policy Press, 51–69

Boddy, M., Lovering, J. and Bassett, K. (1986) *Sunbelt City? A Study of Economic Change in Britain's M4 Growth Corridor*, Oxford: Clarendon Press

Bondi, L. (1991) 'Gender divisions and gentrification: a critique', *Transactions of the Institute of British Geographers* 16: 190–198

Bondi, L. (1999) 'Gender, class, and gentrification: enriching the debate', *Environment and Planning D: Society and Space* 17: 261–282

Booth, P. and Boyle, R. (1993) 'See Glasgow, see culture', in F. Bianchini and M. Parkinson (eds.) *Cultural Policy and Urban Regeneration: The West European Experience*, Manchester: Manchester University Press, 21–47

Bourne, L. (1993) 'The myth and reality of gentrification: a commentary on emerging urban forms', *Urban Studies* 30: 183–189

Boyer, R. (1990) *The Regulation School: A Critical Introduction*, New York: Columbia University Press

Boyle, M. and Hughes, G. (1991) 'The politics of the representation of the "real": discourses from the Left on Glasgow's role as City of Culture, 1990', *Area* 23: 217–228

Boyle, M. and Hughes, G. (1994) 'The politics of urban entrepreneurialism in Glasgow', *Geoforum* 25: 453–470

Boyle, P. and Halfacree, K. (eds.) (1998) *Migration into Rural Areas: Theories and Issues*, Chichester: Wiley

Boyle, P., Halfacree, K. and Robinson, V. (1998) *Exploring Contemporary Migration*, Harlow: Longman

Bradley, A., Hall, T. and Harrison, M. (2002) 'Selling cities: promoting new images for meetings tourism', *Cities* 19: 61–70

Bramley, G. (1999) *The changing physical form of cities: trends, explanations and implications*, Paper presented at ESRC UK/US workshop on Cities, Competitiveness and Cohesion, Glasgow

Bramley, G., Evans, M. and Noble, M. (2005) *Mainstreaming Public Services and their Impact on Neighbourhood Deprivation*, London: ODPM

Bramley, G. and Lambert, C. (2002) 'Managing urban development: land-use planning and city competitiveness', in I. Begg (ed.) *Urban Competitiveness: Policies for Dynamic Cities*, Bristol: Policy Press, 283–210

Breheny, M. (ed.) (1992) *Sustainable Development and Urban Form*, London: Pion

Breheny, M. (1993) 'Planning the sustainable city region', *Town and Country Planning* 62: 71–75

Breheny, M. (1995) 'The compact city and transport energy consumption', *Transactions of the Institute of British Geographers* 20: 81–101

Brennan, A., Rhodes, J. and Tyler, P. (1999) 'The distribution of the SRB Challenge Fund expenditure in relation to local-area need in England', *Urban Studies* 36: 2069–2084

Bridge, G. (1995) 'The space for class? On class analysis in the study of gentrification', *Transactions of the Institute of British Geographers* 20: 236–247

Bridge, G. (2003) 'Time–space trajectories in provincial gentrification', *Urban Studies* 40: 2545–2556

Bristol City Council (1997) *The Development of the Third Sector Economy in Bristol: A Draft for Discussion*, Bristol: BCC

Bristol City Council (2001) *Bristol Social Economy Development Project: Information Brief, October 2001*, Bristol: BCC

British Council of Shopping Centres (1996) *Town Centre Futures: The Long-term Impact of New Developments – Research Report*, London: DTZ Thorpe and Nathaniel Lichfield and Partners

Broadmead BID Bristol (2005) *The Broadmead BID Bristol – Prospectus: Vote for a Better Broadmead*, Bristol: Broadmead BID Bristol

Bromley, R.D.F. (1997) 'The Lower Swansea Valley', in G. Humphrys and S.T. Toole (eds.) *Geographical Excursions from Swansea – Volume 2: Human Landscapes*, Swansea: Department of Geography, University of Wales Swansea, 4th Edition, 29–48

Bromley, R.D.F., Hall, M. and Thomas, C.J. (2003a) 'The impact of environmental improvements on town centre regeneration', *Town Planning Review* 74: 143–164

Bromley, R.D.F. and Humphrys, G. (eds.) (1979) *Dealing with Dereliction: The Redevelopment of the Lower Swansea Valley*, Swansea: University College of Swansea

Bromley, R.D.F. and Jones, G.A. (1996) 'Identifying the inner city in Latin America', *The Geographical Journal* 162: 179–190

Bromley, R.D.F. and Matthews, D.L. (2007) 'Reducing consumer disadvantage: reassessing access in the retail environment', *International Review of Retail, Distribution and Consumer Research* 17: 483–501

Bromley, R.D.F., Matthews, D.L. and Thomas, C.J. (2007a) 'City centre accessibility for wheelchair users: the consumer perspective and the planning implications', *Cities* 24: 229–241

Bromley, R.D.F. and Morgan, R.H. (1985) 'The effects of Enterprise Zone policy: evidence from Swansea', *Regional Studies* 19: 403–413

Bromley, R.D.F. and Nelson, A.L. (2002) 'Alcohol-related crime and disorder across urban space and time: evidence from a British city', *Geoforum* 33: 239–254

Bromley, R.D.F. and Rees, J.C.M. (1988) 'The first five years of the Swansea Enterprise Zone: an assessment of change', *Regional Studies* 22: 263–275

Bromley, R.D.F., Tallon, A.R. and Roberts, A.J. (2007b) 'New populations in the British city centre: evidence of social change from the census and household surveys', *Geoforum* 38: 138–154

Bromley, R.D.F., Tallon, A.R. and Thomas, C.J. (2003b) 'Disaggregating the space-time layers of city-centre activities and their users', *Environment and Planning A* 35: 1831–1851

Bromley, R.D.F., Tallon, A.R. and Thomas, C.J. (2005) 'City centre regeneration through residential development: contributing to sustainability', *Urban Studies* 42: 2407–2429

Bromley, R.D.F. and Thomas, C.J. (1988) 'Retail parks: spatial and functional integration of retail units in the Swansea Enterprise Zone', *Transactions of the Institute of British Geographers* 13: 4–18

Bromley, R.D.F. and Thomas, C.J. (1989a) 'Clustering advantages of out-of-town stores', *International Journal of Retailing* 4: 41–59

Bromley, R.D.F. and Thomas, C.J. (1989b) 'The impact of shop type and spatial structure on shopping linkages in retail parks: planning implications', *Town Planning Review* 60: 45–70

Bromley, R.D.F. and Thomas, C.J. (eds.) (1993a) *Retail Change: Contemporary Issues*, London: UCL Press

Bromley, R.D.F. and Thomas, C.J. (1993b) 'The retail revolution, the carless shopper and disadvantage', *Transactions of the Institute of British Geographers* 18: 222–236

Bromley, R.D.F. and Thomas, C.J. (1995) 'Small town shopping decline: dependence and inconvenience for the disadvantaged', *International Review of Retail, Distribution and Consumer Research* 5: 433–456

Bromley, R.D.F. and Thomas, C.J. (1997) 'Vehicle crime in the city centre: planning for secure parking', *Town Planning Review* 68: 257–278

Bromley, R.D.F. and Thomas, C.J. (2002a) 'Food shopping and town centre vitality: exploring the link', *International Review of Retail, Distribution and Consumer Research* 12: 109–130

Bromley, R.D.F. and Thomas C.J. (2002b) 'Large food stores and town centre revival: evidence from Wales, UK', *European Retail Digest* 35: 35–37

Bromley, R., Thomas, C. and Millie, A. (2000) 'Exploring safety concerns in the night-time city: revitalising the evening economy', *Town Planning Review* 71: 71–96

Brookes, J. (1989) 'Cardiff Bay renewal strategy – another hole in the democratic system', *The Planner* 76: 38–40

Brooks, J. and Young, A. (1993) 'Revitalising the central business district in the face of decline: the case of New Orleans, 1973–1993', *Town Planning Review* 64: 251–271

Brown, A., O'Connor, J. and Cohen, S. (2000) 'Local music policies within a global music industry: cultural quarters in Manchester and Sheffield', *Geoforum* 31: 437–451

Brown, P. (2008) 'Business Improvement Districts: an overview', *Local Economy* 23: 71–75

Brownill, S. (1990) *Developing London's Docklands: Another Great Planning Disaster?*, London: Paul Chapman

Brownill, S. (1993) 'The docklands experience: locality and community in London', in R. Imrie and H. Thomas (eds.) *British Urban Policy and the Urban Development Corporations*, London: Paul Chapman, 41–57

Brownill, S. (1994) 'Selling the inner city: regeneration and place marketing in London's Docklands', in J.R. Gold and S.V. Ward (eds.) *Place Promotion: The Use of Publicity and Marketing to Sell Towns and Regions*, Chichester: Wiley, 133–152

Brownill, S. (1999) 'Turning the East End into West End: the lessons and legacies of the London Docklands Development Corporation', in R. Imrie and H. Thomas (eds.) *British Urban Policy: An Evaluation of the Urban Development Corporations*, London: Sage, 2nd Edition, 43–63

Buchel, P. and Hogervorst, B. (1997) *The Turning Tide*, Amsterdam: IJ Industrial Buildings Guild

Buck, N. and Gordon, I. (2004) 'Does spatial concentration of disadvantage contribute to social exclusion?', in M. Boddy and M. Parkinson (eds.) *City Matters: Competitiveness, Cohesion and Urban Governance*, Bristol: Policy Press, 237–253

Buck, N., Gordon, I., Harding, A. and Turok, I. (eds.) (2005) *Changing Cities: Rethinking Urban Competitiveness, Cohesion and Governance*, Basingstoke: Palgrave Macmillan

Bunce, M. (1994) *The Countryside Ideal: Anglo-American Images of Landscape*, London: Routledge

Burton, E. (2000) 'The compact city: just or just compact? A preliminary analysis', *Urban Studies* 37: 1969–2006

Burwood, S. and Roberts, P. (2002) *Learning from the Past: The BURA Guide to Effective and Lasting Regeneration*, London: BURA

Butler, S.M. (1982) *Enterprise Zones: Greenlining the Inner Cities*, London: Heinemann

Butler, T. (1997) *Gentrification and the Middle Classes*, Aldershot: Ashgate

Butler, T. (2003) 'Living in the bubble: gentrification and its "others" in north London', *Urban Studies* 40: 2469–2486

Butler, T. (2007a) 'For gentrification?', *Environment and Planning A* 39: 162–181

Butler, T. (2007b) 'Re-urbanising London Docklands: gentrification, suburbanization or new urbanism?', *International Journal of Urban and Regional Research* 31: 759–781

Butler, T. and Robson, G. (2001) 'Social capital, gentrification and neighbourhood change in London: a comparison of three south London neighbourhoods', *Urban Studies* 38: 2145–2162

Butler, T. and Robson, G. (2003) 'Negotiating their way in: the middle classes, gentrification and the deployment of capital in a globalising metropolis', *Urban Studies* 40: 1791–1809

Butler, T. with Robson, G. (2003) *London Calling: The Middle Classes and the Remaking of Inner London*, London: Berg

Byrne, D. (1999) 'Tyne and Wear UDC – turning the uses inside out: active de-industrialisation and its consequences', in R. Imrie and H. Thomas (eds.) *British Urban Policy: An Evaluation of the Urban Development Corporations*, London: Sage, 2nd Edition, 128–145

Byrne, D. (2001) *Understanding the Urban*, Basingstoke: Palgrave

Cameron, S. (1992) 'Housing, gentrification and urban regeneration policies', *Urban Studies* 29: 3–14

Cameron, S. (2003) 'Gentrification, housing redifferentiation and urban regeneration: "Going for Growth" in Newcastle upon Tyne', *Urban Studies* 40: 2367–2382

Cameron, S. (2006) 'From low demand to rising aspirations: housing market renewal within regional and neighbourhood regeneration policy', *Housing Studies* 21: 3–16

Cameron, S. and Coaffee, J. (2005) 'Art, gentrification and regeneration: from artist as pioneer to public arts', *European Journal of Housing Studies* 5: 39–58

Cappellin, R. (1991) 'International networks of cities', in R. Camagni (ed.) *Innovation Networks: Spatial Perspectives*, London: Belhaven Press, 230–244

Carley, M. (2000) 'Urban partnerships, governance and the regeneration of Britain's cities', *International Planning Studies* 5: 273–297

Carley, M., Kirk, K. and McIntosh, S. (2001) *Retailing, Sustainability and Neighbourhood Regeneration*, York: York Publishing Services

Carpenter, J. and Lees, L. (1995) 'Gentrification in New York, London and Paris: an international comparison', *International Journal of Urban and Regional Research* 19: 286–303

Cars, G., Healey, P., Madanipour, A. and de Magalhaes, C. (eds.) (2002) *Urban Governance, Institutional Capacity and Social Milieux*, Aldershot: Ashgate

Casado-Diaz, M., Everett, S. and Wilson, J. (eds.) (2007) *Social and Cultural Change: Making Space(s) for Leisure and Tourism*, Eastbourne: Leisure Studies Association

Castells, M. and Hall, P. (1994) *Technopoles of the World: The Making of 21st Century Industrial Complexes*, London: Routledge

Caulfield, J. (1989) 'Gentrification and desire', *Canadian Review of Sociology and Anthropology* 26: 617–632

Caves, R.E. (2000) *Creative Industries: Contracts between Art and Commerce*, London: Harvard University Press

Cebulla, A., Berry, J. and McGreal, S. (2000) 'Evaluation of community based regeneration in Northern Ireland: between social and economic regeneration', *Town Planning Review* 71: 169–187

CEC (Commission of the European Communities) (1990) *Green Paper on the Urban Environment*, Brussels: CEC

Centre for Cities (2008) *Cities Outlook 2008*, London: Centre for Cities

Centre for Local Economy Studies (1996) *Regeneration Through Work: Creating Jobs in the Social Economy*, Manchester: Centre for Local Economy Studies

Champion, T. (2001) 'Urbanization, suburbanization, counterurbanization and reurbanization', in R. Paddison (ed.) *Handbook of Urban Studies*, London: Sage, 143–161

Chanan, G. (2003) *Searching for Solid Foundations: Community Involvement in Urban Policy*, London: ODPM

Chanan, G., West, A., Garratt, C. and Humm, J. (1999) *Regeneration and Sustainable Communities*, London: Community Development Foundation

Chapman, D.W. and Larkham, P.J. (1999) 'Urban design, urban quality and the quality of life: reviewing the Department of the Environment's Urban Design Campaign', *Journal of Urban Design* 4: 211–232

Charney, I. (2007) 'The politics of design: architecture, tall buildings and the skyline of central London', *Area* 39: 195–205

Chatterton, P. (1999) 'University students and city centres – the formation of exclusive geographies: the case of Bristol, UK', *Geoforum* 30: 117–133

Chatterton, P. and Bradley, D. (2000) 'Bringing Britain together? The limitations of area-based regeneration policies in addressing deprivation', *Local Economy* 15: 98–111

Chatterton, P. and Hollands, R. (2002) 'Theorising urban playscapes: producing, regulating and consuming youthful nightlife city spaces', *Urban Studies* 39: 95–116

Chatterton, P. and Hollands, R. (2003) *Urban Nightscapes: Youth Cultures, Pleasure Spaces and Corporate Power*, London: Routledge
Christopherson, S. (1994) 'The fortress city: privatized spaces, consumer citizenship', in A. Amin (ed.) *Post-Fordism: A Reader*, Oxford: Blackwell, 409–427
Church, A. and Frost, M. (1995) 'The Thames Gateway', *Geographical Journal* 161: 199–209
Church, A. and Reid, P. (1996) 'Urban power, international networks and competition: the example of cross-border co-operation', *Urban Studies* 33: 1297–1318
Cities (1995) 'Economic Regeneration Strategies in British Cities', Volume 12, Issue 4
Clarence, E. and Painter, C. (1998) 'Public services under New Labour: collaborative discourses and local networking', *Public Policy and Administration* 13: 8–22
Clarke, G., Eyre, H. and Guy, C. (2002) 'Deriving indicators of access to food retail provision in British cities: studies of Cardiff, Leeds and Bradford', *Urban Studies* 39: 2031–2060
Clement, M. (2007) 'Bristol: "civilising" the inner city', *Race and Class* 48: 97–105
CLG (Communities and Local Government) (2006a) *Planning Policy Statement 3 (PPS3): Housing*, London: CLG
CLG (Communities and Local Government) (2006b) *Strong and Prosperous Communities: The Local Government White Paper*, London: HMSO
CLG (Communities and Local Government) (2006c) *Transferable Lessons from the New Towns*, London: CLG
CLG (Communities and Local Government) (2007a) *Homes for the Future: More Affordable, More Sustainable*, London: Stationery Office
CLG (Communities and Local Government) (2007b) *Regenerating the English Coalfields – Interim Evaluation of the Coalfields Regeneration Programmes*, London: HMSO
CLG (Communities and Local Government) (2007c) *The English Indices of Deprivation*, London: CLG
CLG (Communities and Local Government) (2007d) *The Single Regeneration Budget: Final Evaluation – Urban Research Summary Number 25*, London: CLG
CLG (Communities and Local Government) (2007e) *Strong and Prosperous Communities: The Local Government White Paper – Making It Happen: The Implementation Plan*, London: HMSO
CLG (Communities and Local Government) (2007f) *An Action Plan for Community Engagement: Building on Success*, London: HMSO
CLG (Communities and Local Government) (2008) *Eco-towns: Living a Green Future*, London: Stationery Office
Cochrane, A. (1993) *Whatever Happened to Local Government?*, Buckingham: Open University Press
Cochrane, A. (2007) *Understanding Urban Policy: A Critical Introduction*, Oxford: Blackwell
Cochrane, A., Peck, J. and Tickell, A. (1996) 'Manchester plays games: exploring the local politics of globalisation', *Urban Studies* 33: 1319–1336
Colenutt, B. and Cutten, A. (1994) 'Community empowerment in vogue or vain?', *Local Economy* 9: 236–250
Collis, C., Berkeley, N. and Fletcher, D.R. (2000) 'Retail decline and policy responses in district shopping centres', *Town Planning Review* 71: 149–168
Colomb, C. (2007) 'Unpacking New Labour's "urban renaissance" agenda: towards a socially sustainable reurbanization of British cities?', *Planning Practice and Research* 22: 1–24
Colquhoun, I. (1995) *Urban Regeneration: An International Perspective*, London: Batsford
Comedia (1991) *Out of Hours: A Study of Economic, Social and Cultural Life in Twelve Town Centres in the UK*, London: Comedia
Comedia (2002) *Realising the Cultural Potential of our Core Cities*, London: Comedia
Community Development Foundation (1996) *Regeneration and the Community*, London: Community Development Foundation
Community Development Foundation (1997) *Regeneration and the Community: Guidelines to the Community Involvement Aspect of the SRB Challenge Fund*, London: Community Development Foundation
Conservative Party (2007) *Cities' Renaissance: Creating Local Leadership*, London: Conservative Party
Cook, I.R. (2008) 'Mobilising urban policies: the policy transfer of US Business Improvement Districts to England and Wales', *Urban Studies* 45: 773–795
Cooke, P., Davies, C. and Wilson, R. (2002) 'Urban networks and the new economy: the impact of clusters on planning for growth', in I. Begg (ed.) *Urban Competitiveness: Policies for Dynamic Cities*, Bristol: Policy Press, 233–256
Costello, L. (2005) 'From prisons to penthouses: the changing images of high-rise living in Melbourne', *Housing Studies* 20: 49–62
Couch, C. (1990) *Urban Renewal: Theory and Practice*, Basingstoke: Macmillan

Couch, C. (1999) 'Housing development in the city centre', *Planning Practice and Research* 14: 69–86
Couch, C. (2003) *City of Change and Challenge: Urban Planning and Regeneration in Liverpool*, Aldershot: Ashgate
Couch, C. and Dennemann, A. (2000) 'Urban regeneration and sustainable development in Britain: the example of the Liverpool Ropewalks Partnership', *Cities* 17: 137–147
Couch, C. and Farr, S-J. (2000) 'Museums, galleries, tourism and regeneration: some experiences from Liverpool', *Built Environment* 26: 152–163
Couch, C., Fraser, C. and Percy, S. (eds.) (2003) *Urban Regeneration in Europe*, Oxford: Blackwell
Coulson, A. (1990) 'Flagships and flaws – assessing the UDC decade', *Town and Country Planning* 59: 299–302
Coupland, A. (ed.) (1997) *Reclaiming the City: Mixed Use Development*, London: E & FN Spon
Creative Industries Task Force (1998) *Creative Industries: Mapping Document*, London: DCMS
Crilley, D. (1992a) 'Remaking the image of the Docklands', in P. Ogden (ed.) *London Docklands: The Challenge of Development*, Cambridge: Cambridge University Press, 25–31
Crilley, D. (1992b) 'The great Docklands housing boom', in P. Ogden (ed.) *London Docklands: The Challenge of Development*, Cambridge: Cambridge University Press, 60–66
Crilley, D. (1993) 'Architecture as advertising: constructing the image of redevelopment', in G. Kearns and C. Philo (eds.) *Selling Places: The City as Cultural Capital, Past and Present*, Oxford: Pergamon, 231–252
Crouch, M. (1992) 'Managing town centres', *The Planner* 78: 8–10
Cullingworth, B. and Nadin, V. (2006) *Town and Country Planning in the UK*, London: Routledge, 14th Edition
Davidson, M. and Lees, L. (2005) 'New-build "gentrification" and London's riverside renaissance', *Environment and Planning A* 37: 1165–1190
Davies, L.E. (2008) 'Sport and the local economy: the effects of stadia development on the commercial property market', *Local Economy* 23: 31–46
Davies, R.L. (1984) *Retail and Commercial Planning*, London: Croom Helm
Davies, R.L. (1987) *Help for the High Street: Some New Approaches to Revitalisation*, Cheshunt: Tesco PLC
Davies, R.L. (ed.) (1995) *Retail Planning Policies in Europe*, London: Routledge
Davoudi, S. (1995) 'City Challenge: the three way partnership', *Planning Practice and Research* 10: 333–344
Davoudi, S. and Healey, P. (1995a) 'City Challenge – a sustainable process or temporary gesture?', in R. Hambleton and H. Thomas (eds.) *Urban Policy Evaluation: Challenge and Change*, London: Paul Chapman, 158–174
Davoudi, S. and Healey, P. (1995b) 'City challenge: sustainable process or temporary gesture?', *Environment and Planning C: Government and Policy* 13: 79–95
Dawson, J. (1992) 'European city networks: experiments in trans-national collaboration', *The Planner* 78: 7–9
Dawson, J.A. (1988) 'Futures for the high street', *Geographical Journal* 154: 1–12
DCMS (Department for Culture, Media and Sport) (1998) *Creative Industries Mapping Document 1998*, London: DCMS
DCMS (Department for Culture, Media and Sport) (1999) *Policy Action Team 10: A Report to the Social Exclusion Unit – Arts and Sport*, London: DCMS
DCMS (Department for Culture, Media and Sport) (2000a) *Creative Industries: The Regional Dimension*, London: DCMS
DCMS (Department for Culture, Media and Sport) (2000b) *Culture and Creativity: The Next Ten Years*, London: DCMS
DCMS (Department for Culture, Media and Sport) (2004a) *The Contribution of Culture to Regeneration in the UK: A Review of Evidence*, London: DCMS
DCMS (Department for Culture, Media and Sport) (2004b) *Culture at the Heart of Regeneration*, London: DCMS
DCMS (Department for Culture, Media and Sport) (2004c) *Guidance to the Licensing Act 2003*, London: DCMS
Deakin, N. and Edwards, J. (1993) *The Enterprise Culture and the Inner City*, London: Routledge
Dear, M. (2000) *The Postmodern Urban Condition*, Oxford: Blackwell
Dearlove, J. and Saunders, P. (2000) *Introduction to British Politics*, Cambridge: Polity Press, 3rd Edition
Deas, I. and Giordano, B. (2002) 'Locating the competitive city in England', in I. Begg (ed.) *Urban Competitiveness: Policies for Dynamic Cities*, Bristol: Policy Press, 191–209
Deas, I. and Ward, K. (2000) 'The song has ended but the melody has lingered: Regional Development

Agencies and the lessons of the Urban Development Corporation experiment', *Local Economy* 14: 114–132

de Groot, L. (1992) 'City challenge: competing in the urban regeneration game', *Local Economy* 7: 196–209

Delafons, J. (1996) 'Too many pubs?', *Town and Country Planning* 65: 346–347

DETR (Department of the Environment, Transport and the Regions) (1997) *Involving Communities in Urban and Rural Regeneration: A Guide for Practitioners*, London: HMSO

DETR (Department of the Environment, Transport and the Regions) (1998a) *A New Deal for Transport: Better for Everyone*, London: HMSO

DETR (Department of the Environment, Transport and the Regions) (1998b) *Making the Difference: A New Start for England's Coalfield Communities – Task Force Report*, London: HMSO

DETR (Department of the Environment, Transport and the Regions) (1998c) *The Impact of Large Foodstores on Market Towns and District Centres: Report by CB Hillier Parker and Savell Bird Axon*, London: HMSO

DETR (Department of the Environment, Transport and the Regions) (1999) *Millennium Villages and Sustainable Communities*, London: DETR

DETR (Department of the Environment, Transport and the Regions) (2000a) *Our Towns and Cities: The Future – Delivering an Urban Renaissance*, London: HMSO

DETR (Department of the Environment, Transport and the Regions) (2000b) *The State of English Cities*, London: Stationery Office

DETR (Department of the Environment, Transport and the Regions) (2000c) *Planning Policy Guidance 3: Housing*, London: HMSO

DETR (Department of the Environment, Transport and the Regions) (2000d) *Living in Urban England: Attitudes and Aspirations*, London: HMSO

DETR (Department of the Environment, Transport and the Regions) (2001) *Planning Policy Guidance 13: Transport*, London: HMSO

Diamond, J. and Liddle, J. (2005) *Management of Regeneration: Choices, Challenges and Dilemmas*, London: Routledge

Dicken, P. (2007) *Global Shift: Mapping the Changing Contours of the World Economy*, London: Sage, 5th Edition

Di Gaetano, A. and Strom, E. (2003) 'Comparative urban governance: an integrated approach', *Urban Affairs Review* 38: 356–395

Dixon, T.J. (2005) 'The role of retailing in urban regeneration', *Local Economy* 20: 168–182

Dixon, T. and Adams, D. (2008) 'Housing supply and brownfield regeneration in a post-Barker world: is there enough brownfield land in England and Scotland?', *Urban Studies* 45: 115–139

Dixon, T., Raco, M., Catney, P. and Lerner, D.N. (2007) *Sustainable Brownfield Regeneration: Liveable Places from Problem Spaces*, Oxford: Blackwell

DoE (Department of the Environment) (1977) *Policy for the Inner Cities*, London: HMSO

DoE (Department of the Environment) (1988) *Planning Policy Guidance 6: Major Retail Developments*, London: HMSO

DoE (Department of the Environment) (1992) *The Effects of Major Out of Town Retail Development*, London: HMSO

DoE (Department of the Environment) (1994a) *Planning Policy Guidance 13: Transport*, London: HMSO

DoE (Department of the Environment) (1994b) *Planning Out Crime*, London: HMSO

DoE (Department of the Environment) (1994c) *Sustainable Development: The UK Strategy*, London: HMSO

DoE (Department of the Environment) (1995a) *Projection of Households in England to 2016*, London: HMSO

DoE (Department of the Environment) (1995b) *Our Future Homes: Opportunity, Choice and Responsibility – The Government's Housing Policies for England and Wales*, London: HMSO

DoE (Department of the Environment) (1995c) *Final Evaluation of Enterprise Zones*, London: HMSO

DoE (Department of the Environment) (1996a) *Planning Policy Guidance 6: Town Centres and Retail Developments*, London: HMSO

DoE (Department of the Environment) (1996b) *Household Growth: Where Shall We Live?*, London: HMSO

DoE (Department of the Environment) (1997) *Planning Policy Guidance 1: General Policy and Principles*, London: HMSO

Dorling, D., Rigby, J., Wheeler, B., Ballas, D., Thomas, B., Fahmy, E., Gordon, D. and Lupton, R. (2007) *Poverty, Wealth and Place in Britain, 1968 to 2005*, Bristol: Policy Press

DTLR (Department for Transport, Local Government and the Regions) (2000) *A Review of the Evidence Base for Regeneration Policy and Practice*, London: HMSO

DTLR (Department for Transport, Local Government and the Regions) (2002) *Lessons and Evaluation Evidence from Ten Single Regeneration Budget Case Studies – Mid Term Report*, London: DTLR

Dutton, P. (2003) 'Exporting forms of gentrification: understanding the influence of the global city of London on the gentrification process in Leeds', *Urban Studies* 40: 2557–2572

Dutton, P. (2005) ' "Outside the metropole": gentrification in provincial cities or provincial gentrification?', in R. Atkinson and G. Bridge (eds.) *Gentrification in a Global Context: The New Urban Colonialism*, London: Routledge, 209–224

Dziembowska-Kowalska, J. and Funk, R. (1999) 'Cultural activities: source of competitiveness and prosperity in urban regions', *Urban Studies* 36: 1381–1398

Edgar, B. and Taylor, J. (2000) 'Housing', in P. Roberts and H. Sykes (eds.) *Urban Regeneration: A Handbook*, London: Sage, 153–175

Ellis, G. and McKay, S. (2000) 'City management profile – Belfast', *Cities* 17: 47–54

Emery, J. (2006) 'Bullring: a case study of retail-led urban renewal and its contribution to city centre regeneration', *Journal of Retail and Leisure Property* 5: 121–133

English Heritage (2005) *Low Demand Housing and the Historic Environment*, London: English Heritage

English Heritage (2007a) *Suburbs and the Historic Environment*, London: English Heritage

English Heritage (2007b) *The Heritage of Historic Suburbs*, London: English Heritage

Erikson, R.A. and Friedman, S.W. (1990a) 'Enterprise Zones 1: investment and job creation of state government programs in the USA', *Environment and Planning C: Government and Policy* 8: 251–267

Erikson, R.A. and Friedman, S.W. (1990b) 'Enterprise Zones 2: a comparative analysis of zone performance and state government policies', *Environment and Planning C: Government and Policy* 8: 363–378

Erikson, R.A. and Syms, P.M. (1986) 'The effects of Enterprise Zones on local property markets', *Regional Studies* 20: 1–14

ESRC (Economic and Social Research Council) (2007) *Localism and Local Governance*, ESRC Seminar Series: Mapping the Public Policy Landscape, Swindon: ESRC

Essex, S. and Chalkley, B. (2004) 'Mega-sporting events in urban and regional policy', *Planning Perspectives* 19: 201–232

Etzioni, A. (1997) *The New Golden Rule: Community and Morality in a Democratic Society*, London: Profile Books

Euchner, C. (1999) 'Tourism and sports: the serious competition for play', in D. Judd and S. Fainstein (eds.) *The Tourist City*, London: Yale University Press, 215–232

European Planning Studies (2001) Creativity and Cost in Urban and Regional Development in the 'New Economy', Volume 9, Issue 7

Evans, G. (2001) *Cultural Planning: An Urban Renaissance*, London: Routledge

Evans, G. (2003) 'Hard-branding the cultural city: from Prado to Prada', *International Journal of Urban and Regional Research* 27: 417–440

Evans, G. (2005) 'Measure for measure: evaluating the evidence of culture's contribution to regeneration', *Urban Studies* 42: 959–983

Evans, R. (1997) *Regenerating Town Centres*, Manchester: Manchester University Press

Evans, R.W. (1981) 'The Lower Swansea Valley Enterprise Zone', *Built Environment* 7: 20–30

Fainstein, S.S. (1991) 'Promoting economic development: urban planning in the United States and Great Britain', *Journal of the American Planning Association* 57: 22–33

Fainstein, S. S. (1994) *The City Builders: Property, Planning and Politics in London and New York*, Oxford: Blackwell

Faulkner, B., Chalip, L., Brown, G., Jago, L., March, R. and Woodside, A. (2001) 'Monitoring the tourism impacts of the Sydney 2000 Olympics', *Event Management* 6: 231–246

Fernie, J. (1995) 'The coming of the fourth wave: new forms of out-of-town retail development', *International Journal of Retail and Distribution Management* 23: 4–11

Fernie, J. (1998) 'The breaking of the fourth wave: recent out-of-town retail developments in Britain', *International Review of Retail, Distribution and Consumer Research* 8: 303–307

Filion, P. (1991) 'The gentrification–social structure dialectic: a Toronto case study', *International Journal of Urban and Regional Research* 15: 553–574

Filion, P. (2001) 'Suburban mixed-use centres and urban dispersion: what difference do they make?', *Environment and Planning A* 33: 141–160

Fincher, R. (1992) 'Urban geography in the 1990s', in A. Rogers, H. Viles and A. Goudie (eds.) *The Student's Companion to Geography*, Oxford: Blackwell, 104–110
Fishman, R. (1987) *Bourgeois Utopias: The Rise and Fall of Suburbia*, New York: Basic Books
Florida, R. (2002) *The Rise of the Creative Class and How It's Transforming Work, Leisure, Community and Everyday Life*, New York: Basic Books
Florida, R. (2005) *Cities and the Creative Class*, London: Routledge
Fokkema, T. and Nijkamp, P. (1996) 'Large cities, large problems?', *Urban Studies* 33: 353–377
Foley, P. (1991) 'The impact of the World Student Games on Sheffield', *Environment and Planning C: Government and Policy* 9: 65–78
Foley, P. and Martin, S. (2000) 'A new deal for the community? Public participation in regeneration and local service delivery', *Policy and Politics* 28: 479–492
Fordham, G., Hutchinson, J. and Foley, P. (1998) 'Strategic approaches to local regeneration: the Single Regeneration Budget Challenge Fund', *Regional Studies* 33: 131–141
Forrest, R. and Kearns, A. (2001) 'Social cohesion, social capital and the neighbourhood', *Urban Studies* 38: 2125–2143
Foster, J. (1999) *Docklands: Cultures in Conflict, Worlds in Collision*, London: UCL Press
Francis, A. (1991) 'Private nights in the city centre', *Town and Country Planning* 60: 302–303
Frantz, D. and Collins, C. (1999) *Celebration, U.S.A.: Living in Disney's Brave New Town*, New York: Henry Holt and Company
Frey, H. (1999) *Designing the City: Towards a More Sustainable Urban Form*, London: E & FN Spon
Fulton, W. (1997) *The Reluctant Metropolis: The Politics of Urban Growth in Los Angeles*, Point Arena, CA: Solano Press Books
Furbey, R. (1999) 'Urban "regeneration": reflections on a metaphor', *Critical Social Policy* 19: 419–445
Gaffikin, F. and Morrissey, M. (1999) 'The urban economy and social exclusion: the case of Belfast', in F. Gaffikin and M. Morrissey (eds.) *City Visions: Imagining Place, Enfranchising People*, London: Pluto Press, 34–57
Gaffickin, F. and Warf, B. (1993) 'Urban policy and the post-Keynesian state in the United Kingdom and the United States', *International Journal of Urban and Regional Research* 17: 67–84
García, B. (2004a) 'Cultural policy in European cities: lessons from experience, prospects for the future', *Local Economy* 19: 312–326
García, B. (2004b) 'Urban regeneration, arts programming and major events: Glasgow 1990, Sydney 2000 and Barcelona 2004', *International Journal of Cultural Policy* 10: 103–118
Gardiner, J. (2007) 'Empty promises', *Regeneration and Renewal* 31st August, 18–20
Garreau, J. (1991) *Edge City: Life on the New Frontier*, New York: Doubleday
Garrett, A. (2008) 'Trouble up north as apartment prices slump', *The Observer*, 9th March
Gasson, A. (1992) 'Regeneration in the Lower Swansea Valley', *Geography Review* 5(5): 2–7
Geddes, M. (2006) 'Partnership and the limits to local governance in England: institutionalist analysis and neo-liberalism', *International Journal of Urban and Regional Research* 30: 76–97
Giddens, A. (1990) *The Consequences of Modernity*, Oxford: Blackwell
Giddens, A. (1998) *The Third Way: The Renewal of Social Democracy*, Cambridge: Polity Press
Giddens, A. (2000) *The Third Way and its Critics*, Cambridge: Polity Press
Giddens, A. (2006) *Sociology*, 5th Edition, Cambridge: Polity Press
Giradet, H. (1996) *The Gaia Atlas of Cities' New Directions for Sustainable Urban Living*, London: Gaia Books
Giradet, H. (1999) *Creating Sustainable Cities*, Totnes: Green Books
Giradet, H. (2003) 'Urban design as if people and planet mattered', in *Report of the Sustainability Special Interest Group (Architectural Edition) on Behalf of the Centre for Education in the Built Environment*, Cardiff: Centre for Education in the Built Environment
Giradet, H. (2004) *Cities, People, Planet: Liveable Cities for a Sustainable World*, Chichester: Wiley Academy
GLA (Greater London Authority) (2003) *Spending Time: London's Leisure Economy*, London: GLA
Glass, R. (1964) 'Introduction: aspects of change', in Centre for Urban Studies (ed.) *London: Aspects of Change*, London: MacGibbon and Kee, xiii–xlii
Glasze, G., Webster, C.J. and Frantz, K. (eds.) (2006) *Private Cities: Local and Global Perspectives*, London: Routledge
Gold, J. and Gold, M. (2005) *Cities of Culture: Staging International Festivals and the Urban Agenda, 1851–2000*, Aldershot: Ashgate
Gold, J. and Gold, M. (eds.) (2007) *Olympic Cities: City Agendas, Planning, and the World's Games 1896–2012*, London: Routledge

Gold, J.R. and Ward, S.V. (eds.) (1994) *Place Promotion: The Use of Publicity and Marketing to Sell Towns and Regions*, Chichester: Wiley

Goodlad, R., Burton, P. and Croft, J. (2005) 'Effectiveness at what? The processes and impact of community involvement in area-based initiatives', *Environment and Planning C: Government and Policy* 23: 923–938

Goodwin, M. (1991) 'Replacing a surplus population: the politics of LDDC', in J. Allen and C. Hamnett (eds.) *Housing and Labour Markets*, London: Unwin Hyman, 254–275

Goodwin, M. (1993) 'The city as commodity: the contested spaces of urban development', in G. Kearns and C. Philo (eds.) *Selling Places: The City as Cultural Capital, Past and Present*, Oxford: Pergamon, 145–162

Goodwin, M. (2004) 'The scaling of "urban" policy: neighbourhood, city or region?', in C. Johnstone and M. Whitehead (eds.) *New Horizons in British Urban Policy: Perspectives on New Labour's Urban Renaissance*, Aldershot: Ashgate, 173–184

Goodwin, M. and Painter, J. (1996) 'Local governance, the crises of Fordism and the changing geographies of regulation', *Transactions of the Institute of British Geographers* 21: 635–648

Gordon, I., Brown, B., Buck, N., Hall, P., Harloe, M., Kleinman, M., O'Reilly, K., Potts, G., Smethurst, L. and Sparkes, J. (2004) 'London: competitiveness, cohesion and the policy environment', in M. Boddy and M. Parkinson (eds.) *City Matters: Competitiveness, Cohesion and Urban Governance*, Bristol: Policy Press, 71–90

Gordon, T.M. (2004) 'Moving up by moving out? Planned developments and residential segregation in California', *Urban Studies* 41: 441–461

Gosling, P. (2001) 'The self-help society', *Public Finance* May: 11–17

Goss, J. (1993) 'The "magic of the mall": an analysis of form, function and meaning in the contemporary retail built environment', *Annals of the Association of American Geographers* 83: 18–47

Gotham, K.F. (ed.) (2001) *Critical Perspectives on Urban Redevelopment*, Greenwich, CT: JAI Press

Graham, J. (1992) 'Post-Fordism as politics: the political consequences of narratives on the left', *Environment and Planning D: Society and Space* 10: 393–410

Graham, S. and Marvin, S. (1996) *Telecommunications and the City: Electronic Spaces, Urban Places*, London: Routledge

Grail, J. and Dawkins, G. (2008) 'Business Improvement Districts (BIDs) in London', *Local Economy* 23: 76–80

Grant, J. and Mittelsteadt, L. (2004) 'Types of gated communities', *Environment and Planning B: Planning and Design* 31: 913–930

Greed, C. (2000) *Introducing Planning*, 3rd Edition, London: Athlone Press

Griffiths, R. (1993) 'The politics of cultural policy in urban regeneration strategies', *Policy and Politics* 21: 39–46

Griffiths, R. (1995a) 'Cultural strategies and new modes of urban intervention', *Cities* 12: 253–265

Griffiths, R. (1995b) 'Eurocities', *Planning Practice and Research* 10: 215–221

Griffiths, R. (1998a) 'Making sameness: place marketing and the new urban entrepreneurialism', in N. Oatley (ed.) *Cities, Economic Competition and Urban Policy*, London: Paul Chapman, 41–57

Griffiths, R. (1998b) 'The National Lottery and competitive cities', in N. Oatley (ed.) *Cities, Economic Competition and Urban Policy*, London: Paul Chapman, 181–198

Griffiths, R. (ed.) (1998c) *Social Exclusion in Cities: The Urban Policy Challenge*, Bristol: Faculty of the Built Environment, University of the West of England, Occasional Paper 3

Griffiths, R. (1999) 'Artists organisations and the recycling of urban space', in L. Nystrom (ed.) *City and Culture: Cultural Processes and Urban Sustainability*, Karlskrona: Swedish Urban Environment Council, 460–475

Griffiths, R. (2005) *Creative enterprises and the urban milieu*, paper presented at the AESOP Annual Congress, Vienna, 13–17 July

Griffiths, R. (2006) 'City/culture discourses: evidence from the competition to select the European Capital of Culture 2008', *European Planning Studies* 14: 415–430

Griffiths, R., Bassett, K. and Smith, I. (1999) 'Cultural policy and the cultural economy in Bristol', *Local Economy* 14: 257–264

Griffiths, R., Bassett, K. and Smith, I. (2003) 'Capitalising on culture: cities and the changing landscape of cultural policy', *Policy and Politics* 31: 153–169

Grimshaw, L. and Smith, I. (2007) 'Evaluation, knowledge and learning in neighbourhood governance: the case of the New Deal for Communities', in I. Smith, E. Lepine and M. Taylor (eds.) *Disadvantaged by*

Where You Live? Neighbourhood Governance in Contemporary Urban Policy, Bristol: Policy Press, 185–204

Gripaios, P. (2002) 'The failure of regeneration policy in Britain', *Regional Studies* 36: 568–577

Guy, C. (1994a) 'Whatever happened to regional shopping centres?', *Geography* 76: 291–312

Guy, C. (1994b) *The Retail Development Process: Location, Property and Planning*, London: Routledge

Guy, C. (1998a) ' "High street" retailing in off-centre retail parks: a review of the effectiveness of land use planning policies', *Town Planning Review* 69: 291–313

Guy, C. (1998b) 'Off-centre retailing in the UK: prospects for the future and the implications for town centres', *Built Environment* 24: 16–30

Guy, C. (1998c) 'Alternative-use valuation, open A1 planning consent, and the development of retail parks', *Environment and Planning A* 30: 37–47

Guy, C. (2000) 'From crinkly sheds to fashion parks: the role of financial investment in the transformation of retail parks', *International Review of Retail, Distribution and Consumer Research* 10: 389–400

Guy, C. (2002) 'Is retail planning policy effective? The case of very large store development in the UK', *Planning Theory and Practice* 3: 319–330

Guy, C. (2004) 'Town centres first?', *Town and Country Planning* 73: 48–50

Guy, C. (2007) *Planning for Retail Development: A Critical View of the British Experience*, London: Routledge

Guy, C. (2008) 'Retail-led regeneration: assessing the property outcomes', *Journal of Urban Regeneration and Renewal* 1: 378–388

Guy, C. and Lord, J.D. (1993) 'Transformation and the city centre', in R.D.F. Bromley and C.J. Thomas (eds.) *Retail Change: Contemporary Issues*, London: UCL Press, 88–108

Gwilliam, M., Bourne, C., Swain, C. and Prat, A. (1999) *Sustainable Renewal of Suburban Areas*, Civic Trust and Ove Arup and Partners for the Joseph Rowntree Foundation, York: Joseph Rowntree Foundation

Hackworth, J. (2002) 'Postrecession gentrification in New York City', *Urban Affairs Review* 37: 815–843

Hackworth, J. and Smith, N. (2001) 'The changing state of gentrification', *Tijdschrift voor Economische en Sociale Geografie* 92: 464–477

Hadfield, P., Lister, S., Hobbs, D. and Winlow, S. (2001) 'The "24-hour city" – condition critical', *Town and Country Planning* 70: 300–302

Hall, J. (1992) 'Transport', in P. Ogden (ed.) *London Docklands: The Challenge of Development*, Cambridge: Cambridge University Press, 52–59

Hall, P. (1984) 'Enterprises of great pith and moment?', *Town and Country Planning* 53: 296–297

Hall, P. (1985) 'The people: where will they go?', *The Planner* 71: 3–12

Hall, P. (1997a) 'Modelling the post-industrial city', *Futures* 29: 311–322

Hall, P. (1997b) 'Regeneration policies for peripheral housing estates: inward and outward looking approaches', *Urban Studies* 34: 873–890

Hall, P. (2000) 'Creative cities and economic development', *Urban Studies* 37: 639–649

Hall, R. and Ogden, P. (1992) 'Populations old and new', in P. Ogden (ed.) *London Docklands: The Challenge of Development*, Cambridge: Cambridge University Press, 72–80

Hall, T. (1998) *Urban Geography*, London: Routledge

Hall, T. (2001) *Urban Geography*, London: Routledge, 2nd Edition

Hall, T. (2006) *Urban Geography*, London: Routledge, 3rd Edition

Hall, T. and Hubbard, P. (1996) 'The entrepreneurial city: new urban politics, new urban geographies?', *Progress in Human Geography* 20: 153–174

Hall, T. and Hubbard, P. (eds.) (1998) *The Entrepreneurial City: Geographies of Politics, Regime and Representation*, Chichester: Wiley

Hall, T. and Robertson, I. (2001) 'Public art and regeneration: advocacy, claims and critical debates', *Landscape Research* 26: 5–26

Hallsworth, A.G. (1994) 'Decentralisation in Britain: the breaking of the third wave', *Professional Geographer* 46: 296–307

Hallsworth, A.G. (1997) *The loyal to Leominster initiative: a town centre fightback against out-of-town retailing*, Paper presented at Royal Geographical Society/Institute of British Geographers Annual Conference, Exeter

Hallsworth, A. (1998) 'Loyal to Leominster: a town centre fightback against out-of-town retailing', in *Action for Market Towns: Options for a Share in the Future*, Action for Market Towns, 14–15

Halpern, D. (2004) *Social Capital*, Cambridge: Polity Press

Hambleton, R. (1995) 'The Clinton policy for cities: a transatlantic assessment', *Planning Practice and Research* 10: 359–377

Hambleton, R. (1998) 'Competition and contracting in UK local government', in N. Oatley (ed.) *Cities, Economic Competition and Urban Policy*, London: Paul Chapman, 58–76

Hambleton, R. and Thomas, H. (eds.) (1995) *Urban Policy Evaluation: Challenge and Change*, London: Paul Chapman

Hamnett, C. (1984) 'Gentrification and residential location theory: a review and assessment', in D.T. Herbert and R.J. Johnston (eds.) *Geography and the Urban Environment: Progress in Research and Applications: Volume 6*, Chichester: Wiley, 283–319

Hamnett, C. (1991) 'The blind men and the elephant: the explanation of gentrification', *Transactions of the Institute of British Geographers* 16: 173–189

Hamnett, C. (1996) 'Social polarisation, economic restructuring and welfare state regimes', *Urban Studies* 33: 1407–1430

Hamnett, C. (2000) 'Gentrification, postindustrialism, and industrial and occupational restructuring in global cities', in G. Bridge and S. Watson (eds.) *A Companion to the City*, Oxford: Blackwell, 331–341

Hamnett, C. (2003a) *Unequal City: London in the Global Arena*, London: Routledge

Hamnett, C. (2003b) 'Gentrification and the middle class remaking of Inner London: 1961–2001', *Urban Studies* 40: 2401–2426

Hamnett, C. (2005) 'Urban forms', in P. Cloke, P. Crang and M. Goodwin (eds.) *Introducing Human Geographies*, London: Arnold, 2nd edition, 425–438

Hamnett, C. (2008) 'Buy to let – or to lose?', *The Guardian* 5th May

Hamnett, C. and Whitelegg, D. (2007) 'Loft conversion and gentrification in London: from industrial to postindustrial land use', *Environment and Planning A* 39: 106–124

Hannigan, J. (1998) *Fantasy City: Pleasure and Profit in the Postmodern Metropolis*, London: Routledge

Hannigan, J. (2003) 'Symposium on branding, the entertainment economy and urban place building: introduction', *International Journal of Urban and Regional Research* 27: 352–360

Harcup, T. (2000) 'Re-imaging a post-industrial city: the Leeds St Valentine's Fair as a civic spectacle', *City* 4: 215–231

Harding, A. (1990) 'Public–private partnerships in urban regeneration', in M. Campbell (ed.) *Local Economic Development*, London: Cassell, 108–127

Harding, A. (1991) 'The rise of urban growth coalitions UK style?', *Environment and Planning C: Government and Policy* 9 295–317

Harding, A., Deas, I., Evans, R. and Wilks-Heeg, S. (2004) 'Reinventing cities in a restructuring region? The rhetoric and reality of renaissance in Liverpool and Manchester', in M. Boddy and M. Parkinson (eds.) *City Matters: Competitiveness, Cohesion and Urban Governance*, Bristol: Policy Press, 33–49

Hardoy, J.E., Mitlin, D. and Satterthwaite, D. (2001) *Environmental Problems in an Urbanizing World*, London: Earthscan

Hardy, D. (2004) 'BedZED – eco-footprint in the suburbs', *Town and Country Planning* 73: 29–31

Harrison, T. (2000) 'Urban policy: addressing wicked problems', in H.T.O. Davies, S.M. Nutley and P.C. Smith (eds.) *What Works? Evidence-Based Policy and Practice in Public Services*, Bristol: Policy Press, 207–228

Hartshorn, T.A. and Muller, P. (1989) 'Suburban downtowns and the transformation of Metropolitan Atlanta's business landscape', *Urban Geography* 10: 375–379

Harvey, D. (1982) *The Limits to Capital*, Chicago: University of Chicago Press

Harvey, D. (1985) *Consciousness and the Urban Experience*, Oxford: Blackwell

Harvey, D. (1989) 'From managerialism to entrepreneurialism: the transformation in urban governance in late capitalism', *Geografiska Annaler* 71B: 3–17

Harvey, D. (2000) *Spaces of Hope*, Berkeley: University of California Press

Hastings, A., McArthur, A. and McGregor, A. (1996) *Less Than Equal? Community Organisations and Estate Regeneration Partnerships*, Bristol: Policy Press

Haughton, G. and Hunter, C. (1994) *Sustainable Cities*, London: Regional Studies Association

Hausner, V.A. (1993) 'The future of urban development', *Royal Society of Arts Journal* 141(5441): 523–533

Haywood, I. (1998) 'City management profile – London', *Cities* 15: 381–392

Healey, M.J. and Ilbery, B.W. (1990) *Location and Change: Perspectives on Economic Geography*, Oxford: Oxford University Press

Healey, P. (1991) 'Urban regeneration and the development industry', *Regional Studies* 25: 97–110

Healey, P., Cameron, S., Davoudi, S., Graham, S. and Madanipour, A. (eds.) (1995) *Managing Cities: The New Urban Context*, Chichester: Wiley

Healey, P., Davoudi, S., Tavsanoglu, S., O'Toole, M. and Usher, D. (eds.) (1992) *Rebuilding the City: Property-Led Urban Regeneration*, London: E & FN Spon

Heath, T. (1997) 'The twenty-four hour city concept – a review of initiatives in British cities', *Journal of Urban Design* 2: 193–204

Heath, T. (2001) 'Adaptive re-use of offices for residential use: the experiences of London and Toronto', *Cities* 18: 173–184

Heath, T. and Stickland, R. (1997) 'The twenty-four hour city concept', in T. Oc and S. Tiesdell (eds.) *Safer City Centres: Reviving the Public Realm*, London: Paul Chapman, 170–183

Henderson, S., Bowlby, S. and Raco, M. (2007) 'Refashioning local government and inner-city regeneration: the Salford experience', *Urban Studies* 44: 1441–1463

Henry, I. and Paramio-Salcines, J. (1999) 'Sport and the analysis of symbolic regimes: a case study of Sheffield', *Urban Affairs Review* 34: 641–666

Herbert, D. (2000) 'Towns and cities', in V. Gardiner and H. Matthews (eds.) *The Changing Geography of the United Kingdom*, London: Routledge, 3rd edition, 190–212

Herbert, D.T. and Thomas, C.J. (1997) *Cities in Space: City as Place*, London: David Fulton, 3rd edition

Herrschel, T. and Newman, P. (2002) *Governance of Europe's City Regions: Planning, Policy and Politics*, London: Routledge

Heywood, A. (2007) *Political Ideologies: An Introduction*, Basingstoke: Palgrave Macmillan, 4th edition

Hignell, A. (1996) 'Clarks Village – a new direction for retailing?', *Geography Review* 9: 14–18

Hill, C. (1996) *Olympic Politics: Athens to Atlanta 1896–1996*, Manchester: Manchester University Press

Hillier Parker (2001) *British Shopping Centre Development – Master List Summary: April 2001*, London: Hillier Parker

Hills, J. (1995) *Joseph Rowntree Foundation Inquiry into Income and Wealth*, York: Joseph Rowntree Foundation

Hitters, E. and Richards, G. (2002) 'The creation and management of cultural clusters', *Creativity and Innovation Management* 11: 234–247

HM Government (2005) *Securing the Future: Delivering UK Sustainable Development Strategy*, London: HMSO

HM Treasury (2005a) *Housing Policy: An Overview*, London: HM Treasury

HM Treasury (2005b) *The Government's Response to Kate Barker's Review of Housing Supply*, London: HM Treasury and ODPM

HM Treasury (2006) *Stern Review on the Economics of Climate Change*, London: HMSO

Hobbs, D., Hadfield, P., Lister, S. and Winlow, S. (2003) *Bouncers: Violence and Governance in the Night-time Economy*, Oxford: Oxford University Press

Hobbs, D., Lister, S., Hadfield, P., Winlow, S. and Hall, S. (2000) 'Receiving shadows: governance and liminality in the night-time economy', *British Journal of Sociology* 51: 701–717

Hobbs, D., Winlow, S., Hadfield, P. and Lister, S. (2005) 'Violent hypocrisy: governance and the night-time economy', *European Journal of Criminology* 2: 161–183

Holcomb, B. (1993) 'Revisioning place: de- and re-constructing the image of the industrial city', in G. Kearns and C. Philo (eds.) *Selling Places: The City as Cultural Capital, Past and Present*, Oxford: Pergamon, 133–143

Holcomb, B. (1994) 'City makeovers: marketing the post-industrial city', in J. Gold and S. Ward (eds.) *Place Promotion: the Use of Publicity and Marketing to Sell Towns and Regions*, Chichester: Wiley, 115–132

Holcomb, B. (1999) 'Marketing cities for tourism', in D.R. Judd and S.S. Fainstein (eds.) *The Tourist City*, New Haven: Yale University Press, 54–70

Hollands, R. and Chatterton, P. (2002) 'Changing times for an old industrial city: hard times, hedonism and corporate power in Newcastle's nightlife', *City* 6: 291–315

Hollands, R. and Chatterton, P. (2003) 'Producing nightlife in the new urban entertainment economy: corporatization, branding and market segmentation', *International Journal of Urban and Regional Research* 27: 361–385

Hollins, C., Oc, T. and Tiesdell, S. (1997) 'Safer city centres: the role of town centre management', in T. Oc and S. Tiesdell (eds.) *Safer City Centres: Reviving the Public Realm*, London: Chapman, 89–100

Holloway, L. and Hubbard, P. (2001) *People and Place: The Extraordinary Geographies of Everyday Life*, Harlow: Pearson Education

Home Office/ODPM (Office of the Deputy Prime Minister) (2005) *Citizen Engagement and Public Services: Why Neighbourhoods Matter*, London: HMSO
Hooper, A. and Punter, J. (eds.) (2007) *Capital Cardiff 1975–2020: Regeneration, Competitiveness and the Urban Environment*, Cardiff: University of Wales Press
Hooton, S. (1996) 'Winners and losers: a Bristol perspective on City Challenge', *City* 1/2: 122–128
Hoskins, G. and Tallon, A. (2004) 'Promoting the "urban idyll": policies for city centre living', in C. Johnstone and M. Whitehead (eds.) *New Horizons in British Urban Policy: Perspectives on New Labour's Urban Renaissance*, Aldershot: Ashgate, 25–40
Hoyle, B.S., Pinder, D.A. and Husain, M.S. (eds.) (1988) *Revitalising the Waterfront: International Dimensions of Dockland Redevelopment*, London: Belhaven Press
Hubbard, P. (1996a) 'Re-imaging the city: the transformation of Birmingham's urban landscape', *Geography* 81: 26–36
Hubbard, P. (1996b) 'Urban design and city regeneration: social representations of entrepreneurial landscapes', *Urban Studies* 33: 1441–1461
Hubbard, P. (2000) 'Save our screens?', *Town and Country Planning* 69: 299–301
Hubbard, P. (2002) 'Screen-shifting: consumption, riskless risks and the changing geographies of the cinema', *Environment and Planning A* 34: 1239–1258
Hubbard, P. (2003) 'Going out/staying in: the seductions of new urban leisure spaces', *Leisure Studies* 22: 255–272
Hubbard, P. (2008) 'Regulating the social impacts of studentification: a Loughborough case study', *Environment and Planning A* 40: 323–341
Hughes, G. (1999) 'Urban revitalization: the use of festive time strategies', *Leisure Studies* 18: 119–135
Hutchinson, J. (1994) 'The practice of partnership in local economic development', *Local Government Studies* 20: 41–51
Hutchinson, J. (1997) 'Regenerating the counties: the case of SRB', *Local Economy* 12: 38–54
Imrie, R., Lees, L. and Raco, M. (eds.) (2009) *Regenerating London: Governance, Sustainability and Community in a Global City*, London: Routledge
Imrie, R. and Raco, M. (eds.) (2003a) *Urban Renaissance? New Labour, Community and Urban Policy*, Bristol: Policy Press
Imrie, R. and Raco, M. (2003b) 'Community and the changing nature of urban policy', in R. Imrie and M. Raco (eds.) *Urban Renaissance? New Labour, Community and Urban Policy*, Bristol: Policy Press, 3–31
Imrie, R. and Thomas, H. (1992) '"The wrong side of the tracks": a case study of local economic regeneration in Britain', *Policy and Politics* 20: 213–226
Imrie, R. and Thomas, H. (eds.) (1993a) *British Urban Policy and the Urban Development Corporations*, London: Paul Chapman
Imrie, R. and Thomas, H. (1993b) '"The limits of property-led regeneration", *Environment and Planning C: Government and Policy* 11: 87–102
Imrie, R. and Thomas, H. (1995) 'Urban policy processes and the politics of urban regeneration', *International Journal of Urban and Regional Research* 19: 479–494
Imrie, R. and Thomas, H. (eds.) (1999a) *British Urban Policy: An Evaluation of the Urban Development Corporations*, London: Sage, 2nd edition
Imrie, R. and Thomas, H. (1999b) 'Assessing urban policy and the Urban Development Corporations', in R. Imrie and H. Thomas (eds.) *British Urban Policy: An Evaluation of the Urban Development Corporations*, London: Sage, 2nd edition, 3–39
Imrie, R., Thomas, H. and Marshall, T. (1995) 'Business organisations, local dependence and the politics of urban renewal in Britain', *Urban Studies* 32: 31–46
IPPR (Institute for Public Policy Research) (2007) *Government in Denial Over North–South Divide*, 6 August, London: IPPR
Jackson, P. (1998) 'Domesticating the street: the contested spaces of the high street and the mall', in N.R. Fyfe (ed.) *Images of the Street: Planning, Identity and Control in Public Space*, London: Routledge, 176–191
Jacobs, B.D. (1992) *Fractured Cities: Capitalism, Community and Empowerment in Britain and America*, London: Routledge
Jacobs, J. (1961) *The Death and Life of Great American Cities: The Failure of Town Planning*, New York: Vintage Books
Jacobs, K. (2005) 'Why I don't love Richard Florida', *Metropolismag.com*, 22nd February

Jacobs, K. and Manzi, T. (1998) 'Urban renewal and the culture of conservatism: changing perceptions of the tower block and implications for contemporary renewal initiatives', *Critical Social Policy* 18: 157–174

Jacobs, M. (ed.) (1997) *Greening the Millennium: The New Politics of the Environment*, Oxford: Blackwell

Jacoby, C. (1996) 'The town centre fights back', *Green Futures* 1: 20–24

Jayne, M., Holloway, S.L. and Valentine, G. (2006) 'Drunk and disorderly: alcohol, urban life and public space', *Progress in Human Geography* 30: 451–468

Jayne, M., Valentine, G. and Holloway, S.L. (2008) 'Geographies of alcohol, drinking and drunkenness: a review of progress', *Progress in Human Geography* 32: 247–263

Jensen-Butler, C., Shachar, A. and Van Weesep, J. (eds.) (1997) *European Cities in Competition*, Aldershot: Avebury

Jessop, B. (1995) 'The regulation approach, governance and post-Fordism: alternative perspectives on economic and political change?', *Economy and Society* 24: 307–333

Jessop, B. (2004) 'Hollowing out the "nation-state" and multilevel governance', in P. Kennett (ed.) *A Handbook of Comparative Social Policy*, Cheltenham: Edward Elgar, 11–25

John, P. (2000) 'The Europeanisation of sub-national governance', *Urban Studies* 37: 877–894

Johnson, C. and Osborne, S.P. (2003) 'Local Strategic Partnerships, neighbourhood renewal, and the limits to co-governance', *Public Money and Management* 23: 147–154

Johnston, R.J. (2000a) 'Central business district (CBD)', in R.J. Johnston, D. Gregory, G. Pratt and M. Watts (eds.) *The Dictionary of Human Geography*, Oxford: Blackwell, 4th edition, 71–72

Johnston, R.J. (2000b) 'Inner city', in R.J. Johnston, D. Gregory, G. Pratt and M. Watts (eds.) *The Dictionary of Human Geography*, Oxford: Blackwell, 4th edition, 396

Johnston, R.J., Gregory, D., Pratt, G. and Watts, M. (eds.) (2000) *The Dictionary of Human Geography*, Oxford: Blackwell, 4th edition

Johnstone, C. and Whitehead, M. (eds.) (2004a) *New Horizons in British Urban Policy: Perspectives on New Labour's Urban Renaissance*, Aldershot: Ashgate

Johnstone, C. and Whitehead, M. (2004b) 'Horizons and barriers in British urban policy', in C. Johnstone and M. Whitehead (eds.) *New Horizons in British Urban Policy: Perspectives on New Labour's Urban Renaissance*, Aldershot: Ashgate, 3–21

Jones, C. (2002) 'The stadium and economic development: Cardiff and the Millennium Stadium', *European Planning Studies* 10: 819–829

Jones, C. and Munday, M. (2001) 'Blaenavon and United Nations World Heritage Site status: is conservation of industrial heritage a road to local economic development?', *Regional Studies* 35: 585–590

Jones, K. and Biasiotto, M. (1999) 'Internet retailing: current hype or future reality?', *International Review of Retail, Distribution and Consumer Research* 9: 69–79

Jones, P. (1989) 'The high street fights back', *Town and Country Planning* 58: 43–45

Jones, P. (1990) 'Town centre management schemes in the UK', *International Journal of Retail and Distribution Management* 18: 15–17

Jones, P. (1995) 'Factory outlet shopping developments', *Geography* 80: 277–280

Jones, P. (1996) 'Enter the superpub', *Town and Country Planning* 65: 110–112

Jones, P. (2003) 'Urban regeneration's poisoned chalice: is there an impasse in (community) participation-based policy?', *Urban Studies* 40: 581–601

Jones, P. and Hillier, D. (1997) 'Changes brewing: superpub developments in the UK', *Geography Review* 10(3): 26–28

Jones, P. and Hillier, D. (2000) 'Multi-leisure complexes', *Town and Country Planning* 69: 362–364

Jones, P. and Hillier, D. (2001) 'Multi-leisure parks: the changing geography of entertainment', *Geography Review* 14(5): 8–10

Jones, P., Turner, D. and Hillier, D. (1998) 'Late-night management of town and city centres', *Town and Country Planning* 67: 150–152

Jones, P., Hillier, D. and Turner, D. (1999) 'Towards the "24 hour city" ', *Town and Country Planning* 68: 164–165

Jones, P. and Vignali, C. (1993) 'Factory outlet shopping centres', *Town and Country Planning* 62: 240–241

Jones, P. and Vignali, C. (1994) 'Factory outlet shopping centres: development and planning issues', *Housing and Planning Review* 49: 9–10

Jones, P. and Wilks-Heeg, S. (2004) 'Capitalising culture: Liverpool 2008', *Local Economy* 19: 341–360

Joseph Rowntree Foundation (1999) *Developing Effective Community Involvement Strategies*, York: Joseph Rowntree Foundation

Judd, D. (1994) 'Promoting tourism in US cities', *Tourism Management* 16: 175–187

Judd, D.R. and Fainstein, S.S. (1999) *The Tourist City*, London: Yale University Press
Judge, D., Stoker, G. and Wolman, H. (eds.) (1995) *Theories of Urban Politics*, London: Sage
Karsten, L. (2003) 'Family gentrifiers: challenging the city as a place simultaneously to build a career and to raise children', *Urban Studies* 40: 2573–2584
Kearns, A. (2003) 'Social capital, regeneration and urban policy', in R. Imrie and M. Raco (eds.) *Urban Renaissance? New Labour, Community and Urban Policy*, Bristol: Policy Press, 37–60
Kearns, A. and Forrest, R. (2000) 'Social cohesion and multilevel urban governance', *Urban Studies* 37: 995–1017
Kearns, G. and Philo, C. (eds.) (1993) *Selling Places: The City as Cultural Capital, Past and Present*, Oxford: Pergamon
Keith, M. (2004) 'Knowing the city? 21st century urban policy and the introduction of Local Strategic Partnerships', in C. Johnstone and M. Whitehead (eds.) *New Horizons in British Urban Policy: Perspectives on New Labour's Urban Renaissance*, Aldershot: Ashgate, 185–195
Kintrea, K. and Morgan, J. (2005) *Evaluation of English Housing Policy 1975–2000, Theme 3: Housing Quality and Neighbourhood Quality*, London: ODPM
Kloosterman, R.C. and Musterd, S. (2001) 'The polycentric urban region: towards a research agenda', *Urban Studies* 38: 623–633
Knowles, R. and Wareing J. (1976) *Economic and Social Geography Made Simple*, London: Heinemann
Knutt, E. (2005) 'What the 50 most influential people in regeneration really think about the government, its policies and each other', *Regenerate* March: 12–16
Kochan, B. (2007a) *Achieving a Suburban Renaissance – The Policy Challenges*, York: Joseph Rowntree Foundation
Kochan, B. (2007b) *Achieving a Suburban Renaissance: The Policy Challenges*, London: TCPA
Kotkin, J. (2005) 'Uncool cities', *Prospect* October [www.prospect-magazine.co.uk]
Kreitzman, L. (1999) *The 24-Hour Society*, London: Profile
Krugman, P. (1996a) 'Making sense of the competitiveness debate', *Oxford Review of Economic Policy* 12: 17–25
Krugman, P. (1996b) *Pop Internationalism*, Cambridge, MA: MIT Press
Kunstler, J.H. (1993) *The Geography of Nowhere*, New York: Touchstone
Lambert, C. and Boddy, M. (2002) 'Transforming the city: post-recession gentrification and re-urbanisation', *ESRC Centre for Neighbourhood Research (CNR) Paper 6* [www.neighbourhoodcentre.org.uk]
Lambert, C. and Malpass, P. (1998) 'The rules of the game: competition for housing investment', in N. Oatley (ed.) *Cities, Economic Competition and Urban Policy*, London: Paul Chapman, 93–108
Lambert, C. and Oatley, N. (2002) 'Governance, institutional capacity and planning for growth', in G. Cars, P. Healey, A. Madanipour and C. de Magalhaes (eds.) *Urban Governance, Institutional Capacity and Social Milieux*, Aldershot: Ashgate, 125–141
Lambert, C. and Smith, I. (2003) 'Reshaping the city', in M. Boddy (ed.) *Urban Transformation and Urban Governance: Shaping the Competitive City of the Future*, Bristol: Policy Press, 20–31
Landry, C. (2000) *The Creative City: A Toolkit for Urban Innovators*, London: Earthscan
Landry, C. and Bianchini, F. (1995) *The Creative City*, London: Demos
Landry, C., Greene, L., Matarasso, F. and Bianchini, F. (1996) *The Art of Regeneration: Urban Renewal through Cultural Activity*, London: Comedia
Lang, R.E. (2003) *Edgeless Cities: Exploring the Elusive Metropolis*, Washington, DC: Brookings Institution Press
Lang, R. and LeFurgy, J. (2003) 'Edgeless cities: examining the noncentered metropolis', *Housing Policy Debate* 14: 427–460
Lang, R.E. and LeFurgy, J.B. (2007) *Boomburbs: The Rise of America's Accidental Cities*, Washington D.C.: Brookings Institution Press
Larkham, P.J. (1996) *Conservation and the City*, London: Routledge
Larkham, P.J. (1997) 'Remaking cities: images, control, and postwar replanning in the United Kingdom', *Environment and Planning B: Planning and Design* 24: 741–759
Law, C.M. (1988) 'Public–private partnerships in urban regeneration', *Regional Studies* 22: 446–451
Law, C.M. (1991) 'Tourism and urban revitalisation', *East Midland Geographer* 14: 49–60
Law, C.M. (1992) 'Urban tourism and its contribution to economic regeneration', *Urban Studies* 29: 599–618
Law, C.M. (1993) *Urban Tourism: Attracting Visitors to Large Cities*, London: Mansell
Law, C. (2000) 'Regenerating the city centre through leisure and tourism', *Built Environment* 26: 117–129
Lawless, P. (1981) *Britain's Inner Cities: Problems and Policies*, London: Harper and Row

Lawless, P. (1989) *Britain's Inner Cities*, London: Paul Chapman, 2nd edition
Lawless, P. (1994) 'Partnership in urban regeneration in the UK: the Sheffield central area study', *Urban Studies* 31: 1303–1324
Lawless, P. (1996) 'The inner cities: towards a new agenda', *Town Planning Review* 67: 21–43
Lawless, P. (2004) 'Locating and explaining area-based urban initiatives: New Deal for Communities in England', *Environment and Planning C: Government and Policy* 22: 383–399
Lawless, P. (2006) 'Area-based urban interventions: rationale and outcomes: the New Deal for Communities programme in England', *Urban Studies* 43: 1991–2011
Leadbeater, C. (1999) *Living on Thin Air*, Harmondsworth: Penguin
Lees, L. (1996) 'In the pursuit of difference: representations of gentrification', *Environment and Planning A* 28: 453–470
Lees, L. (2000) 'A reappraisal of gentrification: towards a geography of gentrification', *Progress in Human Geography* 24: 389–408
Lees, L. (2002) 'Urban geography: the "death" of the city?', in A. Rogers and H.A. Viles (eds.) *The Student's Companion to Geography*, Oxford: Blackwell, 2nd edition, 123–128
Lees, L. (2003a) 'Policy (re)turns: gentrification research and urban policy – urban policy and gentrification research', *Environment and Planning A* 35: 571–574
Lees, L. (2003b) 'Visions of "urban renaissance": the Urban Task Force report and the Urban White Paper', in R. Imrie and M. Raco (eds.) *Urban Renaissance? New Labour, Community and Urban Policy*, Bristol: Policy Press, 61–82
Lees, L. (2003c) 'Super-gentrification: the case of Brooklyn Heights, New York City', *Urban Studies* 40: 2487–2509
Lees, L. (ed.) (2004) *The Emancipatory City? Paradoxes and Possibilities*, London: Sage
Lees, L. (2008) 'Gentrification and social mixing: towards an inclusive urban renaissance?', *Urban Studies* 45: 2449–2470
Lees, L. and Bondi, L. (1995) 'De-gentrification and economic recession: the case of New York City', *Urban Geography* 16: 234–253
Lees, L., Slater, T. and Wyly, E. (2008) *Gentrification*, London: Routledge
Le Gales, P. (2002) *European Cities: Social Conflicts and Governance*, Oxford: Oxford University Press
Lepine, E., Smith, I., Sullivan, H. and Taylor, M. (2007) 'Introduction: of neighbourhoods and governance', in I. Smith, E. Lepine and M. Taylor (eds.) *Disadvantaged by Where You Live? Neighbourhood Governance in Contemporary Urban Policy*, Bristol: Policy Press, 1–20
Leunig, T. and Swaffield, J. (2007) *Cities Limited*, London: Policy Exchange
Leunig, T. and Swaffield, J. (2008) *Cities Unlimited: Making Urban Regeneration Work*, London: Policy Exchange
Lever, W.F. (1993) 'Reurbanisation – the policy implications', *Urban Studies* 30: 267–284
Levine, M. (1987) 'Downtown redevelopment as an urban growth strategy: a critical appraisal of the Baltimore renaissance', *Journal of Urban Affairs* 9: 103–123
Levitas, R. (2005) *The Inclusive Society? Social Exclusion and New Labour*, Basingstoke: Palgrave Macmillan, 2nd edition
Lewis, N. (1992) *Inner City Regeneration: The Demise of Regional and Local Government*, Buckingham: Open University Press
Ley, D. (1983) *A Social Geography of the City*, New York: Harper and Row
Ley, D. (1986) 'Alternative explanations for inner-city gentrification: a Canadian assessment', *Annals of the Association of American Geographers* 76: 521–535
Ley, D. (1996) *The New Middle Class and the Remaking of the Central City*, Oxford: Oxford University Press
LGA (Local Government Association) (2005) *The Role of Local Delivery Vehicles in Creating Sustainable Communities*, London: LGA
Lichfield, D. (1992) *Urban Regeneration for the 1990s*, London: London Planning Advisory Committee
Lipietz, A. (1987) *Miracles and Mirages: The Crises of Global Fordism*, London: Verso
Lloyd, G. (1984) 'Policies in search of an opportunity', *Town and Country Planning* 53: 299–301
Local Economy (1994) 'Urban Policy into the 21st Century', Volume 9, Issue 3
Local Economy (2008) 'Sustainable Communities', Volume 23, Issue 3
Local Environment (2006) 'Special Issue: Sustainable Brownfields Redevelopment', Volume 11, Issue 5
Loftman, P. and Nevin, B. (1994) 'Prestige project developments: economic renaissance or economic myth? A case study of Birmingham', *Local Economy* 8: 307–325

Loftman, P. and Nevin, B. (1995) 'Prestige projects and urban regeneration in the 1980s and 1990s: a review of benefits and limitations', *Planning Practice and Research* 10: 299–315

Loftman, P. and Nevin, B. (1996) 'Going for growth: prestige projects in three British cities', *Urban Studies* 33: 991–1019

Loftman, P. and Nevin, B. (1998) 'Pro-growth local economic development strategies: civic promotion and local needs in Britain's second city 1981–1996', in T. Hall and P. Hubbard (eds.) *The Entrepreneurial City: Geographies of Politics, Regime and Representation*, Chichester: Wiley, 129–148

Long, J. and Bramham, P. (2006) 'Joining up policy discourses and fragmented practices: the precarious contribution of cultural projects to social inclusion?', *Policy and Politics* 34: 133–151

LOTS (Living Over The Shop) (1998) *Living Over The Shop: Synopsis and Update*, York: LOTS

Loukaitou-Sideris, A. (1993) 'Privatisation of public space: the Los Angeles experience', *Town Planning Review* 62: 139–167

Lovatt, A. (1996) 'The ecstasy of urban regeneration: regulation of the night-time economy in the transition to a post-Fordist city', in J. O'Connor and D. Wynne (eds.) *From the Margins to the Centre: Cultural Production and Consumption in the Post-Industrial City*, Aldershot: Arena, 141–168

Lovatt, A. (1997) 'Turning up the lights in the cities of the night', *Planning* 1224: 20–21

Lovatt, A. and O'Connor, J. (1995) 'Cities and the night-time economy', *Planning Practice and Research* 10: 127–134

Lovatt, A. with O'Connor, J., Montgomery, J. and Owens, P. (eds.) (1994) *The 24-Hour City: Selected Papers from the First National Conference of the Night-Time Economy*, Manchester: Manchester Institute for Popular Culture, Manchester Metropolitan University

Low, N., Gleeson, B., Elander, I. and Lidskog, R. (eds.) (2000) *Consuming Cities: The Urban Environment in the Global Economy after Rio*, London: Routledge

Low, S. (2003) *Behind the Gates: Life, Security and the Pursuit of Happiness in Fortress America*, London: Routledge

Lowe, M. (1993) 'Local hero! An examination of the role of the regional entrepreneur in the regeneration of Britain's regions', in G. Kearns and C. Philo (eds.) *Selling Places: The City as Cultural Capital, Past and Present*, Oxford: Pergamon Press, 211–230

Lowe, M. (1998) 'The Merry Hill regional shopping centre controversy: PPG6 and new urban geographies', *Built Environment* 24: 57–69

Lowe, M. (2000a) 'From Victor Gruen to Merry Hill: reflections on regional shopping centres and urban development in the US and UK', in P. Jackson, M.S. Lowe, D. Miller and F. Mort (eds.) *Commercial Cultures: Economies, Practices, Spaces*, Oxford: Berg, 245–259

Lowe, M. (2000b) 'Britain's regional shopping centres: new urban forms?', *Urban Studies* 37: 261–274

Lowe, M. (2005) 'The regional shopping centre in the inner city: a study of retail-led urban regeneration', *Urban Studies* 42: 449–470

Lowndes, V., McCabe, A., Nanton, P. and Skelcher, C. (1996) *Community Networks in Urban Regeneration*, Bristol: Policy Press

Lupton, R. (2003) *Neighbourhood Effects: Can We Measure Them and Does it Matter?*, CASE Paper Number 73, London: CASE

McArthur, A. (1995) 'The active involvement of local residents in strategic community partnerships', *Policy and Politics* 23: 61–71

McCarthy, J. (1998a) 'Reconstruction, regeneration and re-imaging: the case of Rotterdam', *Cities* 15: 337–344

McCarthy, J. (1998b) 'Dublin's Temple Bar: a case study of culture-led regeneration', *European Planning Studies* 9: 437–458

McCarthy, J. (2007a) *Partnership, Collaborative Planning and Urban Regeneration*, Aldershot: Ashgate

McCarthy, J. (2007b) 'Quick fix or sustainable solution? Cultural clustering for urban regeneration in the UK', in C. Aitchison, G. Richards and A. Tallon (eds.) *Urban Transformations: Regeneration and Renewal through Leisure and Tourism*, Eastbourne: Leisure Studies Association, 1–17

McDonald, I. and Howick, C. (1981) 'Monitoring the Enterprise Zones', *Built Environment* 7: 31–37

MacDonald, R. (2000) 'Urban tourism: an inventory of ideas and issues', *Built Environment* 26: 90–98

MacFarlane, R. (1993) *Community Involvement in City Challenge: A Good Practice Guide*, Bedford: National Council for Voluntary Organisations

MacGregor, S. and Pimlott, B. (eds.) (1990) *Tackling the Inner Cities: The 1980s Revisited, Prospects for the 1990s*, Oxford: Clarendon Press

Mackinnon, D. and Cumbers, A. (2007) *An Introduction to Economic Geography: Globalization, Uneven Development and Place*, Harlow: Pearson Education

Maclennan, D. (1997) 'The UK housing market: up, down and where next?', in P. Williams (ed.) *Directions in Housing Policy*, London: Paul Chapman, 22–53
MacLeod, D. and Ward, K. (2002) 'Spaces of utopia and dystopia: landscaping the contemporary city', *Geografiska Annaler* 84: 153–170
McNeill, D. (2002) 'The mayor and the world city skyline: London's tall building debate', *International Planning Studies* 7: 325–334
McNeill, D. (2008) 'The hotel and the city', *Progress in Human Geography* 32: 383–398
McQuaid, R.W., Lindsay, C. and Greig, M. (2005) 'Job guarantees, employability training and partnerships in the retail sector', *Local Economy* 20: 22–36
Madanipour, A., Cars, G. and Allen, J. (eds.) (1998) *Social Exclusion in European Cities: Processes, Experiences and Responses*, London: Jessica Kingsley
Madsen, H. (1992) 'Place marketing in Liverpool: a review', *International Journal of Urban and Regional Research* 16: 633–640
Malanga, S. (2004) 'The curse of the creative class', *City Journal* Winter: 36–45
Malbon, B. (1998) 'The club: clubbing: consumption, identity and the spatial practices of every-night life', in T. Skelton and G. Valentine (eds.) *Cool Places: Geographies of Youth Cultures*, London: Routledge, 266–286
Malone, P. (1996) *City, Capital and Water*, London: Routledge
Malpass, P. (1994) 'Policy making and local governance: how Bristol failed to secure City Challenge funding (twice)', *Policy and Politics* 22: 301–312
Manzi, T. and Smith-Bowers, B. (2005) 'Gated communities as club goods: segregation or social cohesion?', *Housing Studies* 20: 345–359
Martin, R. (1988) 'Industrial capitalism in transition: the contemporary reorganisation of the British space economy', in D. Massey and J. Allen (eds.) *Uneven Redevelopment: Cities and Regions in Transition*, London: Hodder & Stoughton, 202–231
Massey, D. (1982) 'Enterprise Zones: a political issue', *International Journal of Urban and Regional Research* 6: 429–434
Mayo, M. (1997) 'Partnerships for regeneration and community development: some opportunities, challenges and constraints', *Critical Social Policy* 17: 3–26
Meegan, R. (1999) 'Urban Development Corporations, urban entrepreneurialism and locality: the Merseyside Development Corporation', in R. Imrie and H. Thomas (eds.) *British Urban Policy: An Evaluation of the Urban Development Corporations*, London: Sage, 2nd edition, 64–105
Merrifield, A. (1993) 'The Canary Wharf debate', *Environment and Planning A* 25: 1247–1265
Miles, M. (2005) 'Interruptions: testing the rhetoric of culturally led urban regeneration', *Urban Studies* 42: 889–911
Miles, S. (2005) ' "Our Tyne": iconic regeneration and the revitalisation of identity in NewcastleGateshead', *Urban Studies* 42: 913–926
Miles, S. and Paddison, R. (2005) 'Introduction: the rise and rise of culture-led urban regeneration', *Urban Studies* 42: 833–839
Mills, C.A. (1988) ' "Life on the upslope": the postmodern landscape of gentrification', *Environment and Planning D: Society and Space* 6: 169–189
Mills, C.A. (1990) 'Creating a Disney world', *Geographical Magazine* December: 40–43
Mingay, G.E. (ed.) (1989) *The Rural Idyll*, London: Routledge
Mohan, J. (1999) *A United Kingdom? Economic, Social and Political Geographies*, London: Arnold
Mole, P. (1996) 'Fordism, post-Fordism and the contemporary city', in J. O'Connor and D. Wynne (eds.) *From the Margins to the Centre: Cultural Production and Consumption in the Post-Industrial City*, Aldershot: Arena, 15–48
Mollenkopf, J.H. and Castells, M. (eds.) (1991) *Dual City: Restructuring New York*, New York: Russell Sage Foundation
Mommaas, H. (2003) 'The creative industries: a brief introduction to an approach', in H. le Blanc (ed.) *The Creative DNA of the Eindhoven Region*, Eindhoven: Alice Foundation, 12–23
Mommaas, H. (2004) 'Cultural clusters and the post-industrial city: towards the remapping of urban cultural policy', *Urban Studies* 41: 507–532
Montgomery, J. (1990) 'Cities and the art of cultural planning', *Planning Practice and Research* 5: 17–24
Montgomery, J. (1994) 'The evening economy of cities', *Town and Country Planning* 63: 302–307
Montgomery, J. (1995) 'The story of Temple Bar: creating Dublin's cultural quarter', *Planning Practice and Research* 10: 135–172

Montgomery, J. (1996) 'Developing the media industries: an overview of strategies and possibilities for the local economic development of the media and cultural industries', *Local Economy* 11: 158–168

Montgomery, J. (2003) 'Cultural quarters as mechanisms for urban regeneration – part 1: conceptualising cultural quarters', *Planning Practice and Research* 18: 293–306

Montgomery, J. (2004) 'Cultural quarters as mechanisms for urban regeneration – part 2: a review of four cultural quarters in the UK, Ireland and Australia', *Planning Practice and Research* 19: 3–31

Mooney, G. (2004) 'Cultural policy as urban transformation? Critical reflections on Glasgow, European City of Culture 1990', *Local Economy* 19: 327–340

Moore, B. and Begg, I. (2004) 'Urban growth and competitiveness in Britain: a long-run perspective', in M. Boddy and M. Parkinson (eds.) *City Matters: Competitiveness, Cohesion and Urban Governance*, Bristol: Policy Press, 93–109

Moore, C. and Richardson, J. (1986) 'Shifting partners: public–private partnerships in local economic regeneration', *Public Policy and Administration* 1: 33–49

Morphet, J. (1991) 'Town centre management – old problems, new solutions', *Town and Country Planning* 60: 202–203

Moulaert, F. and Ailenei, O. (2005) 'Social economy, third sector and solidarity relations: a conceptual synthesis from history to present', *Urban Studies* 42: 2037–2053

Moulaert, F. and Nussbaumer, J. (2005) 'Defining the social economy and its governance at the neighbourhood level: a methodological reflection', *Urban Studies* 42: 2071–2088

Mullins, D. and Murie, A. (2006) *Housing Policy in the UK*, Basingstoke: Palgrave Macmillan

Murray, C. (ed.) (2007) *Working Together: Transformational Leadership for City Growth: Essays from the Leaders of England's Core Cities*, London: The Smith Institute

Myers, D. (2006) *Economics and Property: The Estates Gazette Guide*, London: Estates Gazette, 2nd Edition

Naess, P. (2001) 'Urban planning and sustainable development', *European Planning Studies* 9: 503–524

Nathan, M. and Urwin, C. (2006) *City People: City Centre Living in the UK*, London: Institute for Public Policy Research

National Assembly for Wales (2001) *Communities First Guidance*, Cardiff: National Assembly for Wales

Nelson, A.L. (1997) 'Security shutters: a double-edged sword', *International Journal of Risk, Society and Crime Prevention* 3: 11–19

Nelson, A.L., Bromley, R.D.F. and Thomas, C.J. (2001) 'Identifying micro-spatial and temporal patterns of violent crime and disorder in the British city centre', *Applied Geography* 21: 249–274

Nevin, B., Loftman, P. and Beazley, M. (1997) 'Cities in crisis: is growth the answer?', *Town Planning Review* 68: 145–164

Newby, P. (1993) 'Shopping as leisure', in R.D.F. Bromley and C.J. Thomas (eds.) *Retail Change: Contemporary Issues*, London: UCL Press, 208–228

Newman, O. (1972) *Defensible Space: People and Design in the Violent City*, London: Architectural Press

Nicolaides, B.M. and Wiese, A. (eds.) (2006) *The Suburb Reader*, New York: Routledge

Norcliffe, G.B. and Hoare, A.G. (1982) 'Enterprise Zone policy for the inner city: a review and preliminary assessment', *Area* 14: 265–274

Norris, C. and Armstrong, G. (1999) *The Maximum Surveillance Society: The Rise of CCTV*, Oxford: Berg

Norris, S. (1999) 'Factory outlet centres into the new millennium: "fact and fiction" ', *European Retail Digest* 21: 41–45

Norwood, G. (2005) 'Cities' limits reached', *Estates Gazette* 26th November: 84–85

Oatley, N. (1993) 'Realizing the potential of urban policy: the case of the Bristol Development Corporation', in R. Imrie and H. Thomas (eds.) *British Urban Policy and the Urban Development Corporations*, London: Paul Chapman, 136–153

Oatley, N. (1994) 'Winners and losers in the regeneration game', *Planning* 1068: 24–26

Oatley, N. (1995a) 'Urban regeneration', *Planning Practice and Research* 10: 261–269

Oatley, N. (1995b) 'Competitive urban policy and the regeneration game', *Town Planning Review* 66: 1–14

Oatley, N. (1996) 'Sheffield's cultural industries quarter', *Local Economy* 11: 172–179

Oatley, N. (ed.) (1998a) *Cities, Economic Competition and Urban Policy*, London: Paul Chapman

Oatley, N. (1998b) 'Cities, economic competition and urban policy', in N. Oatley (ed.) *Cities, Economic Competition and Urban Policy*, London: Paul Chapman, 3–20

Oatley, N. (1998c) 'Transitions in urban policy: explaining the emergence of the "challenge fund" model', in N. Oatley (ed.) *Cities, Economic Competition and Urban Policy*, London: Paul Chapman, 21–37

Oatley, N. (1998d) 'Restructuring urban policy: the Single Regeneration Budget and the Challenge Fund', in N. Oatley (ed.) *Cities, Economic Competition and Urban Policy*, London: Paul Chapman, 146–162

Oatley, N. (1999) 'Developing the social economy', *Local Economy* 13: 339–345

Oatley, N. (2000) 'New Labour's approach to age-old problems: renewing and revitalising poor neighbourhoods – the national strategy for neighbourhood renewal', *Local Economy* 15: 86–97

Oatley, N. and Lambert, C. (1995) 'Evaluating competitive urban policy: the City Challenge initiative', in R. Hambleton and H. Thomas (eds.) *Urban Policy Evaluation: Challenge and Change*, London: Paul Chapman, 141–157

Oatley, N. and Lambert, C. (1998) 'Catalyst for change: the City Challenge initiative', in N. Oatley (ed.) *Cities, Economic Competition and Urban Policy*, London: Paul Chapman, 109–126

Oatley, N. and May, A. (1999) ' "Out of touch, out of place, out of time": a valediction for Bristol Development Corporation', in R. Imrie and H. Thomas (eds.) *British Urban Policy: An Evaluation of the Urban Development Corporations*, London: Sage, 2nd Edition, 186–205

O'Brien, L.G. and Harris, F.W. (1991) *Retailing: Shopping, Society, Space*, London: David Fulton

Oc, T. (1991) 'Planning natural surveillance back into city centres', *Town and Country Planning* 60: 237–239

Oc, T. and Tiesdell, S. (1992) 'The London Docklands Development Corporation (LDDC), 1981–1991', *Town Planning Review* 62: 311–330

Oc, T. and Tiesdell, S. (eds.) (1997a) *Safer City Centres: Reviving the Public Realm*, London: Paul Chapman

Oc, T. and Tiesdell, S. (1997b) 'The death and life of city centres', in T. Oc and S. Tiesdell (eds.) *Safer City Centres: Reviving the Public Realm*, London: Paul Chapman, 1–20

Oc, T. and Tiesdell, S. (1997c) 'Housing and safer city centres', in T. Oc and S. Tiesdell (eds.) *Safer City Centres: Reviving the Public Realm*, London: Paul Chapman, 156–169

Oc, T. and Tiesdell, S. (1997d) 'Safer city centres in the USA', in T. Oc and S. Tiesdell (eds.) *Safer City Centres: Reviving the Public Realm*, London: Paul Chapman, 184–197

Oc, T. and Tiesdell, S. (1998) 'City centre management and safer city centres: approaches in Coventry and Nottingham', *Cities* 15: 85–103

Oc, T. and Trench, S. (1993) 'Planning and shopper security', in R.D.F. Bromley and C.J. Thomas (eds.) *Retail Change: Contemporary Issues*, London: UCL Press, 153–169

O'Doherty, R., Durrschmidt, J. and Purdue, D. (1999) 'Local exchange and trading schemes: a useful strand of community economic development policy?', *Environment and Planning A* 31: 1639–1653

ODPM (Office of the Deputy Prime Minister) (2002) *Towns and Cities: Partners in Urban Renaissance*, London: HMSO

ODPM (Office of the Deputy Prime Minister) (2003a) *Sustainable Communities: Building for the Future*, London: HMSO

ODPM (Office of the Deputy Prime Minister) (2003b) *Sustainable Communities: An Urban Development Corporation for the London Thames Gateway – A Consultation Paper*, London: HMSO

ODPM (Office of the Deputy Prime Minister) (2003c) *An Urban Development Corporation for Thurrock – Consultation Paper*, London: HMSO

ODPM (Office of the Deputy Prime Minister) (2003d) *Transferable Lessons from Enterprise Zones*, London: HMSO

ODPM (Office of the Deputy Prime Minister) (2004a) *A Tale of Eight Cities*, London: HMSO

ODPM (Office of the Deputy Prime Minister) (2004b) *Competitive European Cities: Where Do the Core Cities Stand?*, London: HMSO

ODPM (Office of the Deputy Prime Minister) (2004c) *Sustainable Communities: An Urban Development Corporation for West Northamptonshire*, London: HMSO

ODPM (Office of the Deputy Prime Minister) (2004d) *Sustainable Communities: An Urban Development Corporation for the London Thames Gateway – Decision Document*, London: HMSO

ODPM (Office of the Deputy Prime Minister) (2004e) *Urban Regeneration Companies: Guidance and Qualification Criteria*, London: HMSO

ODPM (Office of the Deputy Prime Minister) (2004f) *Urban Regeneration Companies: Policy Stocktake – Final Report*, London: HMSO

ODPM (Office of the Deputy Prime Minister) (2004g) *Urban Regeneration Companies: A Process Evaluation*, London: HMSO

ODPM (Office of the Deputy Prime Minister) (2004h) *The Planning and Compulsory Purchase Act 2004*, London: HMSO

ODPM (Office of the Deputy Prime Minister) (2004i) *Our Cities Are Back: Competitive Cities Make Prosperous Regions and Sustainable Communities*, London: HMSO

ODPM (Office of the Deputy Prime Minister) (2004j) *The English Indices of Deprivation 2004 (Revised)*, London: ODPM

ODPM (Office of the Deputy Prime Minister) (2004k) *Competitive European Cities: Where Do the Core Cities Stand?: Urban Research Summary 13*, London: HMSO

ODPM (Office of the Deputy Prime Minister) (2005a) *Making Connections: An Evaluation of the Community Participation Programmes: Urban Research Summary 15*, London: ODPM

ODPM (Office of the Deputy Prime Minister) (2005b) *Planning Policy Statement 1: Delivering Sustainable Development*, London: HMSO

ODPM (Office of the Deputy Prime Minister (2005c) *Planning Policy Statement 6: Planning for Town Centres*, London: HMSO

ODPM (Office of the Deputy Prime Minister) (2005d) *State of the Cities: A Progress Report to the Delivering Sustainable Communities Summit*, London: HMSO

ODPM (Office of the Deputy Prime Minister) (2005e) *Creating Sustainable Communities: Delivering the Thames Gateway*, London: HMSO

ODPM (Office of the Deputy Prime Minister) (2005f) *Housing Land Availability Assessments: Identifying Appropriate Land for Housing Development – Draft Practice Guidance*, London: ODPM

ODPM (Office of the Deputy Prime Minister) (2005g) *The New Deal for Communities Programme 2001–2005: An Interim Evaluation, Research Report Number 17*, London: ODPM

ODPM (Office of the Deputy Prime Minister) (2005h) *Improving Delivery of Mainstream Services in Deprived Areas: The Role of Community Involvement*, London: ODPM

ODPM (Office of the Deputy Prime Minister) (2006a) *The State of the English Cities*, London: ODPM

ODPM (Office of the Deputy Prime Minister) (2006b) *The State of the English Cities: Urban Research Summary 21*, London: ODPM

Ogden, P. (ed.) (1992) *London Docklands: The Challenge of Development*, Cambridge: Cambridge University Press

Oram, J., Conisbee, M. and Simms, A. (2003) *Ghost Town Britain II: Death on the High Street*, London: New Economics Foundation

Pacione, M. (ed.) (1997) *Britain's Cities: Geographies of Division in Urban Britain*, London: Routledge

Pacione, M. (2001) *Urban Geography: A Global Perspective*, London: Routledge

Pacione, M. (2005) *Urban Geography: A Global Perspective*, London: Routledge, 2nd edition

Paddison, R. (1993) 'City marketing, image reconstruction and urban regeneration', *Urban Studies* 30: 339–350

Page, S. (1995) *Urban Tourism*, London: Routledge

Page, S.J. and Hardyman, R. (1996) 'Place marketing and town centre management: a new tool for urban revitalization', *Cities* 13: 153–164

Pain, R. and Townshend, T. (2002) 'A safer city for all? Senses of "community safety" in Newcastle upon Tyne', *Geoforum* 33: 105–119

Painter, J. (1995) 'Regulation theory, post-Fordism and urban politics', in D. Judge, G. Stoker and H. Wolman (eds.) *Theories of Urban Politics*, London: Sage, 276–295

Painter, J. (1998) 'Entrepreneurs are made, not born: learning and urban regimes in the production of entrepreneurial cities', in T. Hall and P. Hubbard (eds.) *The Entrepreneurial City: Geographies of Politics, Regime and Representation*, Chichester: Wiley, 259–274

Parfect, M. and Power, G. (1997) *Planning for Urban Quality: Urban Design in Towns and Cities*, London: Routledge

Parker, C. and Garnell, C. (2006) 'Regeneration and retail in Liverpool: a new approach', *Journal of Retail and Leisure Property* 5: 292–304

Parkinson, M. (1988) 'Urban regeneration and Development Corporations: Liverpool style', *Local Economy* 3: 109–118

Parkinson, M. (1989) 'The Thatcher government's urban policy, 1979–1989', *Town Planning Review* 60: 421–440

Parkinson, M. (1992) 'City links', *Town and Country Planning* 61: 235–236

Parkinson, M. (1993) 'City Challenge: a new strategy for Britain's cities?', *Policy Studies* 14: 5–13

Parkinson, M. and Bianchini, F. (1993) 'Liverpool: a tale of missed opportunities?', in F. Bianchini and M. Parkinson (eds.) *Cultural Policy and Urban Regeneration: The West European Experience*, Manchester: Manchester University Press, 155–177

Parkinson, M. and Boddy, M. (2004) 'Introduction', in M. Boddy and M. Parkinson (eds.) *City Matters: Competitiveness, Cohesion and Urban Governance*, Bristol: Policy Press, 1–10

Parkinson, M. and Evans, R. (1990) 'Urban Development Corporations', in M. Campbell (ed.) *Local Economic Policy*, London: Cassell, 128–155

Paumier, C.B. (1988) *Designing the Successful Downtown*, Washington, DC: Urban Land Institute

Peabody Trust (2001) *BedZED: A New Low Impact, Zero-Energy Mixed Housing Development in London*, London: Peabody Trust

Peck, J. (2005) 'Struggling with the creative class', *International Journal of Urban and Regional Research* 29: 740–770

Peck, J. and Tickell, A. (1995) 'Business goes local: dissecting the "business agenda" in Manchester', *International Journal of Urban and Regional Research* 19: 55–78

Peck, J. and Ward, K. (eds.) (2002) *City of Revolution: Restructuring Manchester*, Manchester: Manchester University Press

Peel, D. and Lloyd, G. (2005) 'A case for Business Improvement Districts in Scotland: policy transfer in practice?', *Planning Practice and Research* 20: 89–95

Peel, D. and Lloyd, G. (2008) 'Re-generating learning in the public realm: evidence-based policy making and Business Improvement Districts in the UK', *Public Policy and Administration* 23: 189–205

Petherick, A. and Fraser, R. (1992) *Living Over The Shop: A Handbook for Practitioners*, York: University of York

Phillips, M. (2002) 'The production, symbolization and socialization of gentrification: impressions from two Berkshire villages', *Transactions of the Institute of British Geographers* 27: 282–308

Philo, C. and Kearns, G. (1993) 'Culture, history, capital: a critical introduction to the selling of places', in G. Kearns and C. Philo (eds.) *Selling Places: The City as Cultural Capital, Past and Present*, Oxford: Pergamon, 1–32

Pierson, J. and Smith, J. (eds.) (2001) *Rebuilding Community: Policy and Practice in Urban Regeneration*, London: Palgrave

Pilkington, L. (2006) 'From out of the ashes', *Estates Gazette* 15th March: 78–81

Pinch, S. (2002) 'City profile – Southampton', *Cities* 19: 71–78

PIU (Performance and Innovation Unit) (2002) *Social Capital: A Discussion Paper*, London: PIU

Planning Practice and Research (1995) 'Urban Regeneration', Volume 10, Issues 4/5

Plaza, B. (2000) 'Evaluating the influence of a large cultural artefact in the attraction of tourism: the Guggenheim Museum Bilbao case', *Urban Affairs Review* 36: 264–274

Plaza, B. (2006) 'The return on investment of the Guggenheim Museum Bilbao', *International Journal of Urban and Regional Research* 30: 452–467

Porter, L. and Barber, A. (2006) 'Closing time: the meaning of place and state-led gentrification in Birmingham's Eastside', *City* 10: 215–234

Porter, M.E. (1998) *The Competitive Advantage of Nations*, London: Macmillan

Potter, S. (1990) 'Britain's Development Corporations', *Town and Country Planning* 59: 294–298

Potts, G. (2007a) 'From "chains" to partnerships? Supermarkets and regeneration', *Journal of Urban Regeneration and Renewal* 1: 22–36

Potts, G. (2007b) 'Suburban regeneration', *Property Week* 11th May: 56–57

Potts, G. with Besussi, E., Gaus, K., Hassler, J., Lesteven, G., Markovich, J., Munoz-Rojas Oscarsson, O., Parham, H. and Porthe, L. (2008) *Suburban Regeneration: The Real Challenges*, London: BURA

Power, A. (2004) *Sustainable Communities and Sustainable Development: A Review of the Sustainable Communities Plan*, London: Sustainable Development Commission

Prestwich, R. and Taylor, P. (1990) *Introduction to Regional and Urban Policy in the United Kingdom*, Harlow: Longman

Punter, J. (1992) 'Design control and the regeneration of docklands: the example of Bristol', *Journal of Property Research* 9: 49–78

Punter, J. (2007) 'Design-led regeneration? Evaluating the design outcomes of Cardiff Bay and their implications for future regeneration and design', *Journal of Urban Design* 12: 375–405

Purdue, D., Razzaque, K., Hambleton, R., Stewart, M., Huxham, C. and Vangen, S. (2000) *Community Leadership in Area Regeneration*, Bristol: Policy Press

Putnam, R.D. (1993) *Making Democracy Work: Civic Traditions in Modern Italy*, Princeton, NJ: Princeton University Press

Putnam, R.D. (1995) 'Bowling alone: America's declining social capital', *The Journal of Democracy* 6: 65–78

Putnam, R.D. (2000) *Bowling Alone: The Collapse and Revival of American Community*, New York: Simon & Schuster

Putnam, R.D. (ed.) (2002) *Democracies in Flux: The Evolution of Social Capital in Contemporary Society*, Oxford: Oxford University Press

Raco, M. (1997) 'Business association and the politics of urban renewal: the case of the Lower Don Valley, Sheffield', *Urban Studies* 34: 383–402

Raco, M. (2004) 'Urban regeneration in a growing region: the renaissance of England's average town', in C. Johnstone and M. Whitehead (eds.) *New Horizons in British Urban Policy: Perspectives on New Labour's Urban Renaissance*, Aldershot: Ashgate, 41–58

Raco, M. (2005a) 'A step change or a step back? The Thames Gateway and the re-birth of the Urban Development Corporations', *Local Economy* 20: 141–153

Raco, M. (2005b) 'Sustainable development, rolled-out neo-liberalism and sustainable communities', *Antipode* 37: 324–346

Raco, M. (2007a) *Building Sustainable Communities: Spatial Development, Citizenship, and Labour Market Engineering in Post-War Britain*, Bristol: Policy Press

Raco, M. (2007b) 'Securing sustainable communities: citizenship, safety and sustainability in the new urban planning', *European Urban and Regional Studies* 14: 305–320

Raco, M. and Henderson, S. (2006) 'Sustainable urban planning and the brownfield development process in the United Kingdom: lessons from the Thames Gateway', *Local Environment* 11: 499–513

Raemaekers, J. (2000) 'Planning for sustainable development', in P. Allmendinger, A. Prior and J. Raemaekers (eds.) *Introduction to Planning Practice*, Chichester: Wiley, 23–47

Rallings, C., Thrasher, M., Cheal, B. and Borisyuk, G. (2004) 'The New Deal for Communities: assessing procedures and voter turnout at partnership board elections', *Environment and Planning C: Government and Policy* 22: 569–582

Ravenscroft, N. (2000) 'The vitality and viability of town centres', *Urban Studies* 37: 2533–2549

Ravenscroft, N., Reeves, J. and Rowley, M. (2000) 'Leisure, property, and the viability of town centres', *Environment and Planning A* 32: 1359–1374

Ravetz, A. (1996) 'When parts of the town rely on gown', *Town and Country Planning* 65: 72–73

Rees, W. (1997) 'Is "sustainable city" an oxymoron?', *Local Environment* 2: 303–310

Reeve, A. (2004) 'Town centre management: developing a research agenda in an emerging field', *Urban Design International* 9: 133–150

Rhodes, J., Tyler, P. and Brennan, A. (2003) 'New developments in area-based initiatives in England: the experience of the Single Regeneration Budget', *Urban Studies* 40: 1399–1426

Roberts, M. (2004) *Good Practice in the Management of the Evening and the Night-time Economies: A Literature Review from an Environmental Perspective*, London: ODPM

Roberts, M. (2006) 'From "creative city" to "no-go areas" – the expansion of the night-time economy in British town and city centres', *Cities* 23: 331–338

Roberts, M. (2007) 'Sharing space: urban design and social mixing in mixed income new communities', *Planning Theory and Practice* 8: 183–204

Roberts, M. and Eldridge, A. (2007a) 'Quieter, safer, cheaper: planning for a more inclusive evening and night-time economy', *Planning Practice and Research* 22: 253–266

Roberts, M. and Eldridge, A. (2007b) *Expecting 'Great Things'? The Impact of the Licensing Act 2003 on Democratic Involvement, Dispersal and Drinking Cultures*, London: Central Cities Institute, University of Westminster

Roberts, M. and Turner, C. (2005) 'Conflicts of liveability in the 24-hour city: learning from 48 hours in the life of London's Soho', *Journal of Urban Design* 10: 171–193

Roberts, M., Turner, C., Greenfield, S. and Osborn, G. (2006) 'A continental ambiance? Lessons in managing alcohol-related evening and night-time entertainment from four European capitals', *Urban Studies* 43: 1105–1125

Roberts, P. (2000) 'The evolution, definition and purpose of urban regeneration', in P. Roberts and H. Sykes (eds.) *Urban Regeneration: A Handbook*, London: Sage, 9–36

Roberts, P. and Sykes, H. (eds.) (2000) *Urban Regeneration: A Handbook*, London: Sage

Robertson, J. and Fennell, J. (2007) 'The economic effects of regional shopping centres', *Journal of Retail and Leisure Property* 6: 149–170

Robertson, K.A. (1995) 'Downtown redevelopment strategies in the United States: an end-of-the-century assessment', *Journal of the American Planning Association* 61: 429–437

Robinson, F. (1997) *The City Challenge Experience: A Review of the Development and Implementation of Newcastle City Challenge*, Newcastle: Newcastle City Challenge West End Partnership

Robinson, F. and Shaw, K. (1994) 'Urban policy under the Conservatives: in search of the big idea', *Local Economy* 9: 224–235

Robinson, V. (1987) 'The trendy triumvirate: yuppies, gentrification and docklands', *Cambria* 14: 163–175

Robinson, V. and Williams, J.H. (1990) 'The social and demographic transformation of Britain's docklands', in S. Otok (ed.) *Environment in Social Policy of the State*, Warsaw: Warsaw University Press, 187–204

Robson, B.T. (1988) *Those Inner Cities: Reconciling the Social and Economic Aims of Urban Policy*, Oxford: Clarendon Press

Robson, B.T. (1992) 'Competing and collaborating through urban networks', *Town and Country Planning* 61: 236–238

Robson, B.T. (1994) 'Urban policy at the crossroads', *Local Economy* 9: 216–223

Robson, B.T., Bradford, M.G., Deas, I., Hall, E., Harrison, E., Parkinson, M., Evans, R., Harding, A., Garside, P. and Robinson, F. (1994) *Assessing the Impact of Urban Policy*, London: HMSO

Roger Tym and Partners (1982) *Monitoring Enterprise Zones: Year One Report State of the Zones*, London: HMSO

Rogers, R. (1997) *Cities for a Small Planet*, London: Faber & Faber

Rose, G. (1992) 'Local resistance to the LDDC: community attitudes and action', in P. Ogden (ed.) *London Docklands: The Challenge of Development*, Cambridge: Cambridge University Press, 32–42

Ross, A. (2000) *The Celebration Chronicles: Life, Liberty and the Pursuit of Property Value in Disney's New Town*, London: Verso

Rowley, A. (1996) 'Mixed-use development: ambiguous concept, simplistic analysis and wishful thinking?', *Planning Practice and Research* 11: 85–97

Rowley, G. (1993) 'Prospects for the central business district', in R.D.F. Bromley and C.J. Thomas (eds.) *Retail Change: Contemporary Issues*, London: UCL Press, 110–125

Rowley, G. (1994) 'The Cardiff Bay Development Corporation: urban regeneration, local economy and community', *Geoforum* 25: 265–284

Rudlin, D. and Falk, N. (1999) *Building the 21st Century Home: The Sustainable Urban Neighbourhood*, Oxford: Architectural Press

Russell, H., Dawson, J., Garside, P. and Parkinson, M. (1996) *City Challenge Interim National Evaluation*, London: HMSO

Rutheiser, C. (1996) *Imagineering Atlanta*, London: Verso

Rydin, Y. (1998) *Urban and Environmental Planning in the UK*, Basingstoke: Palgrave Macmillan

Sadler, D. (1990) 'The social foundations of planning and the power of capital: Teesside in historical context', *Environment and Planning D: Society and Space* 8: 323–338

Sassen, S. (1994) *Cities in a World Economy*, Thousand Oaks, CA: Pine Forge Press

Schiller, R. (1985) 'Land use controls on UK shopping centres', in J.A. Dawson and J.D. Lord (eds.) *Shopping Centre Development: Policies and Prospects*, Beckenham: Croom Helm, 40–56

Schiller, R.K. (1986) 'Retail decentralization – the coming of the third wave', *The Planner* 72: 13–15

Schopen, F. (2003) 'Urban Development Corporations: special report', *Regeneration and Renewal* 17 October: 18–23

Scott, A. (1999) 'The cultural economy: geography and the creative field', *Media, Culture and Society* 21: 807–817

Scott, A. (2000) *The Cultural Economy of Cities*, London: Sage

Selby, M. (2004) *Understanding Urban Tourism: Image, Culture and Experience*, London: I.B. Tauris

Self, A. and Zealey, L. (eds.) (2007) *Social Trends No. 37 – 2007 Edition*, Basingstoke: Palgrave Macmillan

Senior, M.L., Webster, C.J. and Blank, N.E. (2004) 'Residential preferences versus sustainable cities: quantitative and qualitative evidence from a survey of relocating owner-occupiers', *Town Planning Review* 75: 337–357

Senior, M.L., Webster, C.J. and Blank, N.E. (2006) 'Residential relocation and sustainable urban form: statistical analyses of owner-occupiers' preferences', *International Planning Studies* 11: 41–57

Senn, A. (1999) *Power, Politics and the Olympic Games*, Leeds: Human Kinetics

Seo, J-K. (2002) 'Re-urbanisation in regenerated areas of Manchester and Glasgow: new residents and the problems of sustainability', *Cities* 19: 113–121

SEU (Social Exclusion Unit) (1998) *Bringing Britain Together: A National Strategy for Neighbourhood Renewal*, London: HMSO

SEU (Social Exclusion Unit) (2000) *National Strategy for Neighbourhood Renewal: A Framework for Consultation*, London: HMSO
SEU (Social Exclusion Unit) (2001) *A New Commitment to Neighbourhood Renewal: National Strategy Action Plan*, London: HMSO
Sharp, J., Pollock, V. and Paddison, R. (2005) 'Just art for a just city: public art and social inclusion in urban regeneration', *Urban Studies* 42: 1001–1023
Shaw, K. and Robinson, F. (1998) 'Learning from experience? Reflections on two decades of British urban policy', *Town Planning Review* 69: 49–63
Short, J.R. (1989) 'Yuppies, yuffies and the new urban order', *Transactions of the Institute of British Geographers* 14: 173–188
Short, J.R. (1991) *Imagined Country: Society, Culture and Environment*, London: Routledge
Short, J.R. (1996) *The Urban Order: An Introduction to Cities, Culture and Power*, Oxford: Blackwell
Short, J.R., Benton, L.M., Luce, W.B. and Walton, J. (1993) 'Reconstructing the image of an industrial city', *Annals of the Association of American Geographers* 83: 207–224
Short, J.R. and Kim, Y-H. (1999) *Globalization and the City*, Harlow: Longman
Shorthose, J. (2004) 'The engineered and the vernacular in cultural quarter development', *Capital and Class* 84: 159–178
Shropshire, K.L. (1995) *The Sports Franchise Game: Cities in Pursuit of Sports Franchises, Events, Stadiums, and Arenas*, Philadelphia, PA: University of Pennsylvania Press
Silverman, E., Lupton, R. and Fenton, A. (2005) *A Good Place for Children? Attracting and Retaining Families in Inner Urban Mixed Income Communities*, York: Joseph Rowntree Foundation
Simms, A., Kjell, P. and Potts, R. (2005) *Clone Town Britain: The Survey Results on the Bland State of the Nation*, London: New Economics Foundation
Simms, A., Oram, J., MacGillivray, A. and Drury, J. (2002) *Ghost Town Britain: The Threat from Economic Globalisation to Livelihoods, Liberty and Local Economic Freedom*, London: New Economics Foundation
Slater, T. (2002) 'Looking at the "North American City" through the lens of gentrification discourse', *Urban Geography* 23: 131–153
Slater, T. (2004) 'Municipally-managed gentrification in South Parkdale, Toronto', *The Canadian Geographer* 48: 303–325
Slater, T. (2006) 'The eviction of critical perspectives from gentrification research', *International Journal of Urban and Regional Research* 30: 737–757
SMF (Social Market Foundation) (2007) *Should the Greenbelt be Preserved?*, London: SMF [www.smf.co.uk]
Smith, A. (1989) 'Gentrification and the spatial contribution of the state: the restructuring of London's docklands', *Antipode* 21: 232–260
Smith, A. (1991) 'Political transformation, urban policy and the state in London's Docklands', *GeoJournal* 24: 237–246
Smith, A. (2005) 'Reimaging the city: the value of sports initiatives', *Annals of Tourism Research* 32: 229–248
Smith, A. (2007) 'Large-scale events and sustainable urban regeneration: key principles for host cities', *Journal of Urban Regeneration and Renewal* 1: 178–190
Smith, A. and Sparks, L. (2000) 'The role and function of the independent small shop: the situation in Scotland', *International Review of Retail, Distribution and Consumer Research* 10: 205–226
Smith, D. (2002) 'Patterns and processes of "studentification" in Leeds', *The Regional Review* 12: 15–16
Smith, D.P. (2005) 'Studentification: the gentrification factory?', in R. Atkinson and G. Bridge (eds.) *Gentrification in a Global Context: The New Urban Colonialism*, London: Routledge, 72–89
Smith, D.P. and Holt, L. (2007) 'Studentification and "apprentice" gentrifiers within Britain's provincial towns and cities: extending the meaning of gentrification', *Environment and Planning A* 39: 142–161
Smith, I., Lepine, E. and Taylor, M. (eds.) (2007a) *Disadvantaged by Where You Live? Neighbourhood Governance in Contemporary Urban Policy*, Bristol: Policy Press
Smith, I., Howard, J. and Evans, L. (2007b) 'Mainstreaming and neighbourhood governance: the importance of process, power and partnership', in I. Smith, E. Lepine and M. Taylor (eds.) *Disadvantaged By Where You Live? Neighbourhood Governance in Contemporary Urban Policy*, Bristol: Policy Press, 165–183
Smith, N. (1996) *The New Urban Frontier: Gentrification and the Revanchist City*, London: Routledge
Smith, N. (2000) 'Gentrification', in R.J. Johnston, D. Gregory, G. Pratt and M. Watts (eds.) *The Dictionary of Human Geography*, Oxford: Blackwell, 4th Edition, 295–296
Smith, N. (2002) 'New globalism, new urbanism: gentrification as global urban strategy', *Antipode* 34: 428–450

Smith, N. and Williams, P. (1986) 'Alternatives to orthodoxy: invitation to a debate', in N. Smith and P. Williams (eds.) *Gentrification of the City*, London: Allen and Unwin, 1–12

Smyth, H. (1994) *Marketing the City: The Role of Flagship Developments in Urban Regeneration*, London: E & FN Spon

Somerville, P. (2005) 'Community governance and democracy', *Policy and Politics* 33: 117–144

Sorkin, M. (ed.) (1992) *Variations on a Theme Park: The New American City and the End of Public Space*, New York: Hill and Wang

Sparks, L. (1987) 'Retailing in Enterprise Zones: the example of Swansea', *Regional Studies* 21: 37–42

Spear, R. (2001) *Tackling Social Exclusion in Europe: The Contribution of the Social Economy*, Aldershot: Ashgate

Stanton, R. (1996) 'The retreat from social need: competitive bidding and local public investment', *Local Economy* 11: 194–201

Steel, M. and Symes, M. (2005) 'The privatisation of public space? The American experience of Business Improvement Districts and their relationship to local governance', *Local Government Studies* 31: 321–334

Stegman, M.A. (1995) 'Recent US urban change and policy initiatives', *Urban Studies* 32: 1601–1607

Stewart, M. (1987) 'Ten years of inner cities policy', *Town Planning Review* 58: 129–145

Stewart, M. (1990) 'Urban policy in Thatcher's England', *School for Advanced Urban Studies, Working Paper Number 90*, Bristol: University of Bristol

Stewart, M. (1994) 'Between Whitehall and Town Hall: the realignment of urban policy in England', *Policy and Politics* 22: 133–146

Stewart, M. (1998) 'Partnership, leadership and competition in urban policy', in N. Oatley (ed.) *Cities, Economic Competition and Urban Policy*, London: Paul Chapman, 77–90

Stewart, M. (2003) 'Towards collaborative capacity', in M. Boddy (ed.) *Urban Transformation and Urban Governance: Shaping the Competitive City of the Future*, Bristol: Policy Press, 76–89

Stewart, M. and Taylor, M. (1995) *Empowerment and Estate Regeneration*, Bristol: Policy Press

Stimson, R.J. and Taylor, S.P. (1999) 'City profile – Brisbane', *Cities* 16: 285–296

Stoker, G. (1989) 'Urban Development Corporations: a review', *Regional Studies* 23: 159–173

Stoker, G. (1998) 'Governance as theory: 5 propositions', *International Social Science Journal* 155: 17–28

Stoker, G. (ed.) (2000) *The New Politics of British Local Governance*, Basingstoke: Macmillan

Stoker, G. (2004) *British Local Government into the 21st Century*, Basingstoke: Palgrave Macmillan

Stoker, G. and Mossberger, K. (1995) 'The post-Fordist local state: the dynamics of its development', in J. Stewart and G. Stoker (eds.) *Local Government in the 1990s*, London: Macmillan, 210–227

Summerfield, C. and Gill, B. (2005) *Social Trends No. 35 – 2005 Edition*, Basingstoke: Palgrave Macmillan

Sustainable Development Commission (2003) *Mainstreaming Sustainable Regeneration: A Call to Action*, London: Sustainable Development Commission

Symes, M. and Steel, M. (2003) 'Lessons from America: the role of Business Improvement Districts as an agent of urban regeneration', *Town Planning Review* 74: 301–313

Tallon, A.R. (2001) 'Towards the "twenty-four hour city": housing and city centre revitalisation', *Swansea Geographer* 36: 35–52

Tallon, A.R. (2002) *Living in the British City Centre: Revitalisation and the Urban Renaissance*, Unpublished PhD Thesis, University of Wales Swansea

Tallon, A.R. (2003) 'Residential transformation and the promotion of city centre living', *Town and Country Planning* 72: 190–193

Tallon, A.R. (2006) 'Regenerating Bristol's Harbourside', *Town and Country Planning* 75: 278–282

Tallon, A.R. (2007a) 'City profile – Bristol', *Cities* 24: 74–88

Tallon, A.R. (2007b) 'Carnivals and regeneration', *Town and Country Planning* 76: 256–260

Tallon, A.R. (2008) 'Mega-retail-led regeneration', *Town and Country Planning* 77: 131–137

Tallon, A.R. and Bromley, R.D.F. (2002) 'Living in the 24-hour city', *Town and Country Planning* 71: 282–285

Tallon, A.R. and Bromley, R.D.F. (2004) 'Exploring the attractions of city centre living: evidence and policy implications in British cities', *Geoforum* 35: 771–787

Tallon, A.R., Bromley, R.D.F., Reynolds, B. and Thomas, C.J. (2006) 'Developing leisure and cultural attractions in the regional city centre: a policy perspective', *Environment and Planning C: Government and Policy* 24: 351–370

Tallon, A.R., Bromley, R.D.F. and Thomas, C.J. (2005) 'City profile – Swansea', *Cities* 22: 65–76

Tang, Z. and Batey, P.W.J. (1996) 'Intra-urban spatial analysis of housing-related urban policies: the case of Liverpool, 1981–1991', *Urban Studies* 33: 911–936

Taylor, M. (1998) 'Combating the social exclusion of housing estates', *Housing Studies* 13: 819–832

Taylor, M. (2000) 'Communities in the lead: power, organisational capacity and social capital', *Urban Studies* 37: 1019–1035

Taylor, M. (2002) 'The new public management and social exclusion: cause or response?', in K. McLaughlin, S. Osborne and E. Ferlie (eds.) *New Public Management: Current Trends and Future Prospects*, London: Routledge, 109–128

Taylor, M. (2003) *Public Policy in the Community*, Basingstoke: Palgrave Macmillan

Taylor, M. (2007) 'Community participation in the real world: opportunities and pitfalls in new governance spaces', *Urban Studies* 44: 291–317

Thomas, C.J. (1989) 'Retail change in Greater Swansea: evolution or revolution?', *Geography* 71: 201–213

Thomas, C.J. and Bromley, R.D.F. (1987) 'The growth and functioning of an unplanned retail park: the Swansea Enterprise Zone', *Regional Studies* 21: 287–300

Thomas, C.J. and Bromley, R.D.F. (1993) 'The impact of out-of-centre retailing', in R.D.F. Bromley and C.J. Thomas (eds.) *Retail Change: Contemporary Issues*, London: UCL Press, 126–152

Thomas, C.J. and Bromley, R.D.F. (1995) 'Retail decline and the opportunities for commercial revitalisation of small shopping centres: a case study in South Wales', *Town Planning Review* 66: 431–452

Thomas, C.J. and Bromley, R.D.F. (1996) 'Safety and shopping: peripherality and shopper anxiety in the city centre', *Environment and Planning C: Government and Policy* 14: 469–488

Thomas, C.J. and Bromley, R.D.F. (2000) 'City-centre revitalisation: problems of fragmentation and fear in the evening and night-time city', *Urban Studies* 37: 1407–1433

Thomas, C.J. and Bromley, R.D.F. (2002) 'The changing competitive relationship between small town centres and out-of-town retailing: town revival in South Wales', *Urban Studies* 39: 791–817

Thomas, C.J. and Bromley, R.D.F. (2003) 'Retail revitalization and small town centres: the contribution of shopping linkages', *Applied Geography* 23: 47–71

Thomas, C.J., Bromley, R.D.F. and Tallon, A.R. (2004) 'Retail parks revisited: a growing competitive threat to traditional shopping centres?', *Environment and Planning A* 36: 647–666

Thomas, C.J., Bromley, R.D.F. and Tallon, A.R. (2006) 'New "high streets" in the suburbs? The growing competitive impact of evolving retail parks', *International Review of Retail, Distribution and Consumer Research* 16: 43–68

Thomas, H. and Imrie, R. (1993) 'Cardiff Bay and the project of modernisation', in R. Imrie and H. Thomas (eds.) *British Urban Policy and the Urban Development Corporations*, London: Paul Chapman, 74–88

Thomas, H. and Imrie, R. (1999) 'Urban policy, modernisation, and the regeneration of Cardiff Bay', in R. Imrie and H. Thomas (eds.) *British Urban Policy: An Evaluation of the Urban Development Corporations*, London: Sage, 2nd edition, 106–127

Thornley, A. (1991) *Urban Planning under Thatcherism: The Challenge of the Market*, London: Routledge

Thornley, A. (2002) 'Urban regeneration and sports stadia', *European Planning Studies* 10: 813–818

Tiesdell, S. (1995) 'Tensions between revitalization and conservation: Nottingham's lace market', *Cities* 12: 231–241

Tiesdell, S. and Allmendinger, P. (2001) 'Neighbourhood regeneration and New Labour's third way', *Environment and Planning C: Government and Policy* 19: 903–926

Tiesdell, S. and Allmendinger, P. (2004) 'City profile – Aberdeen', *Cities* 21: 167–179

Tiesdell, S., Oc, T. and Heath, T. (1996) *Revitalising Historic Urban Quarters*, London: Butterworth–Heineman

Townshend, T. (2006) 'From inner city to inner suburb? Addressing housing aspirations in low demand areas in NewcastleGateshead, UK', *Housing Studies* 21: 501–521

Trench, S. (1991) 'Reclaiming the night', *Town and Country Planning* 60: 235–237

Trench, S., Oc, T. and Tiesdell, S. (1992) 'Safer city centres for women: perceived risks and planning measures', *Town Planning Review* 63: 279–296

Trott, T. (2002) *Best Practice in Regeneration: Because it Works*, Bristol: Policy Press

Turok, I. (1992) 'Property-led urban regeneration: panacea or placebo?', *Environment and Planning A* 24: 361–379

Turok, I. (2005) 'Urban regeneration: what can be done and what should be avoided?', in *Istanbul 2004 International Urban Regeneration Symposium: Workshop of Kucukcekmece District*, Istanbul: Kucukcekmece Municipality Publication, 57–62

Turok, I., Bailey, N., Atkinson, R., Bramley, G., Docherty, I., Gibb, K., Goodlad, R., Hastings, A., Kintrea, K., Kirk, K., Leibovitz, J., Lever, B., Morgan, J. and Paddison, R. (2004) 'Sources of city prosperity and cohesion: the case of Glasgow and Edinburgh', in M. Boddy and M. Parkinson (eds.) *City Matters: Competitiveness, Cohesion and Urban Governance*, Bristol: Policy Press, 13–31

Turok, I. and Edge, N. (1999) *The Jobs Gap in Britain's Cities: Employment Loss and Labour Market Consequences*, Bristol: Policy Press

Uitermark, J., Duyvendak, J.W. and Kleinhans, R. (2007) 'Gentrification as a governmental strategy: social control and social cohesion in Hoogvliet, Rotterdam', *Environment and Planning A* 39: 125–141

Unsworth, R. and Nathan, M. (2006) 'Beyond city living: remaking the inner suburbs', *Built Environment* 32: 235–249

Unsworth, R. and Stillwell, J. (eds.) (2004) *Twenty-First Century Leeds: Geographies of a Regional City*, Leeds: Leeds University Press

Urban Studies (1999) 'Competition Cities', Volume 36, Number 5–6

Urban Task Force (1999) *Towards an Urban Renaissance*, London: E & FN Spon

Urban Task Force (2005) *Towards a Strong Urban Renaissance*, London: Urban Task Force [www.urbantaskforce.org]

Urban Villages Group (1992) *Urban Villages: A Concept for Creating Mixed Use Urban Development on a Sustainable Scale*, London: Urban Villages Group

URBED (Urban and Economic Development Group) (1994) *Vital and Viable Town Centres: Meeting the Challenge*, Report of a study undertaken in association with Comedia, Hillier Parker, Bartlett School of Planning, University College London, and Environmental and Transport Planning, Department of the Environment, London: HMSO

URBED (Urban and Economic Development Group) (1997) *Town Centre Partnerships: A Survey of Good Practice and Report of an Action Research Project*, London: HMSO

Urry, J. (1995) *Consuming Places*, London: Sage

Valentine, G. (1989) 'The geography of women's fear', *Area* 21: 385–390

van den Berg, L. and Braun, E. (1999) 'Urban competitiveness, marketing and the need for organising capacity', *Urban Studies* 36: 987–1000

van den Berg, L., Braun, E. and van der Meer, J. (1997) 'The organising capacity of metropolitan regions', *Environment and Planning C: Government and Policy* 15: 253–272

Van Kempen, R. and Van Weesep, J. (1994) 'Gentrification and the urban poor: restructuring and housing policy in Utrecht', *Urban Studies* 31: 1043–1056

Van Weesep, J. (1994) 'Gentrification as a research frontier', *Progress in Human Geography* 18: 74–83

Vicario, L. and Manuel Martinez Monje, P. (2003) 'Another "Guggenheim effect"? The generation of a potentially gentrifiable neighbourhood in Bilbao', *Urban Studies* 40: 2383–2400

Wackernagel, M. and Rees, W. (1995) *Our Ecological Footprint: Reducing Human Impact on the Earth*, Gabriola Island, BC: New Society Publishers

Waitt, G. (1999) 'Playing games with Sydney: marketing Sydney for the 2000 Olympics', *Urban Studies* 36: 1055–1077

Walburn, D. (2008) 'Business Improvement Districts and local economic development', *Local Economy* 23: 69–70

Walker, G. (1996) 'Retailing development: in town or out of town?', in C. Greed (ed.) *Investigating Town Planning: Changing Perspectives and Agendas*, Harlow: Longman, 155–180

Walton, D.S. (1990) 'Cardiff Bay development', *The Planner* 76: 10–18

Ward, K. (2003a) 'Entrepreneurial urbanism, state restructuring and civilizing "New" East Manchester', *Area* 35: 116–127

Ward, K. (2003b) 'The limits to contemporary urban redevelopment: "doing" entrepreneurial urbanism in Birmingham, Leeds and Manchester', *City* 7: 199–211

Ward, K. (2006) ' "Policies in motion", urban management and state restructuring: the trans-local expansion of Business Improvement Districts', *International Journal of Urban and Regional Research* 30: 54–75

Ward, S.V. (1998) *Selling Places: The Marketing and Promotion of Towns and Cities, 1850–2000*, London: E & FN Spon

Ward, S.V. (2004) *Planning and Urban Change*, London: Sage, 2nd edition

Ward, S.V. and Gold, J.R. (1994) 'Introduction', in J.R. Gold and S.V. Ward (eds.) *Place Promotion: The Use of Publicity and Marketing to Sell Towns and Regions*, Chichester: Wiley, 1–17

Warde, A. (1991) 'Gentrification as consumption: issues of class and gender', *Environment and Planning D: Society and Space* 9: 223–232

Warnaby, G., Alexander, A. and Medway, D. (1998) 'Town centre management in the UK: a review, synthesis and research agenda', *International Review of Retail, Distribution and Consumer Research* 8: 15–31

Waterman, S. (1998) 'Carnivals for elites? The cultural politics of arts festivals', *Progress in Human Geography* 22: 54–74

Watkins, H. and Herbert, D. (2003) 'Cultural policy and place promotion: Swansea and Dylan Thomas', *Geoforum* 34: 249–266

Weaver, M. (2003) 'Maze of initiatives "like spaghetti" ', *The Guardian*, 14 January

Webster, C.J. (2001) 'Gated cities of tomorrow', *Town Planning Review* 72: 149–170

Webster, C.J. (2002) 'Property rights and the public realm: gates, green belts and gemeinschaft', *Environment and Planning B: Planning and Design* 29: 397–412

Webster, C., Glasze, G. and Frantz, K. (2002) 'The global spread of gated communities', *Environment and Planning B: Planning and Design* 29: 315–320

Whitelegg, D. (2000) 'Going for gold: Atlanta's bid for fame', *International Journal of Urban and Regional Research* 24: 801–817

Whitson, D. (2004) 'Bringing the world to Canada: "the periphery of the centre" ', *Third World Quarterly* 25: 1215–1232

Whysall, P. (1995) 'Regenerating inner city shopping centres: the British experience', *Journal of Retailing and Consumer Services* 2: 3–13

Wilks-Heeg, S. (1996) 'Urban experiments limited revisited: urban policy comes full circle?', *Urban Studies* 33: 1263–1279

Wilks-Heeg, S. and North, P. (2004) 'Cultural policy and urban regeneration', *Local Economy* 19: 305–311

Williams, C. (1996a) 'An appraisal of local exchange and trading systems in the UK', *Local Economy* 11: 259–266

Williams, C. (1996b) 'The new barter economy: an appraisal of LETS', *Journal of Public Policy* 16: 85–101

Williams, C. (1996c) 'Informal sector responses to unemployment: an evaluation of the potential of Local Exchange Trading Systems (LETS)', *Work, Employment and Society* 10: 341–359

Williams, C.C. (1997) 'Rethinking the role of the retail sector in economic development', *Service Industries Journal* 17: 205–220

Williams, C. (2000a) 'Paving the Third Way? Evaluating the potential of LETS', *Town and Country Planning* 69: 89–91

Williams, C. (2000b) 'Are local currencies an effective tool for tackling social exclusion?', *Town and Country Planning* 69: 323–325

Williams, C. and Windebank, J. (2000a) 'Rebuilding social capital in deprived urban neighbourhoods', *Town and Country Planning* 69: 352–353

Williams, C. and Windebank, J. (2000b) 'Helping people to help themselves: policy lessons from a study of deprived urban neighbourhoods in Southampton', *Journal of Social Policy* 29: 355–373

Williams, G. (1995a) 'Local governance and urban prospects: the potential of City Pride', *Local Economy* 10: 1–8

Williams, G. (1995b) 'Prospecting for gold: Manchester's City Pride experience', *Planning Practice and Research* 10: 345–358

Williams, G. (1996) 'City profile – Manchester', *Cities* 13: 203–212

Williams, G. (1998) 'City vision and strategic regeneration – the role of City Pride', in N. Oatley (ed.) *Cities, Economic Competition and Urban Policy*, London: Paul Chapman, 163–180

Williams, G. (2003) *The Enterprising City Centre: Manchester's Development Challenge*, London: Routledge

Williams, K., Burton, E. and Jenks, M. (eds.) (2000) *Achieving Sustainable Urban Form*, London: Spon

Williams, K. and Johnstone, C. (2000) 'The politics of the selective gaze: closed circuit television and the policing of public space', *Crime, Law and Social Change* 34: 183–210

Williams, R.H. (1996) 'Hands across the border: networking, lobbying and cross border planning', in R.H. Williams (ed.) *European Urban Spatial Policy and Planning*, London: Paul Chapman, 145–166

Williams, S. (1998) *Tourism Geography*, London: Routledge

Wood, A. (1988) 'Making sense of urban entrepreneurialism', *Scottish Geographical Magazine* 114: 120–123

Worpole, K. (1991) 'The age of leisure', in J. Corner and S. Harvey (eds.) *Enterprise and Heritage: Crosscurrents of National Culture*, London: Routledge, 137–150

Worpole, K. (1992) *Towns for People: Transforming Urban Life*, Buckingham: Open University Press

Worpole, K. (1998) 'Time and the city', *Town and Country Planning* 67: 114–115

Worthington, S. (1998) 'Loyalty cards as an icon for revitalisation of the urban centre', in CIRM

(Contemporary Issues in Retailing and Marketing) *Marketing and Managing Urban Centres*, Conference Proceedings, Manchester Metropolitan University, 37–45

Wray, I. (1987) 'The Merseyside Development Corporation: progress versus objectives', *Regional Studies* 21: 763–767

Wrigley, N. (2002) 'Food deserts in British cities: policy context and research priorities', *Urban Studies* 39: 2029–2040

Wrigley, N., Guy, C. and Lowe, M. (2002) 'Urban regeneration, social exclusion and large store development: the Seacroft development in context', *Urban Studies* 39: 2102–2114

Wrigley, N. and Lowe, M. (2002) *Reading Retail: A Geographical Perspective on Retailing and Consumption Spaces*, London: Arnold

Wrigley, N., Warm, D.L. and Margetts, B.M. (2003) 'Deprivation, diet and food retail access: findings from the Leeds "food deserts" study', *Environment and Planning A* 35: 151–188

Wrigley, N., Warm, D.L., Margetts, B.M. and Lowe, M.S. (2004) 'The Leeds "food deserts" intervention study: what the focus groups reveal', *International Journal of Retail and Distribution Management* 31: 123–136

Wyly, E.K. and Hammel, D.J. (1999) 'Islands of decay in seas of renewal: housing policy and the resurgence of gentrification', *Housing Policy Debate* 10: 711–771

Young, C. and Lever, J. (1997) 'Place promotion, economic location and the consumption of city image', *Tijdschrift voor Economische en Sociale Geografie* 88: 332–341

Young, C., Diep, M. and Drabble, S. (2006) 'Living with difference? The "cosmopolitan city" and urban re-imaging in Manchester, UK', *Urban Studies* 43: 1687–1714

Zukin, S. (1989) *Loft Living: Culture and Capital in Urban Change*, New Brunswick, NJ: Rutgers University Press

Index

References to Figures or Tables will be in *italic* print